VII Brazilian Congress on Metrology

(Metrologia 2013)

Journal of Physics: Conference Series Volume 575

Ouro Preto, Brazil
24-27 November 2013

Editors:

Rodrigo Costa-Félix
José Carlos Valente de Oliveira
José Mauro Granjeiro

ISBN: 978-1-63439-908-1
ISSN: 1742-6588

Printed from e-media with permission by:

Curran Associates, Inc.
57 Morehouse Lane
Red Hook, NY 12571

Some format issues inherent in the e-media version may also appear in this print version.

Printed by Curran Associates, Inc. (2015)

For permission requests, please contact the Institute of Physics
at the address below.

Institute of Physics
Dirac House, Temple Back
Bristol BS1 6BE UK

Phone: 44 1 17 929 7481
Fax: 44 1 17 920 0979

techtracking@iop.org

Additional copies of this publication are available from:

Curran Associates, Inc.
57 Morehouse Lane
Red Hook, NY 12571 USA
Phone: 845-758-0400
Fax: 845-758-2634
Email: curran@proceedings.com
Web: www.proceedings.com

TABLE OF CONTENTS

OPTICS

IONIZING RADIATION

TIME AND FREQUENCY

CHEMISTRY

SEVENTH BRAZILIAN CONGRESS ON METROLOGY (METROLOGIA 2013)
Metrology and Quality for a Sustainable Development

From November 24th to 27th 2013 was issued the Seventh Brazilian Congress on Metrology (*Metrologia 2013*), which is a biannual conference organized and sponsored by the Brazilian Society of Metrology (SBM) and the Brazilian National Institute of Metrology, Quality and Technology (Inmetro). This edition was held in the charming and historical city of Ouro Preto, MG, Brazil, and aimed to join people and institutions devoted to the dissemination of the metrology and conformity assessment. The *Metrologia 2013* Conference consisted of Keynote Speeches (7) and regular papers (204). Among the regular papers, the 47 most outstanding ones, comprising a high quality content on Metrology and Conformity Assessment, were selected to be published in this issue of the *Journal of Physics: Conference Series*.

The topics of the conference covered all important areas of Metrology, which were agglutinated in the following sessions in the present issue:

- Physical Metrology (Acoustics, Vibration and Ultrasound; Electricity and Magnetism; Mechanics; Optics);
- Metrology on Ionizing Radiations;
- Time and Frequency
- Chemistry Metrology;
- Materials Metrology;
- Biotechnology;
- Uncertainty, Statistics and Mathematics;
- Legal Metrology;
- Conformity Assessment.

It is our great pleasure to present this volume of IOP *Journal of Physics: Conference Series* (JPCS) to the scientific community to promote further research in Metrology and related areas. We believe that this volume will be both an excellent source of scientific material in the fast evolving fields that were covered by *Metrologia 2013*.

President of the congress
Americo Bernardes
Federal University of Ouro Preto
atb@iceb.ufop.br

Editor-in-chief
Rodrigo Costa-Félix
Brazilian National Institute of Metrology, Quality and Technology
rpfelix@inmetro.gov.br

Editors
José Carlos Valente de Oliveira (Editor on Mechanical Metrology)
Brazilian National Institute of Metrology, Quality and Technology
jcoliveira@inmetro.gov.br

José Mauro Granjeiro (Editor on Biotechnology)
Brazilian National Institute of Metrology, Quality and Technology
jmgranjeiro@inmetro.gov.br

Ruth Epsztejn (Editor on Conformity Assessment)
Brazilian National Institute of Metrology, Quality and Technology
repsztejn@inmetro.gov.br

Waldemar Ihlenfeld (Editor on Electrical Metrology)
Brazilian National Institute of Metrology, Quality and Technology
wgihlenfeld-pronametro@inmetro.gov.br

Valnei Smarçaro da Cunha (Editor on Chemistry Metrology)
Brazilian National Institute of Metrology, Quality and Technology
vscunha@inmetro.gov.br

Technical and Scientific Committee for *Metrologia 2013*

Peer review statement

All papers published in this volume of *Journal of Physics: Conference Series* have been peer reviewed through processes administered by the proceedings Editors. Reviews were conducted by expert referees to the professional and scientific standards expected of a proceedings journal published by IOP Publishing.

Journal of Physics: Conference Series **575** (2015) 012001 doi:10.1088/1742-6596/575/1/012001

Monitoring biodiesel reactions of soybean oil and sunflower oil using ultrasonic parameters

MK K Figueiredo[1],CE R Silva[1], AV Alvarenga[1], RP B Costa-Félix[1]

[1]Laboratory of Ultrasound (Labus), Directory of Scientific and Industrial Metrology (Dimci), National Institute of Metrology, Quality and Technology (Inmetro), Av. Nossa Sra. das Graças 50, Duque de Caxias, RJ, Brazil, ZIP 25250-020
E-mail: mkfigueiredo@inmetro.gov.br

Abstract. Biodiesel is an innovation that attempts to substitute diesel oil with biomass. The aim of this paper is to show the development of a real-time method to monitor transesterification reactions by using low-power ultrasound and pulse/echo techniques. The results showed that it is possible to identify different events during the transesterification process by using the proposed parameters, showing that the proposed method is a feasible way to monitor the reactions of biodiesel during its fabrication, in real time, and with relatively low-cost equipment.

1. Introduction

Meeting current energy demand is a challenging issue. Environmental concerns associated with fossil fuel burning have brought in focus the necessity of renewable energy resources [1]. Biodiesel is an example of the employment of (renewable) biomass to produce energy. Biomass is an alternative to diesel that is biodegradable, nontoxic, and has a low-emission profile [2].

It is important ensure the quality of the fuel and to understand the stages of the transesterification reaction. Monitoring reaction kinetics is of extreme importance for the optimization of industrial processes. A robust method that performs accurate non-destructive monitoring and ensures the quality of biofuel is in demand. In this direction, ultrasound has emerged as a feasible tool. Koc in 2009, for example, has shown the feasibility of using ultrasound to monitor the deposition of glycerol during the transesterification reaction [3].

This paper presents a study that evaluates the usefulness of ultrasound parameters as tools for real-time monitoring of transesterification reactions by using low-power ultrasound and pulse/echo techniques. Time of flight and amplitude of ultrasonic signals were the parameters used in the study.

2. Method

The materials used in this study were soybean and sunflower oil, methanol and ethanol (purity > 95 %), and potassium hydroxide (KOH) (purity > 95 %).The alcohol to oil proportion was 6:1 and 1 % (m/m) of catalyser.

The ultrasonic measurement system consisted of a Beaker (vessel) in which the reactions were carried out, a 1 MHz ultrasonic transducer with a diameter of 12.7 mm (A303-S, Panametrics-NDT Olympus Corporation, Japan), a waveform generator (33250A, Agilent Technologies, CA, USA), an oscilloscope (TDS3012B, Tektronix, Beaverton, OR, USA), a data acquisition unit (34970A, Agilent Technologies, CA, USA) and a computer with a program developed in LabView 11, (National Instruments, Austin, TX, USA). It is noteworthy that the ultrasonic power delivered to this system was low, i.e., less than 100 mW. Typically, ultrasonic power delivered in an ultrasonic-assisted process is in the range of hundreds of watts [4, 5], much greater than that delivered in this work.

A waveform generator was used to excite the transducer by driving a 4-cycle sine wave burst at 20 Vpp of amplitude. The echo signal was acquired and digitalized by using the oscilloscope. Temperature measurement was carried out by using a thermocouple and a data acquisition system. All signals and acquired data were recorded every 30 seconds and analysed by using the software developed in LabView. In Figure 1, a scheme of the experimental setup is presented. It allows identifying the variation that occurs during the reaction through the ultrasonic time of flight and amplitude of signal. Amplitude, time of flight of the ultrasonic waveform, and temperature of the reaction are displayed in real time and recorded in all the experiments.

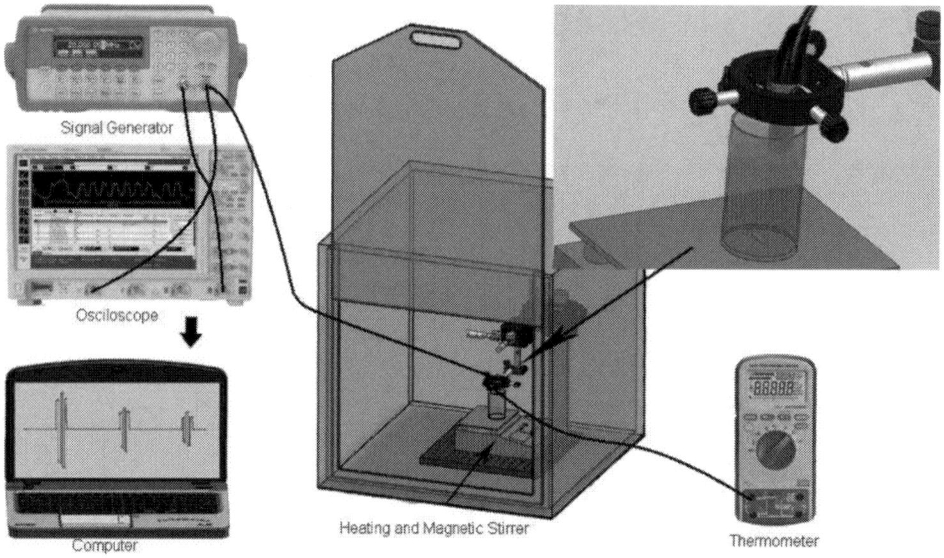

Figure 1. Experimental setup that uses ultrasound for monitoring the transesterification reaction.

3. Results

The results of the transesterification reaction of soybean oil with methanol are presented in Figure 2. In Figure 2a, one can observe the ultrasonic waveform, where the signal with the highest amplitude represents the echo signal from the beaker bottom (the particular waveform disclosed is a sample picked up after the stirring was finished).

The temperature variation during the transesterification process is shown in Figure 3b. The amplitude (Figure 2c) and time of flight (Figure 2d) calculated from the echo signal also vary correspondently. After the end of the stirring, the glycerol deposition starts, and a two-phase sample (biodiesel/glycerol) is formed. The ultrasound reflection on that interface appears as an echo signal before the beaker bottom reflection (arrow in Figure 2a). As the layer of glycerol increases, the distance between the glycerol/biodiesel interface and the beaker bottom increases. Figure 2a was obtained when total glycerol deposition has already occurred.

Once the stirring starts, temperature, time of flight, and amplitude decrease quickly, and after approximately one minute, all the parameters start to increase. The time of flight achieves its maximum in 30 seconds and immediately starts to decrease presenting three stages of decay: till the 5th minute, the time of flight decays very fast (negative steepest part of the curve); between the 5th and 30th minutes, the rate of decay reduces; and finally, after the stirring is turned off, the rate of decay reduces a little bit more. The temperature keeps increasing until approximately the 10th minute, and then it starts to decrease until the 30th minute, when the stirring is turned off. After that, the temperature increases until the 50th minute and then it decreases rapidly. Contrarily, the amplitude stays increasing fast until the 10th minute, and after that, it presents a lower rate of increase.

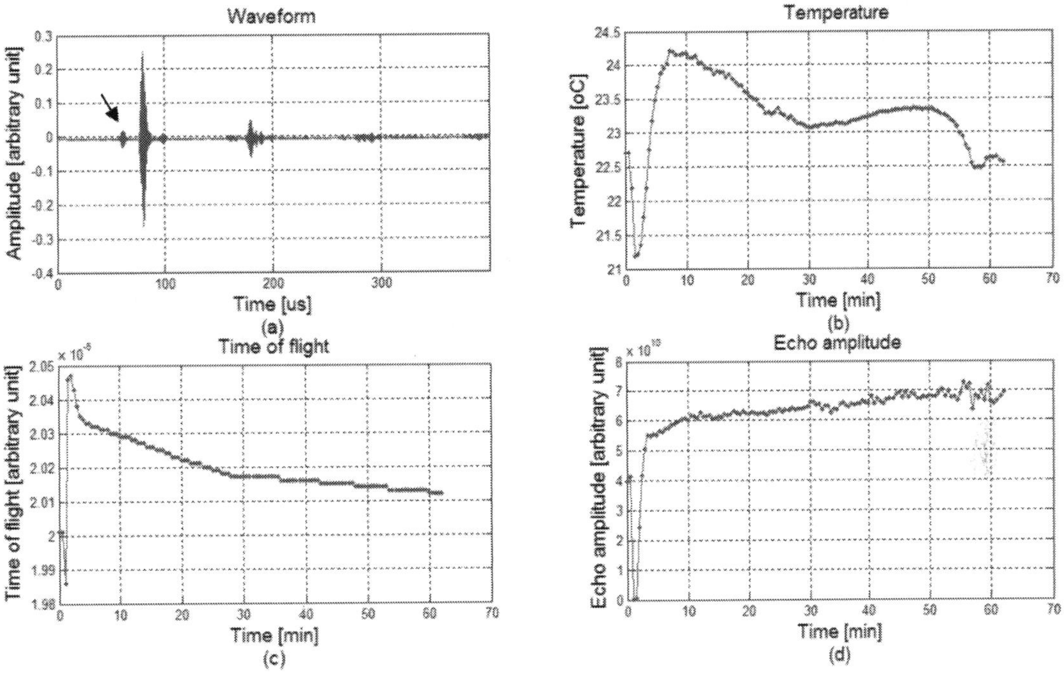

Figure 2. Monitoring of the transesterification reaction of soybean oil with methanol. One can observe (a) the ultrasonic waveform at the end of the experiment, (b) the temperature variation during the transesterification process, (c) the time of flight, and (d) the amplitude calculated from the echo signal.

An example of the parameters obtained during the transesterification reaction of soybean oil with ethanol is presented in Figure 3. Figure 3a shows the waveform at the end of the transesterification, where the signal with the highest amplitude represents the echo signal from the beaker bottom. Glycerol deposition was not observed immediately after the end of the reaction. Correspondingly, no echo signal concerning glycerol deposition was observed in the waveform. Similarly, that reaction using methanol, once the stirring starts, temperature, time of flight, and amplitude decrease quickly. After approximately one minute, temperature and time of flight quickly increase in the 10 first minutes and tend to stabilize after reaching the maximum temperature, while the amplitude decreases until the stirring is turned off. Thereafter, the amplitude increased.

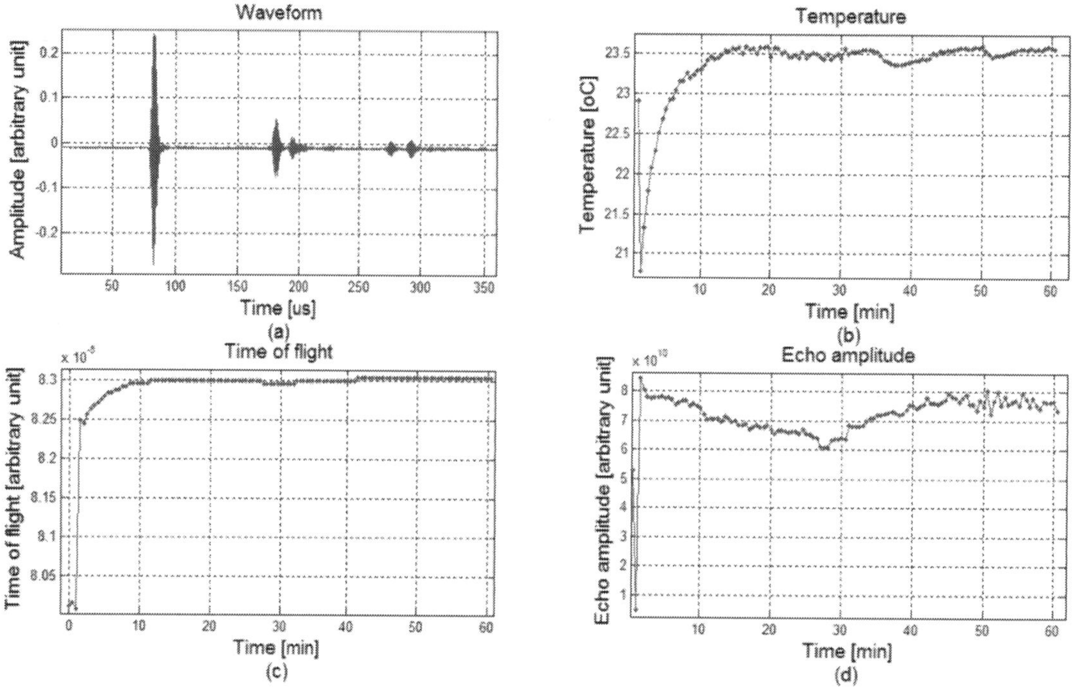

Figure 3. Monitoring of the transesterification reaction of soybean oil with ethanol. One can observe (a) the ultrasonic waveform at the end of the experiment, (b) the temperature variation during the transesterification process, (c) the time of flight, and (d) the amplitude calculated from the echo signal.

An example of the parameters obtained during the transesterification reaction of sunflower oil with methanol is presented in Figure 4. Figure 4a shows the waveform at the end of the transesterification. Similarly, that reaction using soybean oil with methanol, after the end of the stirring, the glycerol deposition starts, and a two-phase sample (biodiesel/glycerol) is formed. The ultrasound reflection on that interface appears as an echo signal before the beaker bottom reflection (Figure 4a).

The same way, that using the soybean oil, once the stirring starts, temperature, time of flight and amplitude decrease quickly. Then, the temperature keeps increasing until approximately the 7th minute, and then it starts to decrease until the 30th minute, when the stirring is turned off. The time of flight increase quickly in one minute and after increasing slow until 15 th minute and hereafter tend to stabilize.The amplitude stays increasing fast until the 4th minute, and after that, it presents a lower rate of decrease and after 6th minute tend to stabilize.

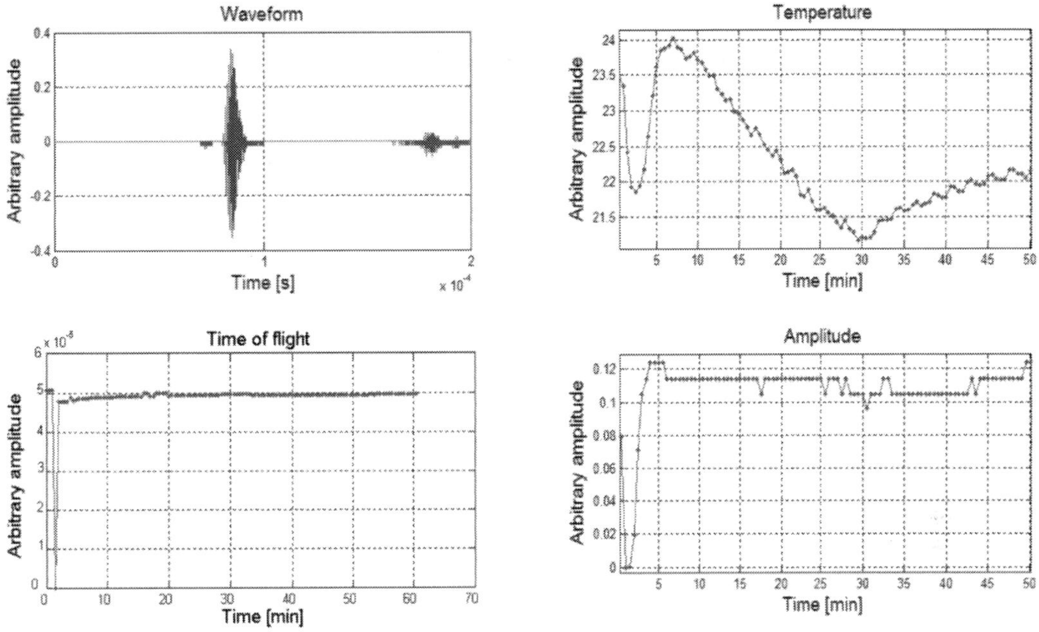

Figure 4. Monitoring of the transesterification reaction of sunflower oil with methanol. One can observe (a) the ultrasonic waveform at the end of the experiment, (b) the temperature variation during the transesterification process, (c) the time of flight, and (d) the amplitude calculated from the echo signal.

Analysing the transesterification reaction of sunflower oil with ethanol. It is observed that glycerol deposition was not observed immediately after the end of the reaction. Furthermore once the stirring starts, temperature and amplitude decrease quickly, however the time of flight increase speedily and after approximately one minute decrease for one minute and then increase until the 5th minute, and after that, it presents a lower rate of decrease.

The temperature increases until approximately the 8th minute, and then it starts to decrease until the 30th minute, when the stirring is turned off. The amplitude increases for one minute, remains stable for two minutes, and after that, it presents decrease and after 5th minute tend to stabilize, when the stirring is turned off.

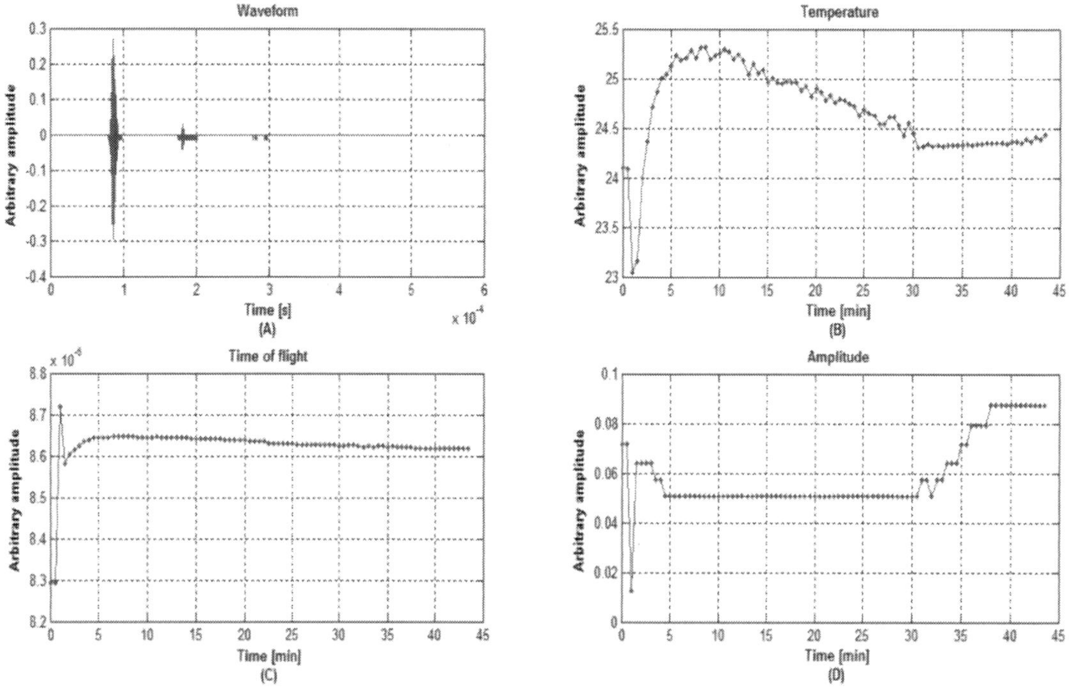

Figure 5. Monitoring of the transesterification reaction of sunflower oil with ethanol. One can observe (a) the ultrasonic waveform at the end of the experiment, (b) the temperature variation during the transesterification process, (c) the time of flight, and (d) the amplitude calculated from the echo signal.

4. Discussion

While assessing the methanol biodiesel reactions, in all cases, glycerol deposition was observed after turning off the stirring. Consequently, all ultrasonic waveforms presented the echo signal from the interface of glycerol and biodiesel. On the contrary, none of the ethanol biodiesel reactions presented glycerol deposition after turning off the stirring presented the same only sometime later when the sample was placed to rest in the funnel. According to literature, separation of glycerol is not spontaneous when ethanol is employed to obtain biodiesel owing to the higher water presence in its composition [1].

In this study, three parameters were studied during the transesterification reactions: the time of flight and amplitude, both calculated from the ultrasonic waveform, and temperature. With the graphics shows in this work, it note that the behaviour of all the parameters studied were sensitive to the type of alcohol and the type of feedstock used, which proves to be a selective method.

Observing the same feedstock(soybean oil) and different alcohols, note that the parameters have different behaviours, temperature and time of flight for example, when the reaction occurred via methanolic, had a tendency to increase up to a certain point and then decreased until the moment that the stirring is turned off. However, when the reaction occurred via ethanolic, these same parameters had only a tendency to increase.

Analysing the graphs of the same alcohol, such as ethanol, and different feedstock is also observed significant differences, thus showing the efficiency of the method developed.

5. Conclusion

With the proposed procedure, it was possible to identify different events during the transesterification process by using the proposed parameters. Through the method developed was also possible to identify differences between the feedstock and alcohols used, showing to be a selective method. The preliminary results suggest that the proposed method may be feasible for monitoring the reactions of biodiesel during its fabrication, in real time, and with relatively low-cost equipment.

Acknowledgement
The authors thank the CNPq and FAPERJfor the financial support.

References

[1] Lôbo, I. P., Ferreira, S.L.C., Cruz, R.S., 2009, Biodiesel: Quality parameters and analytical methods, *Quím. Nova*, v. 32, n^0. 6, 1596-1608 [in Portuguese].

[2] Corrêa, S. M., Arbilla, G.; 2006, Aromatic hydrocarbons emissions in diesel and biodiesel exhaust, *Atmos. Environ.*, 40, 6821-6826.

[3] Koc, A. B., 2009, Ultrasonic monitoring of glycerol settling during transesterification of soybean oil, *Bioresource Technology*, v.100, 19-24.

[4] Ji, J.; Wang, J.; Li, Y.; Yu, Y.; Xu, Z.; 2006, Preparation of biodiesel with the help of ultrasonic and hydrodynamic cavitation, *Ultrasonics*, 44, e411-e414.

[5] Badday, A. S.; Abdullah, A. Z.; Lee, K. T.; Khayoon, M. S.; 2012, Intensification of biodiesel production via ultrasonic-assisted process: A critical review on fundamentals and recent development, *Renewable and Sustainable*,16, pp. 4574-4587.

Journal of Physics: Conference Series **575** (2015) 012002 doi:10.1088/1742-6596/575/1/012002

Uncertainty assessment in measurement of myotube thickness in cells culture treated with and without therapeutic ultrasound

V M Abrunhosa[1,2], M L. Costa[2], C Mermelstein[2], A V Alvarenga[1], R P B Costa-Félix[1]

[1]Laboratory of Ultrasound, Directory of Scientific and Industrial Metrology (DIMCI), National Institute of Metrology, Quality and Technology (Inmetro), Av. Nossa Sra das Graças, 50 (Prédio 1), Duque de Caxias, RJ, Brazil, ZIP 25250-020

[2]Laboratory of Muscle Differentiation and Cytoskeleton, Biomedical Sciences Institute, Federal University of Rio de Janeiro (UFRJ), Cidade Universitária, Rio de Janeiro, RJ, Brazil, ZIP 21949-590

E-mail: [1,2]vmabrunhosa@inmetro.gov.br

Abstract. Effectiveness of an ultrasound treatment shall be assessed by experiments. A reliable cell culture protocol is available and spatial discrepancies could arise. To assure if the spatial difference are relevant or not, and how they should be dealt with, an uncertainty model for the treatment result is a metrological reliable solution. The present work reports a metrological approach to assess myotube thickness and to validate a primary cell culture of muscle after a therapeutic ultrasound treatment, comparing it with a control group. The results reinforced the importance of such approach and show an efficacy of treatment on myotube differentiation.

1. Introduction

Muscle injuries are common traumas in athletic and non-athletic population. For these lesions, some treatments are recommended. One of the most usual treatments is ultrasound wave (Speed, 2001). Ultrasound is defined as acoustic sound of a frequency above 20 kHz. To evidence the benefits of therapeutic ultrasound (TU) in the treatment of soft tissue injuries, experimental studies are necessary. Research about the development of myogenesis with muscle culture system is potentially capable of revealing the molecular events which occur as cell prepares for differentiation (Morris & Cole, 1972). The chick myoblast primary culture is a robust in vitro model of myogenesis. Skeletal myogenesis proceeds in these cell cultures through the following main sequential stages: myoblast proliferation, cell cycle withdrawal, myoblast alignment and fusion, and their subsequent differentiation into mature and striated multinucleated myotubes. Even under appropriated in vitro conditions, isolated embryonic muscle cells fuse and differentiation to form myotubes.

As these events describe above do not occur uniformly on petri dishes, it was decided to study the importance of analyzing the whole dish and evaluated the interference of TU in this type of cell culture. Uncertainties (type A) of thickness of myotubes were assessed to prove the demand of evaluated the whole dish.

2. Materials and Methods

2.1. Ultrasonic Pressure Field Mapping System
A water bath measuring 1700 mm x 1000 mm x 800 mm from the Laboratory of Ultrasound (National Institute of Metrology, Quality and Technology, Inmetro, Rio de Janeiro, Brasil) was used for mapping the ultrasound pressure field of the transducer. A needle hydrophone having an active element of 0.2 mm was applied in the mapping procedure (Precision Acoustics Ltd, Dorchester, Dorset, UK). To integrate all system components, and also to provide a user-friendly interface, a virtual software was developed in LabView platform (National Instruments Corporation, Austin, TX, USA). The software automatically performs the raster scan necessary to calculate the effective radiating area (A_{ER}) of the physiotherapeutic ultrasound transducer (Alvarenga & Costa-Felix, 2009).

2.2. Primary myogenic cell cultures
Primary cultures of myogenic cells were prepared from breast muscles of 11-day-old chick embryos based in Mermelstein et al., 2005. This study using chick embryos was approved by the Ethics Committee for Animal Care and Use in Scientific Research from the Federal University of Rio de Janeiro and received the approval number: DAHEICB004.

2.3. Treatment with ultrasound wave
An ultrasound apparatus (Sonopulse, Ibramed, Brazil) was used with one mode of operation (continuous), effective intensity (0.5 W/cm2) and duration of treatment (5 minutes). The frequency of the TU was 1 MHz. The ultrasound was transmitted through the bottom of the culture dish with a coupling gel between the ultrasound transducer and the dish (Figure 1). All 24-h cell cultures were exposed to the ultrasound wave only once. Control cultures were subjected to the same experimental procedures as the treated cultured, except that the ultrasound equipment was not turned on (sham ultrasound). All analysis were performed after 48-h of treatment (with cells grown for 72 hours in culture). The ultrasound apparatus was fully tested according to IEC 61689:2007 before its use in this study.

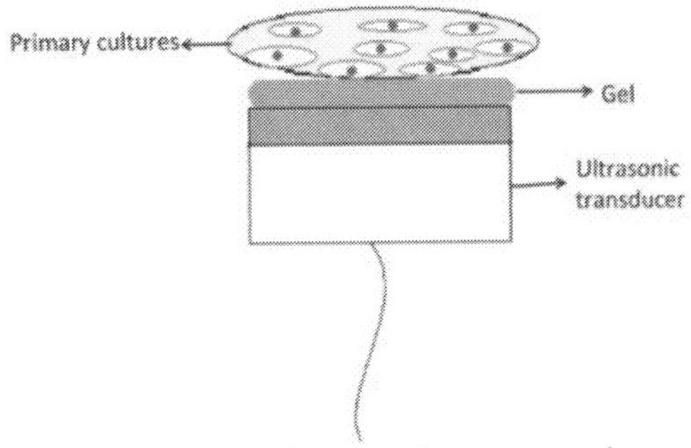

Figure 1 – Experimental setup showing how ultrasound wave transducer was applied to cultures cells.

2.4. Quantification of cultured images
Phase contrast microscopy images of live cultured cells were acquired with an Axiovert 100 microscope (Carl Zeiss, Germany). Images of 72 h-myogenic cultures were acquired from several equally apart microscopic fields, representative of whole dish for each culture condition. The coordinates in referring microscope are position 1 (136x45 mm), position 2 (136x52 mm), position 3 (136x59 mm), position 4 (143x45 mm), position 5 (143x52 mm), position 6 (143x59 mm), position 7 (150x45 mm), position 8 (150x52 mm), position 9 (150x59 mm) (Figure 2). The diameter of myotubes was measured using the public domain software ImageJ (http://rsb.info.nih.gov/ij/). The measurements were performed in at least three independent experiments.

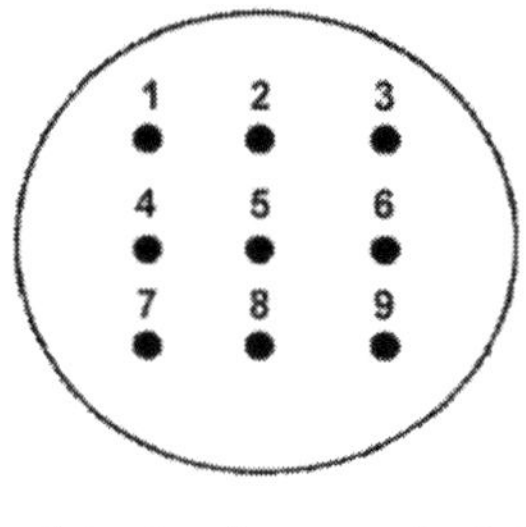

Figure 2 – Schematic representation of the 9 microscopic fields used for the acquisition of digital images of cultured cultures.

2.5. Determination of standard uncertainty of type A

Data were presented as the mean ± uncertainty of type A. The analyses were made with 2 different forms. The first, it was decided to calculate the myotube thickness in each position (mean ± uncertainty of type A) and the second we evaluated the thickness of myotube of the whole dish (mean ± uncertainty of type A).

3. Results

The myotube thicknesses measured over the 9 positions of the dish are presented in Figure 3, while the results considering the whole dish are presented in Figure 4.

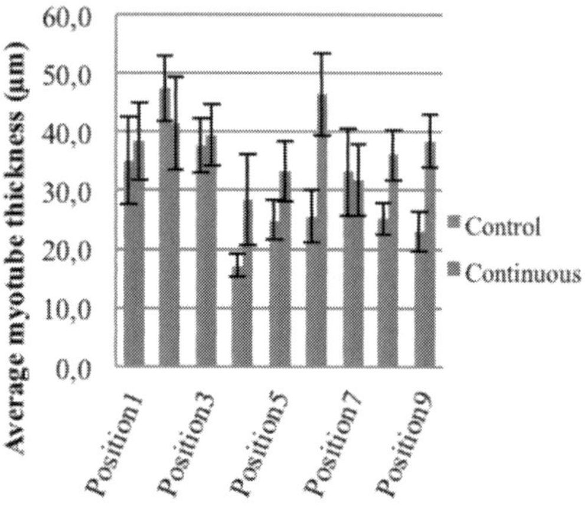

Figure 3 – Myotube thickness: mean value (depicted by the high of the bars) for each of the 9 positions in the dish used to collect the data, and the Type A uncertainty (bar errors).

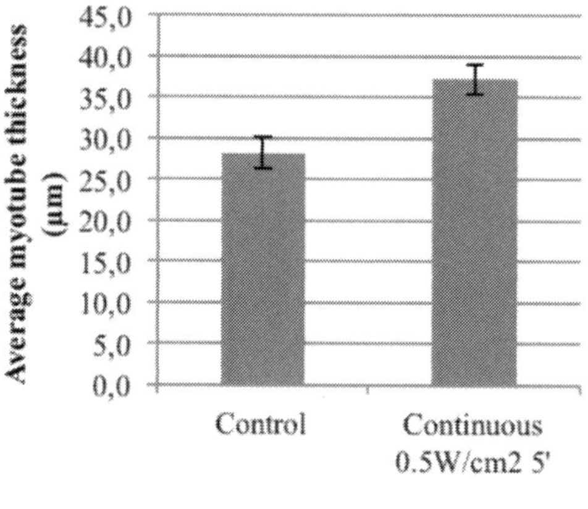

Figure 4 – Overall average myotube thickness for both the control and the treated cell cultures, represented with each Type A uncertainty.

Acknowledgment

The authors thank CNPq and Faperj for the financial support.

References

[1] Alvarenga A.V. e Costa-Felix R. 2009 Uncertainty assessment of effective radiation area and beam non-uniformity ratio of ultrasound transducers determined according to IEC 61689:2007. *Metrologia*.; **46** 367-374.

Journal of Physics: Conference Series **575** (2015) 012002 doi:10.1088/1742-6596/575/1/012002

[2] Mermelstein C.S., Portilho D.M., Medeiros R.B., Matos A.R., Torteloti G.G., Lamas M.E., Vieyra A.R., Costa M.L. 2005 Cholesterol depletion by methyl-b-cyclodextrin enhances myoblast fusion and induces the formation of myotubes with disorganized nuclei. *Cell Tissue Res.*, **319(2)**: 289-297.
[3] Speed C.A. 2001 Therapeutic ultrasound in soft tissue lesions. *Rheumatology.* **40**: 1331-1336.

Primary calibration system for vibration transducers from 0.4 Hz to 160 Hz

C D Ferreira, G P Ripper, R S Dias, D B Teixeira

National Institute of Metrology, Quality and Technology – Inmetro
Division of acoustics and vibrations metrology – Diavi
Vibration Laboratory – Lavib
Av. N. Sra. das Graças 50, Xerém – Duque de Caxias – RJ – CEP 25250-020 – Brazil

lavib@inmetro.gov.br

Abstract. This paper presents a system developed at the Vibration Laboratory of Inmetro, which is used for primary calibration of vibration transducers by the fringe counting method. This system includes a vibration exciter, a Michelson interferometer, a data acquisition board, a band-pass filter, a universal counter and a software for measurement automation. It allows the laboratory to perform calibrations in accordance with the international standard ISO 16063-11 in the frequency range from 0.4 Hz to 160 Hz. Some experimental results are presented herein.

1. Introduction

Vibration measurements at low frequencies are used in several applications, as for example: environmental monitoring, seismic measurement, human-body vibration assessment, mechanical tests, transportation of goods, large structures vibration monitoring, inertial navigation and so on. Since the demand for low-frequency vibration calibration services has grown up during the last years, the Vibration Laboratory (Lavib) of Inmetro has started the development of specific systems to attend this field.

A comparison calibration system requires a calibrated reference standard to quantify the measurand with high accuracy. In comparison accelerometer calibrations the sensitivity of the device under test (DUT) is determined by direct comparison of the voltage outputs given by the reference and DUT measuring chains when both transducers are submitted to a given motion input [1].

Reference vibration standards shall preferably be calibrated by a primary method capable of providing metrological traceability to SI units with low measurement uncertainties. For the purpose of mutual recognition and participation in interlaboratory comparison activities, it is important that the uncertainties are comparable with the ones claimed by the leading National Metrology Institutes (NMI) in the field.

In order to achieve this goal, a new primary interferometric system based on a Michelson interferometer was developed at Lavib for performing automated calibrations of reference vibration transducers at low frequencies.

2. Calibration Method

The primary calibration method used in this new system is in compliance with the requirements of the Method 1, which is described in the international standard ISO 16063-11 [2] and known as the fringe counting method. This method is based on primary measurement by laser interferometry of the displacement amplitude of a sinusoidal harmonic motion, generated by a vibration exciter and applied to the transducer under calibration.

The sensitivity of the transducer is determined as a function of excitation frequency by calculating the ratio between the electrical output quantity provided by the transducer under calibration and the corresponding motion input quantity.

A two-arm Michelson interferometer is used to measure the displacement amplitude of the vibration by the well known fringe counting method. The ratio R_f is determined by measuring the number of optical interference fringes per period of sinusoidal vibration motion. In practical terms this is done by measuring the ratio of two frequencies: the frequency of the optical interference signal captured by a Silicon photodetector and the frequency of motion given by a signal generator, which drives the vibration exciter.

The magnitude of the displacement is directly proportional to the ratio R_f. Assuming a sinusoidal motion of the moving table with a frequency f, then the magnitude of corresponding velocity and acceleration can be calculated using the angular frequency $\omega = 2\pi f$.

Measuring the voltage output V_{obj} furnished by the device under test (DUT), it is possible to calculate its sensitivity S_{obj} at each frequency by

$$S_{obj}(f) = \frac{8\sqrt{2}}{(2\pi f)^2 \lambda}\left(\frac{V_{obj}(f)}{R_f}\right) \tag{1}$$

where λ is the wavelength of the light emitted by a He-Ne laser.

3. Calibration System

The basic block diagram of the primary low-frequency accelerometer calibration system is presented in figure 1.

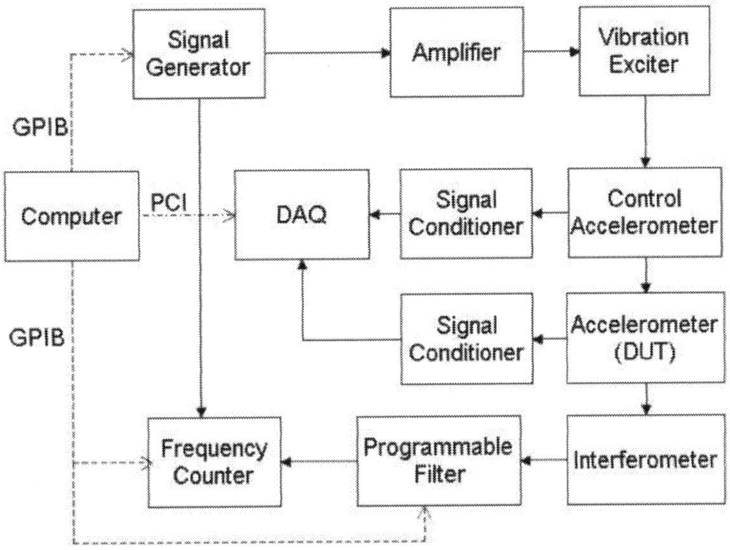

Figure 1. Block diagram of the calibration system.

In order to obtain a transducer output with a good signal-to-noise ratio at low frequencies it is important to use a vibration exciter capable of generating a motion with large displacement amplitudes. This occurs because the acceleration level a_{pk} is proportional to the displacement amplitude d_{pk} multiplied by the angular frequency squared ($a_{pk} = d_{pk}\,\omega^2$). Considering that most commonly used transducers respond to acceleration, then larger displacements are needed for calibrations at lower frequencies for a given acceleration level.

In the system depicted in figure 1, the mechanical vibration is generated by an electrodynamic exciter APS Dynamics model 129, which has a 25 x 25 cm moving table guided by a linear air bearing. This exciter can generate up to 75 mm displacement amplitude motion and offers a payload capacity up to 23 kg. The exciter base is rigidly mounted on the top of a 1200 kg concrete inertial mass, which lies on rubber sheets. The air bearing linear guide model was selected because it produces low-friction and low noise levels, also helping to generate a linear sinusoidal motion with reduced effect of transverse and rocking motions.

A function generator 33120A and a power amplifier APS model 124 provide the sinusoidal signals necessary for driving the exciter with at the desired amplitudes and frequencies. The default calibration frequencies are the nominal 1/3-octave frequencies in the range from 0.4 Hz to 160 Hz.

An acceleration measuring set, which includes a B&K 4370 piezoelectric accelerometer and a B&K 2626 charge amplifier, is used to control the acceleration levels of the moving table.

The measurement of the displacement of the table is made by a Michelson interferometer with a non-stabilized 15 mW laser. Retroreflectors are used in both arms of the interferometer in order to facilitate the optical alignment along the large displacements of the moving table. The interferometer is mounted on an optical breadboard, which lies on helical steel springs and an independent seismic block. The use of passive isolation and two independent massive blocks are important to avoid any mechanical short-circuit and isolate the interferometer setup from any undesired vibrations.

A programmable filter Krohn-Hite is used for pre-conditioning of the frequency modulated signal obtained by the photodetector. The filtered signal is then directed to a universal counter configured in frequency ratio mode for measuring of the ratio R_f.

An analog to digital conversion board is applied for measurement of the electrical output signals provided by the DUT and the control accelerometer. A 12 bit 4-channel, simultaneous sampling data acquisition board (DAQ) National Instruments model PCI-6115 is used. This DAQ can be configured with a maximum sampling rate of 10 MS/s per channel.

A computational program was developed in LabVIEW environment for automation of the calibration process. This software configures and controls all the equipments via an IEEE 488 interface, also controlling the acceleration level provided by the vibration exciter. Beyond this, it also configures the DAQ for the measurement of the electrical output voltage signals of the DUT and control accelerometer chains. The program front panel includes graphical indicators of the measured signal, both in time and frequency domain, and presents the calibrated sensitivity results graphically in real time. The measured data is saved in an ASCII file for further post-processing and analysis.

The voltage values of the DUT are measured at the fundamental frequency of the vibration excitation by a Fast Fourier Transform (FFT) routine. The Total Harmonic Distortion is also measured for assessment of the quality of the movement during the calibration.

Figure 2 shows the moving table of the exciter (A) with a mounting adaptor (C) used to hold both the DUT (D) and the moving retroreflector (B) of the interferometer. In addition, it shows the control accelerometer (E) mounted on a bracket.

In figure 3, the whole interferometric setup is presented, including the adaptor mounted on the exciter moving table (B) and the interferometer mounted on the optical breadboard. The photodetector and the fixed retroreflector are identified in this figure as B and A, respectively. The same mechanical adaptor presented here is also used for comparison calibrations of accelerometers. In this case, the retroreflector is just replaced by a standard reference accelerometer.

In order to minimize the effects of any transverse vibrations acting on the transducer under test, two measurements are usually carried out in sequence. Initially, the transducer is first calibrated in a certain position. In sequence, it is rotated 180° degrees relatively to its own axis and is calibrated again. The arithmetic mean of the sensitivities measured at these two positions is reported as the final calibration result. This procedure minimizes the effect of any coupling between the transverse sensitivity of the DUT and the transverse accelerations generated by the exciter or even to the earth gravity field.

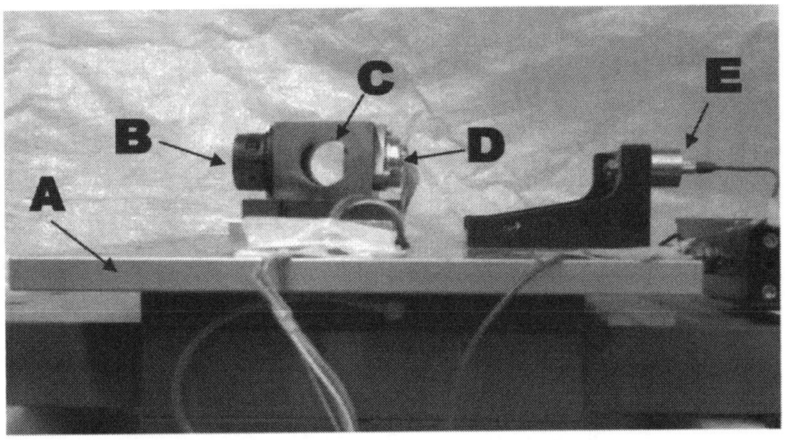

Figure 2. Moving table of the exciter (A), retroreflector (B), mounting adaptor (C), accelerometer under test (D) and control accelerometer (E).

Figure 3. Michelson interferometer view: Retroreflectors (A and B) and photodetector (C).

4. Experimental Results

This section presents some results obtained by the primary calibration system. Figure 4 presents the results of sixteen sensitivity calibrations of a servo-accelerometer Allied Signal QA3000, used as a reference standard in the comparison calibration system. This kind of transducer is applied for inertial navigation and has a high temporal stability. Therefore, this test basically evaluates the repeatability level achievable with the calibration system developed.

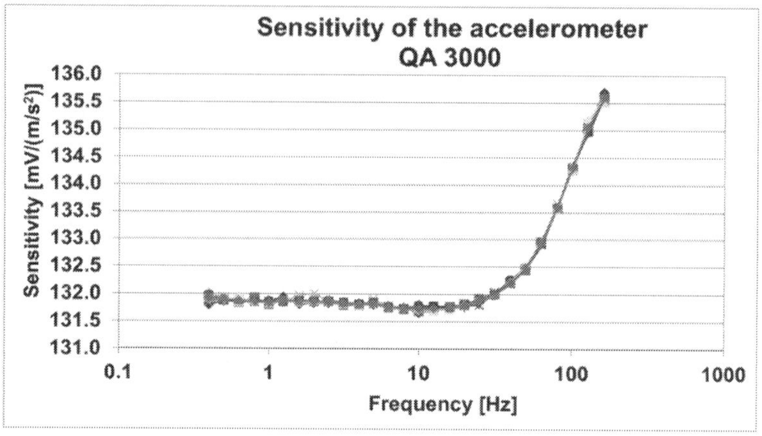

Figure 4. Results of sensitivity obtained for a servo-accelerometer QA-3000.

Figure 5 presents the deviation (%) of the same measurements shown in figure 4 relative to the mean sensitivity calculated for the servo-accelerometer. It can be observed that most of the individual results lie within 0.1 % of the mean sensitivity value.

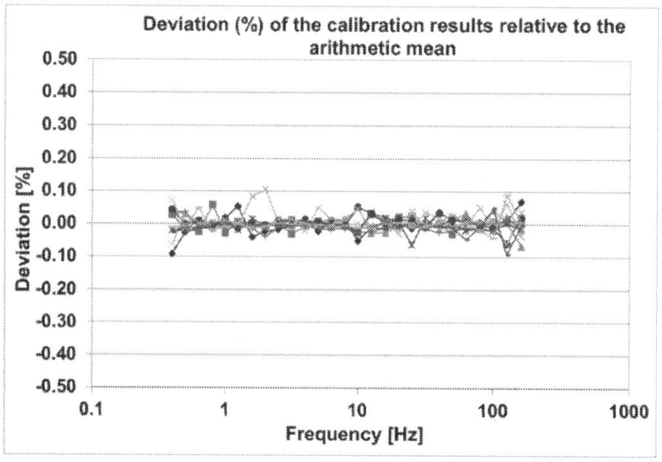

Figure 5. Deviation of the sensitivity values from the arithmetic mean (in %) for sixteen calibrations of the servo-accelerometer QA-3000.

5. Measurement Uncertainty

The expanded measurement uncertainty (U) estimated for the results of magnitude of the sensitivity obtained by the primary calibration system for vibration transducers at low frequencies was 0.35%, under normal calibration conditions and for the whole range of frequencies. This uncertainty was calculated in accordance with the Guide to the Expression of Uncertainty in Measurement – GUM [3] and the Annex A of the international standard ISO 16063-11[2]. This value is given by the multiplication of the combined standard uncertainty with the coverage factor $k = 2$, which defines an interval having a level of confidence of approximately 95%.

6. Interlaboratory Comparison

Figure 6 shows the results of the interlaboratory bilateral comparison AFRIMETS.AUV.V-S2, which was developed between Inmetro and the National Metrology Institute of South Africa (NMISA) [4].

The transducer used for this comparison was an IEPE type accelerometer, PCB 301M26, in the frequency range from 0.4 Hz to 50 Hz. The expanded uncertainty reported by both institutes was 0.3%. This reduced value is possible due the especial conditions dedicated to the comparison, including more measurements than the usual for a normal calibration.

The good agreement obtained between the results demonstrates the capability of Inmetro's system and provides the technical support to the typical uncertainty of 0.35% reported to general customers.

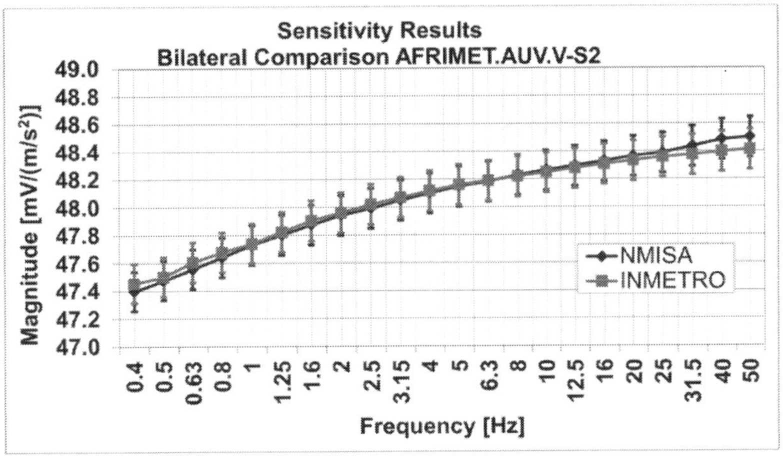

Figure 6. Results of the bilateral interlaboratory comparison carried out between Inmetro and NMISA.

7. Conclusion

The system presented in this article allowed the Vibration Laboratory of Inmetro to extend and improve its capability to realize primary laser calibrations of vibration transducers at low frequencies.

This system is currently able to execute calibrations of accelerometers and vibration measuring systems in the range from 0.4 Hz to 160 Hz with a 0.35% expanded uncertainty of measurement, considering a level of confidence of approximately 95%.

The calibration process is fully automated, minimizing the influence of the operator. Some results were presented, including the results obtained in an international bilateral interlaboratory comparison, which validates and supports the claimed uncertainties.

Therefore, the system implemented at Lavib will enable Inmetro to provide the needed traceability to the International System of Units to several economic sectors in Brazil.

Lavib will continue to develop researches at low frequencies, and intends to implement in the near future a quadrature interferometric system. This further improvement will provide the additional capability to measure the phase shift of complex sensitivity of vibration transducers.

References

[1] TEIXEIRA, D B *et al* 2009 *Sistema para calibração comparativa de transdutores em baixa frequência* (Salvador, Bahia, Brazil: V CONGRESSO BRASILEIRO DE METROLOGIA).

[2] International Organization for Standardization 1999 *ISO 16063-11, Methods for the calibration of vibration and shock transducers – Part 11: Primary vibration calibration by laser interferometry* (Genéve, Switzerland).

[3] *Evaluation of Measurement – Guide to the expression of uncertainty in measurement (GUM)*, (JCGM 100:2008, BIPM, IEC, IFCC, ILAC, ISO, IUPAC, IUPAP and OIML).

[4] Veldman C S, Ripper G P 2012 *Final report on supplementary comparison AFRIMETS. AUV.V-S2* Tech. Suppl. 09001 (*Metrologia*) p 49.

Hydrophone's sensitivity calibration based on its complex transfer function

E G Oliveira[1], J C Machado[2], R P B Costa-Felix[1]
[1]Laboratory of Ultrasound (Labus), Directory of Scientific and Industrial Metrology (Dimci), National Institute of Metrology, Quality, and Technology (Inmetro)
[2] Coordination of the Pos-Graduation Programs on Engineering (COPPE), Federal University of Rio de Janeiro (UFRJ)

E-mail: egobira@inmetro.gov.br

Abstract. A novel approach to calibrate ultrasonic hydrofone's sensitivity magnitude has been developed. The primary calibration method known as self-reciprocity with an auxiliary transducer was improved with a measurement technique based on the assessment of complex transfer function of the system. The protocol was experimented from 1.0 MHz to 7.0 MHz, and thereafter compared with a previous calibration realized at the National Physical Laboratory (NPL, Teddington, UK). Within the frequency range of interest, the novel calibration results were statistically identical to the previous calibration data. The 95% confidence level uncertainty varied from 6.6% to 7.5%.

1. Introduction

For ultrasound hydrophones, calibration means to determine their sensitivity, that is, its voltage response (output) to a dynamic pressure that reaches its active element. The ratio between the amplitude of the voltage measured at the output of the hydrophone and the amplitude of the acoustic pressure wave incident on the active element of the hydrophone defines the reception sensitivity (V/Pa). Self-reciprocity method is a traditional, long time validated and well known on ultrasonics [1]. Reciprocity refers to a reciprocal transducer that acts as a projector and receptor. The reciprocity coefficient (J) is defined as the ratio of the free-field voltage sensitivity of the transducer, M, to its transmiting response to current, S, [1], [2]. Typically, measurements are carried out at discrete frequencies. A relatively simple measurement setup is used, as far as reciprocity procedure demands electrical measurements and mechanical quantities in the reciprocal coefficient. The self-reciprocity method requires a measurement of an open circuit voltage appearing at the terminals of the transducer when it is operating as a receiver in a pulse-echo mode.

A novel technique to calibrate the magnitude of hydrophone´s sensitivity is introduced, using the transducer transfer function. The method employed in the present work is based on auxiliary transducer acting in a pulse-echo setup and on electrical measurements such as, voltage and current.

An advantage with the proposed system and method, it is possible to calibrate hydrophones from common equipment in a laboratory of ultrasound. The obtained hydrophone sensitivity was compared with their sensitivity as calibrated at NPL (UK's National Physical Laboratory). To establish a metrological base of comparison, an uncertainty model was developed, and the results are the kernel of the present document.

2. Theoretical approach

2.1. Self-reciprocity

The theoretical principles of self-reciprocity calibration are described in [2] and [3], and standardized by [1]. As it is an absolute calibration method, the pressure on the transducer face is obtained directly from the electrical voltage and current, and the current obtained indirectly by measuring the impedance of the probe, and mechanical quantities present in the parameter plane wave reciprocity $J=2A/\rho c$, where A is the effective area of the auxiliary transducer, ρ is the density of the propagation medium and c is the propagation velocity propagation in medium. The echo is generated by a reflector (flat face and the reflection coefficient calibrated in advance) placed in a specific field position, as shown in Figure 1. Knowing the transducer transfer function, as a function of frequency ω, it is possible to determine the pressure field at the point where the target reflector was positioned.

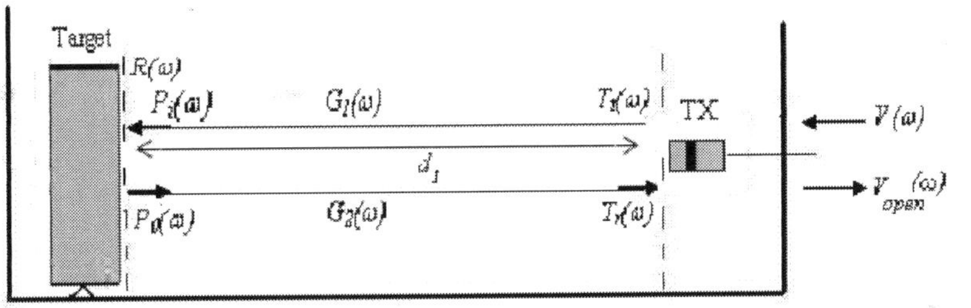

Figure 1. Self-reciprocity pulse-echo scheme.

2.2. Determination of the transfer function

Traditional transfer function assessment is based in the complex frequency spectrum analyses, and it has been published in the literature [4, 5]. The novel approach used in the present work deals with the transfer function in the time domain, instead, as presented by [6]. Despite the technique uses the waveforms in the time domain, as they are used single frequency tone burst as excitation, it is straightforward to compute all parameters as function of ω, i.e, the angular frequency.

2.3. Magnitude of the sensitivity

Joining both methods, i.e., self-reciprocity and determination of the transfer function, the hydrophone sensitivity (M) can be expressed as function of ω by:

$$M_h(\omega)=\frac{\mathrm{V_h}(\omega)\sqrt{2A\cdot R(\omega)Z(\omega)}\sqrt{G_1(\omega)}}{\sqrt{V_{open}{}^{*}(\omega)\frac{I_k(\omega)}{I_0(\omega)}V(\omega)\rho_0 c_0\cdot G_2(\omega)}} \tag{1}$$

where $V_h(\omega)$ is the output signal from the hydrophone placed at the same position that the transducer previously calibrated by self-reciprocity; $G_1(\omega)$ and $G_2(\omega)$ and are the correction due diffraction in either way from the transducer to the target and vice-verse, respectively; $V_{open}{}^{*}(\omega)$ is the apparent open circuit voltage measured in the reception activity of the transducer when loaded by a finite impedance; $I_k(\omega)/I_0(\omega)$ is a correction factor to compute correctly the open circuit voltage, being $I_0(\omega)$ the measured current and $I_k(\omega)$ is computed to correct the measurement to the open circuit situation;

$V(\omega)$ is the driven voltage to the transducer in the output mode; ρ_0 and c_0 are the density and the speed of sound in the medium, respectively; A is the effective radiation area of the transducer; $R(\omega)$ is the reflection coefficient of the reflecting target; and $Z(\omega)$ is the complex impedance of the transducer. The complete derivation of (1) is omitted for simplification in this text.

2.4. Measurement uncertainty

The approach depicted in [7] was applied in (1), deriving the Type B uncertainties for the hydrophone sensitivity. Type A source of uncertainty was calculated as the standard deviation of the mean after a repeatability experiment.

3. Material and method

A schematic drawing of the experimental setup of the system to perform calibration of a hydrophone is presented in Figures 2.

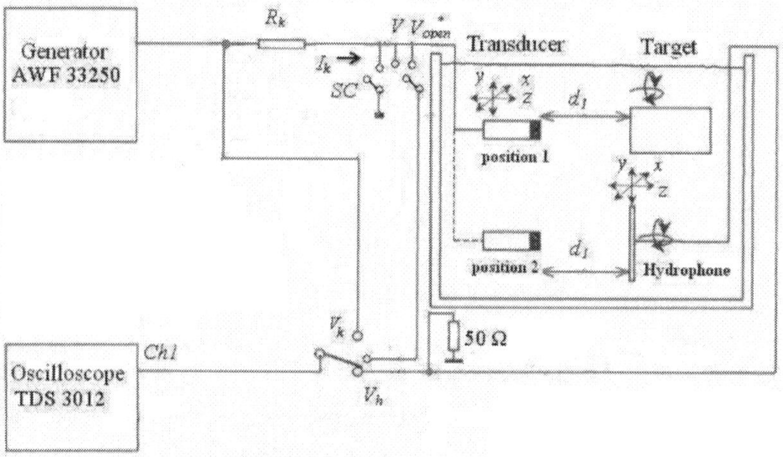

Figure 2. Experimental setup

The following parameter are recorded in the during a hydrophone calibration (see (1)): frequency (MHz), $V(\omega)$, $V_{open}^*(\omega)$, $V_h(\omega)$, water temperature (°C), pulse echo distance d_1, $R(\omega)$, $Z(\omega)$, $R_k(\omega)$, $G_1(\omega)$. The correction factor $G_2(\omega)$ should be applied to the pressure reaching the surface of the sensor element hydrophone as a function of normalized distance and the ratio of the effective radius of the probe emission and hydrophone. The short-circuit voltage $V_k(\omega)$ and short-circuit resistance $R_k(\omega)$ both used to assess $I_k(\omega)$. The transducer and the target reflector should be fully submerged in water, at least 100 mm below the water level in the tank.The transducer should be aligned with the target reflector. The hydrophone under test used must be aligned with the transducer. The water temperature should be measured with the measuring end of the probe located at least 100 mm below the water level in the tank. The central region of the face of the hydrophone's active element must be located approximately 180 mm from the geometric center of the face of the target reflector toward the horizontal axis (x-axis). The amplitude measured generator to 20 V peak to peak with a "burst" of 20 cycles. The transducer was moved in the Z axis to position d_1 away from the target reflector. The distance d_1 is determined using the formula $Z_n = d^2/4\lambda$, where d is the diameter of the transducer, and λ is given by $\lambda = c/f$, where c is the propagation speed of ultrasound in water and f is operating frequency of the transducer. The two way pulse echo distance $2d_1$ should be within the following range: $1.5Z_n \leq 2d_l \leq 3.0Z_n$ [1]. Type A uncertainty is assessed as the precision under repeatability conditions after 4 repetitions. The high frequency signal was generated using an arbitrary wave generator model AWF 33250 (Agilent Technologies, USA), and voltage measurements are undertaken with an oscilloscope model TDS 3012 (Tektronix Inc., USA). R_k (in Ω) and the 50 Ω shunt are in series and parallel to the circuit, respectively. The reflecting target is a stainless steel cylinder (Ø = 80 mm diam × 80 mm long).

4. Results

Table 1 disclose the results, comprising the calibrated values using the present method $M_h(\omega)$ and its expanded uncertainty (with 95% confidence level), the calibration values from the NPL with their uncertainty, and the normalized error (E_{norm}).

Table 1. Results of magnitude sensitivity calibration of membrane hydrophone (Precision Acoustics) calibrated by self-reciprocity and calibrated by the NPL.

Inmetro			National Physical Laboratory - NPL		
$F\,[MHz]$	$M_h\,[nV/Pa]$	$U_{M_h}\,[nV/Pa]$	$M_h\,[nV/Pa]$	$U_{M_h}\,[nV/Pa]$	E_{norm}
1	66,7	4,9	71	4,3	0,66
2	67,7	4,6	74	4,4	0,99
3	70,5	4,7	75	4,5	0,69
5	78,4	5,2	76	4,6	0,35
7	80,4	5,3	76	4,6	0,63

5. Conclusion

Table 1 summarize the results of this work. For all frequencies tested, the difference between the calibration from NPL and the results with the proposed method are statistically equal, using both uncertainties as parameter. It can be affirmed as the differences in all cases are lower than the addition of the respective uncertainties. From that point of view, the present method could be considered validated. Further, the uncertainties obtained are numerically not much greater than those provided by NPL. Some improvements particularly in the voltage measurements should be applied, the uncertainty would be even lower.

Acknowledgments
The authors would like to acknowledge the financial support from the CNPq and Faperj.

6. References

[1] IEC International Electrotechnical Commission 2013 *62127-2:2013 Ultrasonics – Hydrophones – Part 2: Calibrations for Ultrasonics Fields up to 40 MHz*, ed IEC Central Office (Geneva)
[2] Foldy and Primakoff H 1945 *The Journal of the Acoustical Society of America* **17** 109-120
[3] Carstensen E L 1947 *The Journal of the Acoustical Society of America* **19(4)** 961-965
[4] Ludwig G and Brendel K 1988 *Transactions on Ultrasonics Ferroelectrics Frequency Control* **35(2)** 168-174

[5] Raduslescu E G, Lewin P A, Wojcik J and et al 2003 *Ultrasonics* **41** 247-254
[6] Neer P L M J V, Vos H J and Jong N D 2011 *Ultrasonics* **51** 1-6
[7] BIPM Bureau International des Poids et Measures 2008 *100:2008 Evaluation of Measurement – Guide to the Expression of Uncertainty in Measurement, (First Edition September 2008)*, available in www.bipm.org accessed in February 2014.

Effects of the concentration of emulsion of oil-in-water on the propagation velocity and attenuation

L S F Silva[1], D S Bibiano[1], M K K Figueiredo[1], R P B Costa-Félix[1]
[1]Laboratory of Ultrasound (Labus), Directory of Scientific and Industrial Metrology (Dimci), National Institute of Metrology, Quality and Technology (Inmetro), Av. Nossa Sra. das Graças 50, Duque de Caxias, RJ, Brazil, ZIP 25250-020

E-mail: lsegne@inmetro.gov.br

Abstract. Soybean oil is an important feedstock for production of biodiesel that generates about 20 % of oily effluents. This paper studied the effect of concentration of soybean oil-in-water emulsions, in the range from 6 000 to 14 000 ppm, on the propagation velocity and ultrasonic attenuation. The Emission-Reception method has shown that the propagation velocity depends linearly on the concentration. The behavior of attenuation is similar to the velocity. Thus, both parameters can be used to measure oils and greases content in water.

1. Introduction

Large wastes produced mainly by the branches of energy and biofuel industries are water polluted by high concentrations of oil and grease. Extractions processes, improvement and oil recovery of several sources lead to concern that oil wastes must be treated and measured, before they were disposed on nature.

The National Council for the Environment (CONAMA) delimits quantities of oil-in-water and greases wastes, in order to reduce potential impacts on the natural receiving bodies. According to the CONAMA Resolution nr. 430/2011, oil and grease content in water, daily disposed, have to be limited to 50 ppm for vegetable oils and 20 ppm for oils of mineral sources.[1]

Measurement of oil and greases in water are taken into account like parameters to release wastes of water on natural receiving bodies. It is constituted by hydrocarbons dissolved and dispersed in water and others compounds like phenols and carboxylic acids [2].

There are many techniques used to oil and greases measurement, among them, analyzers of total organic carbon (TOC), a destructive technique and infrared spectroscopy, a nondestructive technique. The infrared technique is widely used. Nevertheless, it requires extraction by tetrachloroethylene (TCE) solvent on a large volume.TCE is an expensive solvent used in sample preparation for infrared analysis too.Losing on the sampling process and high cost on the solvent are disadvantages thereof [3].

Generally, oil contains potential surfactants that act on the thermodynamic balance and affect phase behavior on a system. In addition, oil phase could lay itself on water like a film. Otherwise, it could be dissolved in water medium and dispersed forming an emulsion. The figure 1 illustrates phase behavior.

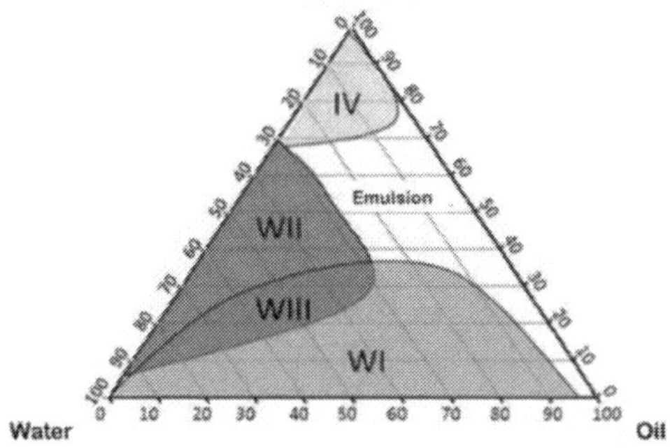

Figure 1. Representation of pseudo ternary phase diagram. WI – emulsion balanced with oil phase, WII – emulsion balanced with water phase, WIII – emulsion balanced with oil and water phases, IV – microemulsion.

Due to multiple equilibrium, a study of phase behavior must be taken into account in order to delimit one region, where ultrasound parameters must be acquired and assessed.[4]

Ultrasound comprises acoustic waves with frequencies above 20 kHz, it does mean, over the upper limit of the range of human audition (20 Hz to 20 kHz). These waves are formed due the piezoelectric properties of materials, which exhibit response to electric potential applied, with elastic deformations of its shapes and vice versa. [5,6]

Piezoelectric properties are fundamentals to ultrasonic transducer because of its use like vibrant source. Once ultrasonic transducer is in contact with a liquid, its vibration is able to generate areas with pressures different at all around. The particles (molecules, ions or atoms), in the structural arrangement of the medium, go through processes of compressing due to high pressure areas followed by relaxation on low pressure areas in alternating way. That is the physical phenomenon by which the propagation of ultrasonic waves takes place. It is illustrated in figure 2. [7]

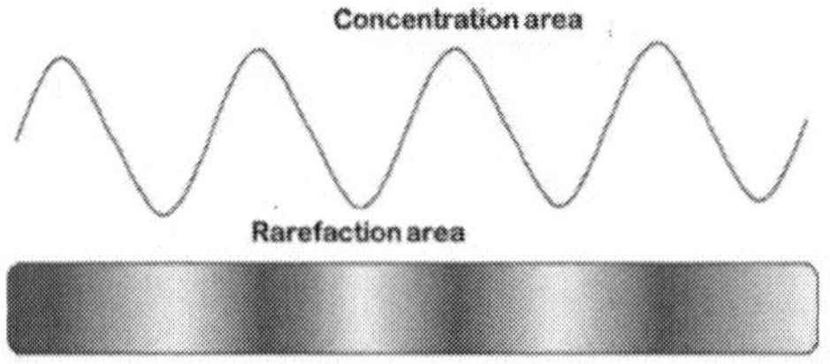

Figure 2. Propagation of ultrasonic waves.

Ultrasonic parameters allow us to make assessment through techniques which have fundamentals on scattering, propagation velocity, vibration potential and attenuation in order to characterize dispersion or emulsions systems. Ultrasonic velocity depends on the hardness of the material and the specific weight thereof. Attenuation is a response to loss of energy that occurs to a wave travelling along a medium.[8]

2. Materials and method

This paper describes the method named Emission-Reception. In the Emission-Reception method, two ultrasonic transducers which have nominal frequency of 15 MHz, 12.5 mm diameter (Panametric-NDT-Olympus Co. Japan) were accurately aligned on vertical axis, one on top other on bottom of the sample, like figure 3 illustrates.

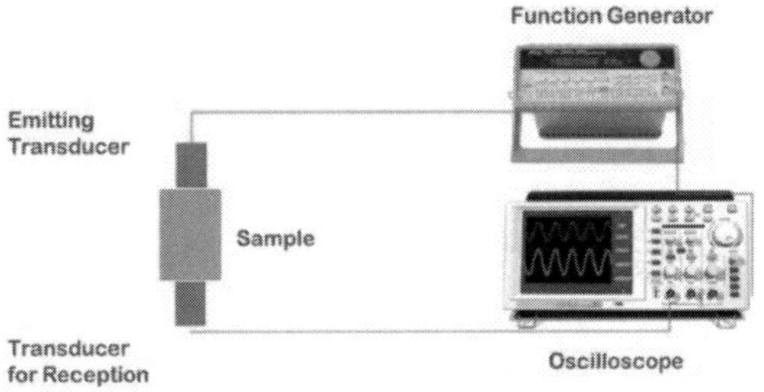

Figure 3. Emission-Reception method Set-up.

The sample concentrations were prepared within emulsion region and, in sequence, diluted 25% one after another. Attenuation parameters and the time of fly in a burst were acquiredthrough manual mode. Measurement had been made for deionized water and it was taken like reference. The measurement was made on 5 replica for each concentration level. Velocity and attenuation on samples were obtained by comparing to those of the water in repeatability condition of measurement.

3. Results and discussion

3.1 Phases diagram

Mixtures of Sodium Lauryl Sulfate (SLS) 0.5 % at aqueous medium and soybean oil at organic phase were carried out at 10 levels of proportion. Aliquots of Sodium Chloride (NaCl) 0.5 % in aqueous solutions were used like titrant for each level of mixtures. It resulted on the phases equilibrium demonstrated in figure 4.

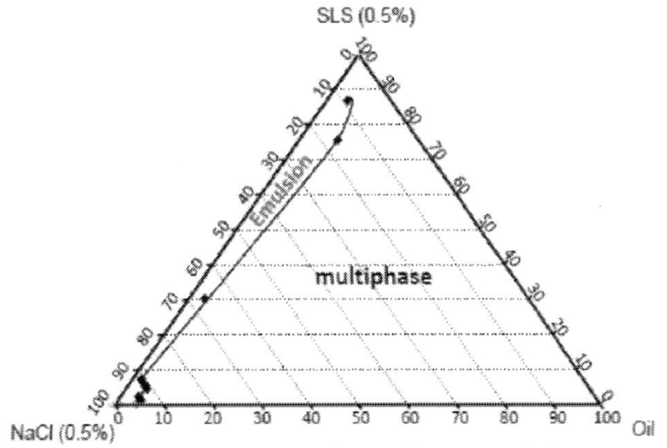

Figure 4. Pseudo ternary phase diagram of the system saline water (NaCl), surfactant (SLS) and soybean oil.

Despite the emulsion region seems to be thin, it's homogeneous and stable enough to allow us to perform the experiments.

3.2 Propagation velocity of ultrasonic waves through emulsion

Ultrasonic velocity was obtained by the measurement of five replicas of the mixture of oil in water for each level of concentration and comparing with reference liquid. The figures 5 and 6 show the measurement precision.

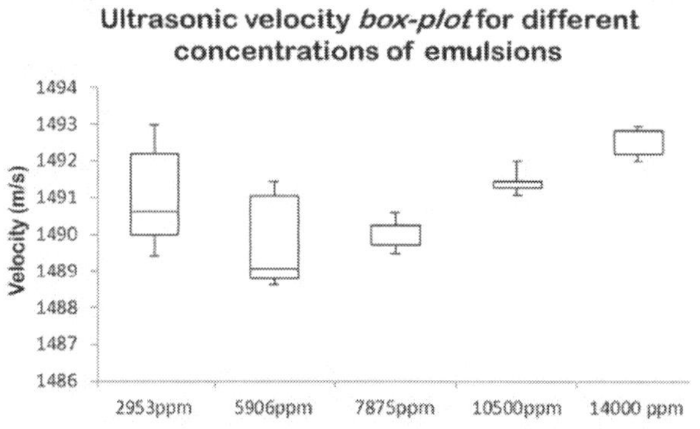

Figure 5. Ultrasonic velocity *Box-plot* of 5 levels of concentration of oil-in-water emulsions.

According to the figure 6, the samples with concentrations 2 953 ppm and 5 906 ppm have high repeatability estimative values (4,17 e 3,71, respectively).

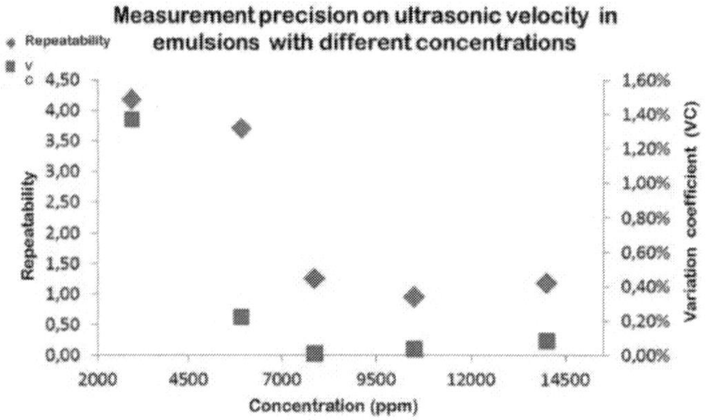

Figure 6. Measurement precision on ultrasonic velocity.

It was established that the criteria of acceptance of repeatability estimative values had to be less than 1.5. Likewise, the variation coefficient, VC values, had to be less than 0.15 %. Therefore, it was decided to discard measurements of 2 953 ppm e 5 906 ppm of the linear regression, shown in figure 7.

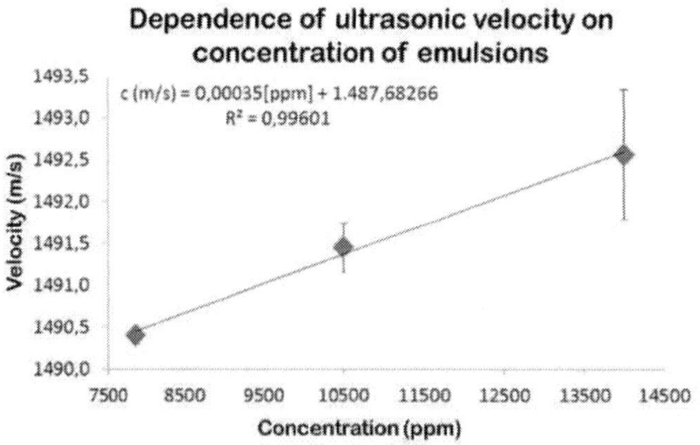

Figure 7. Effect of oil-in-water emulsions on ultrasonic velocity

3.3 Oil-in-water emulsions ultrasonic attenuation

Attenuation had been assessed and shown low measurement precision, observed in figure8. It may be attributed to undesired dispersion on measures, due to manual mode of acquisition or measurement bias.

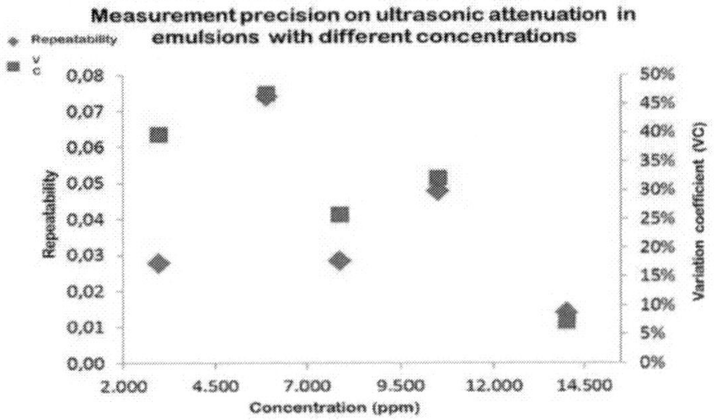

Figure 8. Measurement precision on ultrasonic attenuation

Nevertheless, once the median value can be taken as most likely value which is unaffected by extremes, we may assume that a linear curve could match that dispersion within a certain reasonability. It's illustrated in both figures 9 and 10.

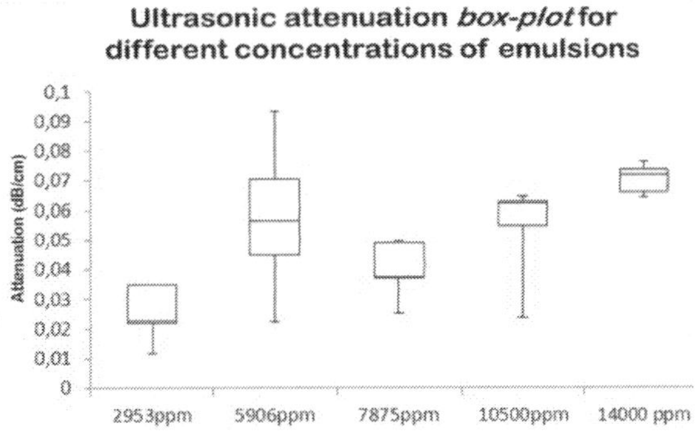

Figure 9. Ultrasonic attenuation *Box-plot* of 5 levels of concentration of oil-in-water emulsions

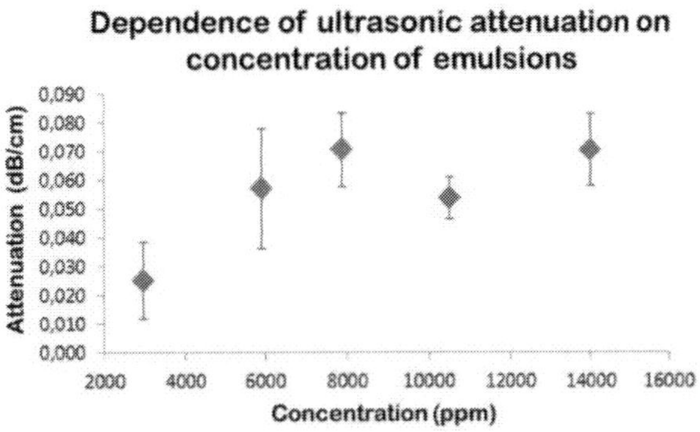

Figure 10. Effect of oil-in-water emulsions on ultrasonic attenuation.

4. Conclusion

The dependence of oi-in-water emulsions on ultrasonic velocity and attenuationhavebeen studying after varying five levels of concentration.

The linear correlation between ultrasonic velocity and concentration proved to be useful to obtain an analytical curve that should be need to measure oil and grease content in water within the range from 6 000 to 14 000 ppm.

The attenuation measurement allows us to assume that there is a linear bias on data, which should be assessed with improvements of the method.

Acknowledgements

We thank financial support by the National Council for Scientific and Technological Development (CNPq): PROMETRO (Training Program of Science and Technology for Scientific and Industrial Metrology of INMETRO.

References

[1] BRASIL. Ministry of Environment.*National Council for the Environment Resolution nr. 430, 2011May 13*.available at: http://www.mma.gov.br/port/conama> Accessed at: 2013 Feb 04.

[2] Duhan, H 2012. *Oil and GasFacilities – SPE*, **21**.

[3] Farmaki, E, Kaloudis, T at al. 2007. *Desalination*, **210** 52–60.

[4] Pal, R 1994. Colloids and Surfaces A: Physicochem. Eng. Aspects, **84** 141-93.

[5] Richter, A, Voigt, T, Ripperger, S. 2007. *Journal of Colloid and Interface Science*, **315**, 482–92.

[6] Kinsler, L E, Frey, A R, Coppens, A B, Sanders, J.V. 2000. Fundamentals of acoustic. 4th Ed., (New York:John Wiley6 SonsInc.)

[7] Figueiredo, M K K, Costa-Felix, R P B, Maggi, L E, Alvarenga, A E, Romeiro, G A. 2012. *Fuel*, **91** 209–12.

[8] Richter, A, Babick, F, Stintz, M. 2006. *Ultrasonics*, **44**, 483–90.

Journal of Physics: Conference Series **575** (2015) 012006 doi:10.1088/1742-6596/575/1/012006

Ultrasonic transverse velocity calibration of standard blocks for use in non-destructive testing

C E R Silva, D S Braz, L E Maggi and R P B Costa Felix
Laboratory of Ultrasound (Labus), Directory of Scientific and Industrial Metrology (Dimci), National Institute of Metrology, Quality, and Technology (Inmetro), Av. Nossa Sra. das Graças, 50 (Building 1), Duque de Caxias, RJ, Brazil, ZIP 25250-020

E-mail: cesilva@inmetro.gov.br

Abstract. Standard blocks are employed in the verification of the equipment used in Ultrasound Non-Destructive Testing. To assure the metrology reliability of all the measurement process, it is necessary to calibrate or certify these Standard blocks. In this work, the transverse wave velocity and main dimensions were assessed according to the specifications ISO Standards. For transverse wave velocity measurement, a 5 MHz transverse wave transducer, a waveform generator, an oscilloscope and a computer with a program developed in LabVIEW™ were used. Concerning the transverse wave velocity calibration, only two Standard blocks of the 4 tested is in accordance with the standard.

1. Introduction

Standard blocks named V1 and V2 are employed in the calibration and verification of equipment used in Ultrasonic Non-destructive testing (US NDT). To assure the metrology reliability of all the measurement process, it is necessary that standard blocks are used, preferentially if they were certified such as a CRM (Certified Reference Material) [1]. These follow the principle of CRM and the technical requisites of ISO 2400 [2] and ISO 7963 [3] standards. The transverse velocity is the main characteristic to be evaluated.

The aim this work was to calibrate 4 reference blocks, evaluating its main dimensions and acoustic properties. Therefore, four reference blocks, being two V1 blocks and two V2 blocks were used. The main dimensions and the transverse velocity were assessed.

2. Theorical Approach

In V1 blocks, to dimensional calibration, the specified maximum admissible error is of ± 0.10 mm, in accordance with ISO 2400 standard. In V2 blocks, maximum permissible measurement error is of ± 0.10 mm, in accordance ISO 7963 standard.

The measurement transverse wave velocity shall be 3255 m/s ± 15m/s. The velocity shall be measured within a maximum permissible measurement error of ± 0.2%, i.e. with uncertainty of ± 6 m/s for transverse waves.

3. Methods

The dimensional calibrations of two V1 blocks (Nº 121 e Nº176) and two V2 blocks (Nº 255 and Nº182) were carried out in the Laboratory of Dimensional Metrology (Lamed)/INMETRO. A system composed by laser interferometer (reference value), a table with a system of horizontal micrometric displacement (maximum range: 300 mm), moved by c.c. motor and a microscope with camera CCD coupled to a computer were used. Specific software to determine the measurements was used. The scale is aligned to the laser and to the microscope in this system.

For transverse wave velocity assessment, the ultrasonic measurement system consisted of 5 MHz transverse wave transducer with diameter of 12,7 mm (Olympus, V155), a waveform generator (Agilent Technologies, USA, 33250A), an oscilloscope (Agilent Technologies, USA, DSO-X 3012A), and a computer with a program (BCNDT_rev00.vi) developed in LabVIEW™ 11, (National Instruments, Austin, TX, USA).

The waveform generator was used to excite the transducer using a 3-cycle sine wave burst with 20 Vpp of amplitude. The signal from echo was acquired and digitalized using the oscilloscope. All signals and acquired data were recorded analysed using the software developed in LabVIEW™.
This system is used to measure the time of flight difference between the first and second back wall echoes. Velocities were then calculated (i.e. path/time/2). The time of flight was measured in different directions, i.e. at two well separated positions through the 25 mm thickness and one position through the 100 mm thickness for V1 standard blocks. For V2 standard blocks, three different positions through the 12.5 mm thickness were carried out. In each position, five measurements, under repeatability conditions, were realized. Because shear waves are polarized, in each location of the probe was made two measurements with the plane of polarization in the second measurement perpendicular to the first measurement and parallel to one side of the block.

Each measurement series was carried out in three consecutive days.

3.1. Method validation

To BCNDT_rev00.vi software validation, at hand measurement (HM) using the cursors oscilloscope to determine the time of flight difference between the first and second back wall echoes was carried out.

Therefore, the results were statistically compared using the normalized error, in order to know the measurement performance according equation (1) [4][5]:

$$E_n = \frac{x - X}{\sqrt{U_{HM}^2 + U_S^2}} \qquad (1)$$

Where:

x - At Hand measurement data.
X - Software measurement data;
U_{HM} - Expanded uncertainty from Hand measurement data;
U_S - Expanded uncertainty from software measurement data

For E_n numbers:

- $|E_n| \leq 1.0$ - indicates "satisfactory" performance;
- $|E_n| > 1.0$ – indicates "unsatisfactory" performance.

4. Determination of standard uncertainty of Type A and Type B.

All uncertainties were assessed according to GUM [4]. The time base resolution of the oscilloscope is assumed to present a rectangular distribution; hence the Type B uncertainty is estimated by dividing the resolution by $2\sqrt{3}$. The distance is obtained from calibration certified and thus its Type B uncertainty is defined as expanded uncertainty divided by K.

The Type A uncertainties were estimated as the standard deviation from 5 repeated measurements, divided by $\sqrt{5}$. Type B uncertainties were estimated, as following in Table 1:

Table 1. Uncertainty components of Type B.

Uncertainty components	Probability Distribution	Sensitivity coefficients
Time base resolution of the oscilloscope	Rectangular	-d .t -2[a]
Distance from calibration certified	Rectangular	1.d-1[a]

[a] Where: d distance of sound propagation (m);
t time (s).

Finally, the expanded uncertainty was calculated considering a coverage factor k = 2 (95.45% confidence level).

The combination between all calculated uncertainties to determine the transverse velocity uncertainty was obtained using the equation (2):

$$U = \sqrt{\left(\frac{s}{\sqrt{n}}\right)^2 + m\acute{a}x\left(\frac{U_i}{k_i}\right)} \qquad (2)$$

Where:
s - Standard deviation between the average the found values.
n - Number of independent repeated observations
$m\acute{a}x\left(\frac{U_i}{k_i}\right)$ - Maximum found uncertainty divided by respective coverage factor.

5. Results and Discussion

The main dimensions to calculate V1 and V2 blocks transverse velocity are given in Table 2. In V2 and V1 blocks it could be observed that the systematic measurement error of the main dimensions were lower than the maximum permissible measurement error established by ISO 7963 and 2400 standards, which is ± 0.10 mm.

The transverse velocity measurements results and its respective uncertainty realized for different modes (HM and S) are shown in Table 3. The software developed in LabVIEW™ was validated. Consequently, the software can be used to calibrate the standard blocks.

The transverse velocity values of V1 and V2 standard blocks and its respective uncertainties are in Table 4.

The nominal transverse wave velocity values (according the ISO Standards) were compared with the average obtained values from 03 (three) days of measurements. The systematic measurements errors were compared with of maximum permissible measurement error as specified ISO 2400 and 7963 standards (15 m/s). As it can be seen in all measurements only V1 0176/09 and V2 182/09 Standard blocks showed transverse wave velocity average values below the maximum error admissible and measurement uncertainty (0.2% or ± 6 m / s).

Table 2. Nominal value and systematic measurement error of V1 and V2 standard blocks main dimensions.

Block	Serial number	quota	Nominal value	Measured quantity value [mm]	Measurement uncertainty [mm]	*k*	Systematic error [6].
V1	121/06	a	100	100.0020	0.0030	2.0	0.005
		b	25	25.0360	0.0030	2.0	0.039
	0176/09	a	100	99.9950	0.0020	2.0	0.007
		b	25	25.0000	0.0020	2.0	0.002
V2	0182/09	c	12.5	12.5190	0.0030	2.0	0.022
	225	c	12.5	12.5030	0.0020	2.0	0.005

Table 3. Software validation results.

Mode			Velocity [m/s]	Expanded Uncertainty [m/s]	EN number	Performance
V1	121/06	HM	3242.4	1.3	0.09	Satisfactory
		S	3242.8	4.3		
	0176/09	HM	3244	31	0.13	Satisfactory
		S	3254	69		
V2	0182/09	HM	3240.40	0.90	0.98	Satisfactory
		S	3242.8	2.3		
	225	HM	3251	15	0.04	Satisfactory

Table 4. Transverse wave velocity and measurement uncertainty of V1 and V2 Standard blocks.

Standard Block/ Serial number	Nominal value/ maximum permissible measurement error [m/s]	Average transverse wave velocity [m/s]	Expanded uncertainty [m/s]	*k*	%	Systematic error [6].
V1 121/06	3255 ± 15	3254	69	1.96	2.09	1
V1 0176/09	3255 ± 15	3242.8	4.3	1.96	0.13	12.2
V2 0182/09	3255 ± 15	3242.8	2.3	1.96	0.07	12.2
V2 225	3255 ± 15	3252	22	1.96	0.67	2

6. Conclusion

This work concludes that V1 and V2 blocks in calibrated main dimensions are in agreement with the technical specifications of ISO 7963 and ISO 2004 standards. Concerning the transverse wave velocity calibration only V1 0176/09 and V2 182/09 Standard blocks are in accordance with these standards.

Acknowledge
The authors would like to acknowledge the financial support from the CNPq and Faperj. Especial thanks to Lamed/Dimec/Inmetro.

References

[1] Instituto Nacional de Metrologia, Normalização e Qualidade Industrial – (Inmetro). URL: http://www.inmetro.gov.br/ [accessed on 16/May/2013].
[2] ISO 2400, Non-destructive testing. Ultrasonic testing. Specification for calibration block No. 1. 2012.
[3] ISO 7963, Non-destructive testing. Ultrasonic testing. Specification for calibration block No. 2. 2010.
[4] Evaluation of measurement data – Guide to the expression of uncertainty in measurement (GUM) – JCGM 100:2008, First Edition, Sep – 2008.
[5] ISO/IEC17043, Conformity assessment - General requirements for proficiency testing, 2010.
[6] International vocabulary of metrology - Basic and general concepts and associated terms (VIM) – JCGM 200:2012.

Journal of Physics: Conference Series **575** (2015) 012007 doi:10.1088/1742-6596/575/1/012007

Hydro acoustic transducer's calibration by the reciprocity

S A Miqueleti[1], C A O Martins[2], C E P Ribeiro[2], R P B Costa-Felix

[1]Laboratório de Ultrassom (Labus), Diretoria de Metrologia Científica (Dimci), Instituto Nacional de Metrologia, Qualidade e Tecnologia (Inmetro); [2]Laboratório de Instrumentação Oceanográfica(LIOc), Universidade Federal do Rio de Janeiro (UFRJ).

E-mail: samiqueleti@inmetro.gov.br

Abstract. This paper presents a calibration technique of underwater acoustic transducers, hydroacoustics, known as three-transducer spherical-wave reciprocity and performs an assessment of the type B standard uncertainty of the results obtained for the frequency measurement used.

1. Introduction

In the various activities of ocean exploration as oceanographic research, hydrographic surveys, underwater communications and sea surveillance, all components of hydroacoustic systems and their combinations should be capable of providing data according with acceptance criteria that ensure that the measuring instrument or standard has the accuracy required for the intended use. Metrologically, the way to ensure the required accuracy is, in general, through proper and periodic calibration according to specific technical standards, such as IEC 60565 [1].

2. Theoretical basis

The transducers used in underwater acoustics can be divided into three groups:

- Projectors (P) are built to be transmitters of acoustic waves and transmit sound power to the middle.
- Hydrophones (H) are constructed to be receivers of acoustic waves and have the characteristic high sensitivity of detection.
- Reciprocal transducers (T) can be used as transmitters or receivers.

The calibration of transducers can be performed (primary) absolutely, as known through calibration by reciprocity with spherical waves, based on reciprocity technique.

2.1. Reciprocity

In order for a transducer be reciprocal it must have the following properties: being a linear, reversible and passive element. The parameter of reciprocity J is constant for a given frequency [2], and relates the reception sensitivity (M) with the transmission sensitivity (S) of a piezoelectric element, defined as:

$$J(f) = \frac{M(f)}{S(f)} = \frac{2r}{\rho f} \tag{1}$$

where:

- r is distance from the spherical wave source
- ρ density of the medium (water)
- f the vibration frequency

2.2. *Three-transducer spherical-wave reciprocity*

Three transducers are needed in this calibration technique by reciprocity to determine the sensitivity of a free-field hydrophone: hydrophone (H), a projector (P) and a reciprocal transducer (T). Three measurements are also needed, as shown in Figure 1.

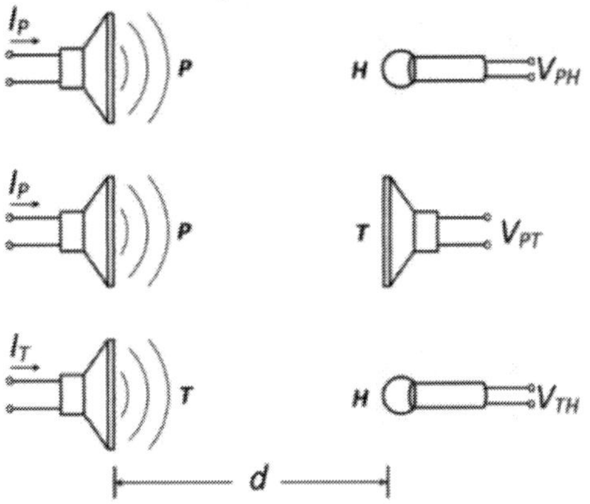

Figure1 – Three steps of calibration by reciprocity

For a pair of transducers, the acoustic transfer impedance (Z) is defined as the ratio of the open circuit voltage (V) at the receiver, induced by the excitation current (I) through the transmitter device [3]:

$$Z_{PH} = \frac{V_{PH}}{I_P} \qquad Z_{PT} = \frac{V_{PT}}{I_P} \qquad Z_{TH} = \frac{V_{TH}}{I_T} \tag{2}$$

Voltage induced on the hydrophone (H) is the product of its reception sensitivity (M_H) by the pressure received in his face (p). As the pressure generated by the projector (P) at a separation distance (d), the product of the sensitivity of the transmission (S_P) by the excitation current (I_P) divided by (d) in case where the spherical wave acoustic pressure decays to half with twice the distance, we have:

$$V_{PH} = pM_H = \frac{S_P I_P M_H}{d} \tag{3}$$

In an analogous way we obtain V_{PT} and V_{TH}. Thus we have a system with three equations and four unknowns variables (M_H, S_P, M_T, S_T). Using the reciprocity parameter for the reciprocal transducer (T), the equation (1), is obtained the fourth equation and can determine all parameters of the transducers, as in this case:

$$M_H = \left(Jd\frac{Z_{PH}Z_{TH}}{Z_{PT}}\right)^{\frac{1}{2}} = \left(\frac{2d^2}{\rho f}\frac{V_{PH}V_{TH}}{V_{PT}I_T}\right)^{\frac{1}{2}} \tag{4}$$

Rewriting the equation (4), so as to visualize the measurement variables, we have:

$$M_H = \sqrt{2}\cdot d\cdot \rho^{-\frac{1}{2}} f^{-\frac{1}{2}} V_{PH}^{\frac{1}{2}} V_{TH}^{\frac{1}{2}} V_{PT}^{-\frac{1}{2}} I_T^{-\frac{1}{2}} \tag{5}$$

From equation (5) evaluation of the standard uncertainty of type B was taken.

3. Materials and methods

The experiments were performed in hydroacoustic tank tests at the Institute of Marine Research (IPqM), in Rio de Janeiro, RJ, Brazil. That non-anechoic freshwater tank has dimensions of 10 meters in length and 10 meters wide by 5 feet deep (Figure 2). The following measuring instruments and accessories have been used: two reciprocal transducers Model 1001, one as a projector (P) and another as reciprocal (T) and a hydrophone (H) Model 1032, all manufactured by International Transducer Corporation (ITC); preamplifier and amplifier dedicated (Scientific-Atlanta, USA); model DSO6032A oscilloscope and signal generator model 3325A, both from Agilent Technologies. For control, data acquisition and processing a routine done in LabView™ (National Instruments, USA) was used.

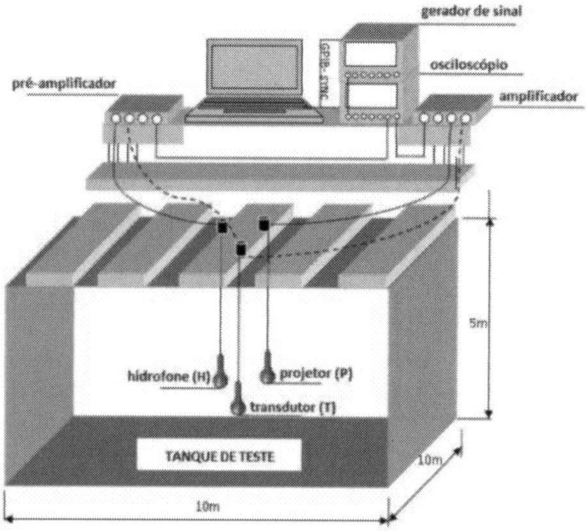

Figure2 – Tank testing and instrumentation used

The details of the method used in this work can be found in [4]. The transducers were prepared, and their surfaces were clean and free of air bubbles, so that it ensures the perfect coupling between the acoustic transducer and the medium (water). The transducers were positioned symmetrically at 96 cm from each other and the half depth of the tank to minimize the effects of reflections on the walls.

The excitation currents were measured by the voltage drop across a resistor shunt in the amount of $1.82 \pm 0.01\Omega$.

4. Results

The speed of sound was measured obtaining 1.488 ± 15 m/s under the experimental conditions. According to [5], the density of fresh water is 997.77 ± 0.96 kg/m^3 for the range of temperatures observed during the experiment (21°C e 23°C). The separation distance between the transducers, as measured by time of flight of the signal was 0.96 ± 0.02m. We carried out a calibration of the reception sensitivity (M_H) of the transducer ITC1032 in the range of 3 kHz to 15 kHz in steps of 1 kHz, obtaining the result shown in Figure 3. In this figure the sensitivity of the hydrophone is also presented, according to the manufacturer data.

4.1. *Evaluation of standard uncertainty type B*

As non-repetitions of the same experiment were performed, an assessment of uncertainty type B from the specifications of the instruments used and the uncertainties in the variables in (5) was taken, using the Guide to the Expression of Uncertainty in Measurement, GUM [6].

Table 1. Uncertainty (M_H) of the transducer ITC1032

freq.[Hz]	sup [dB]	M_H [dB]	inf [dB]
3000	-192.1	-192.5	-192.9
4000	-192.5	-192.9	-193.3
5000	-192.6	-192.9	-193.3
6000	-193.1	-193.4	-193.8
7000	-193.3	-193.7	-194.0
8000	-193.7	-194.0	-194.4
9000	-194.0	-194.3	-194.7
10000	-194.3	-194.6	-195.0
11000	-194.6	-195.0	-195.4
12000	-195.0	-195.4	-195.8
13000	-195.2	-195.6	-196.0
14000	-195.4	-195.8	-196.1
15000	-195.5	-195.8	-196.2

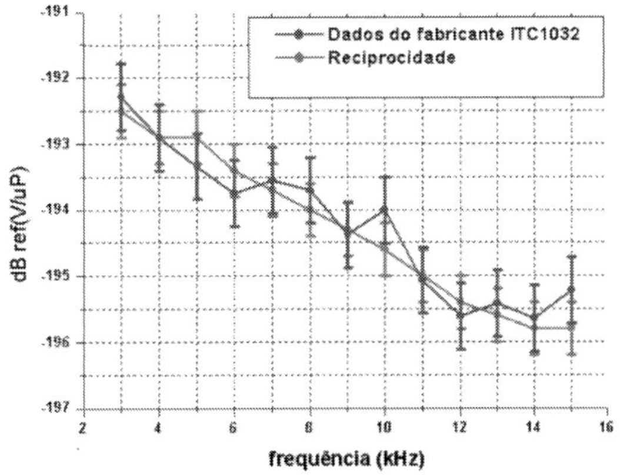

Figure3 – Curve (M_H) of the transducer ITC1032

5. Analisys

The IEC 60565:2007 [1] tells that the standard calibrations for reciprocity in free field is carried out by observing the requirements of the standard and taking into account the parameters described in Annex D have uncertainties typically less than 1 dB. Already in reference [7] the authors performed calibrations on 2 methods, independent laboratories obtaining uncertainties typically less than 0.5 dB. Our findings are similar to [7] and are well below the typical uncertainties [1].

This is justified by the fact that this work was not done study of uncertainty of type A. Additionally the system used has adequate accuracy for the method studied since it results in a low natural component of Type B uncertainty.

Regarding the results, Figure 3 shows the results obtained with the calibration comparing them with the specifications of the manufacturer. It is observed that in all the frequencies results differed less than the respective uncertainties Type B. This shows that the method was consistent with the expected results for the calibration of the hydrophone.

A number of factors can influence the calibration results, such as:

- The separation between the transducers must be controlled;
- The reflections on the walls of the tank;
- Electrical calibrations of instruments;
- Monitoring of water temperature.

These parameters were considered most relevant in the uncertainty model developed in this study and were considered in this model.

6. Conclusion

The method of absolute calibration of hydroacoustic transducers was implemented successfully. This is a validated method of long standing, including an international standard for their implementations have been used. The newness of this work lies in the fact that this implementation had never been held in Brazil and open space for future developments in underwater acoustics.

ACKNOWLEDGMENT
The authors thank the Research Institute of the Navy of Brazil (IPqM).

References

[1] IEC60565:2007. Underwater Acoustics Calibration in the Frequency Range 0.01 Hz to 1 MHz. International Eletrotechnical Comission. Geneva.
[2] Bobber R J. General Reciprocity Parameter. The Journal of Acoustical Society of America. Vol. 39. No 4. 680-687. April 1966
[3] Bobber R J. Underwater Electroacoustic Measurements. 1ed. Washington. Naval Research Laboratory. 1970.
[4] Martins C A O. Calibração de transdutores hidroacústicos pelo método da reciprocidade empregando salvas de sinais de excitação monotônicos e de frequência modulada. Dissertação de Mestrado. Programa de Engenharia Naval e Oceanica. COPPE/UFRJ. 2013.
[5] Lide D R (Ed.). CRC Handbook of Chemistry and Physics (70th Edn.). Boca Raton (FL):CRC Press. 1990.
[6] Guia para a expressão de incerteza de medição – GUM 2008. Duque de Caxias. RJ: INMETRO/ CICMA/ SEPIN. 2012.
[7] Robinson S P. Theobald P D. Validation of primary hydrophone calibrations by inter-laboratory comparisons and by independent calibration methods. annals of Acoustics'08. June 29-July 4. Paris. 2008.

Primary calibration in acoustics metrology

T A Bacelar Milhomem and Z M Defilippo Soares

Inmetro, Duque de Caxias, Rio de Janeiro, Brazil

E-mail: tbmilhomem@inmetro.gov.br

Abstract. SI unit in acoustics is realized by the reciprocity calibrations of laboratory standard microphones in pressure field, free field and diffuse field. Calibrations in pressure field and in free field are already consolidated and the Inmetro already done them. Calibration in diffuse field is not yet consolidated, however, some national metrology institutes, including Inmetro, are conducting researches on this subject. This paper presents the reciprocity calibration, the results of Inmetro in recent key comparisons and the research that is being developed for the implementation of reciprocity calibration in diffuse field.

1. Introduction

Confidence of sound pressure level measurements is guaranteed when its traceability is evidenced. Traceability which passes through the accredited calibration laboratories, the national metrology institutes (NMI) and the International Bureau of Weights and Measures (BIPM, abbreviation in French) [1].

BIPM, the highest body in the traceability hierarchy, determines that SI unit in acoustics, pascal, is realized by the reciprocity calibration of one inch and half inch laboratory standard microphones (LS1 and LS2) [1]. Figure 1 shows pictures of LS1 and LS2 microphones.

Figure 1. Pictures of LS1 (left) and LS2 microphones.

Calibration of a measurement microphone consists in determine its sensitivity and its frequency response. Sensitivity is the ratio between the output signal and the input signal. In the case of a microphone, input signal is the variation of dynamic pressure and output signal is the voltage. Thus, the sensitivity of a microphone is measured in volts per pascal and the sensitivity level in decibels referenced to 1 V/Pa.

A microphone calibration can be made in three distinct idealized sound fields: the free field, the diffuse field and the pressure field. The free field is a field with only one sound wave (there is no reflections). The diffuse field is a field such any position receives sound waves from all directions with equal probability and equal intensity. The pressure field is a field such in any section the sound wave is uniformly distributed.

The objective of this paper is to present the reciprocity calibration of microphones, the Inmetro's participation in recent key comparisons of microphones in pressure field and in free field, and the Inmetro's research for implementation of reciprocity calibration in diffuse field.

2. Reciprocity calibration

Calibration by reciprocity is traditionally obtained using three microphones. In the first step two microphones, microphones 1 and 2, are acoustically coupled. Using the microphone 1 as a transmitter and the microphone 2 as a receiver the product of sensitivities is determined (1):

$$M_1 M_2 = \frac{Z_{e,12}}{Z_{a,12}} \tag{1}$$

Where M_1, M_2, $Z_{e,12}$, and $Z_{a,12}$ are the sensitivities of the microphones 1 and 2, the electrical transfer impedance between the microphones 1 and 2 and the acoustic transfer impedance between microphones 1 and 2. The electrical transfer impedance is the quotient of the open circuit voltage of the microphone used as a receiver by the input current through the electrical terminals of the microphone used as a transmitter. The acoustical transfer impedance is function of the acoustic coupling between the microphones [1].

In the second step the microphone 2 is replaced by the third microphone, the microphone 3, and the procedure is repeated. In the third step the microphone 1 is replaced by the microphone 2 and again the procedure is repeated. In the end, there is a system of three equations and three unknowns and each sensitivity can be calculated. [1] Thus, the sensitivity of the microphone 1 is obtained by:

$$M_1 = \left(\frac{Z_{e,13} Z_{e,12}}{Z_{e,23}} \frac{Z_{a,23}}{Z_{a,12} Z_{a,13}} \right)^{1/2} \tag{2}$$

3. Reciprocity calibration in pressure field

Reciprocity calibration in pressure field provides traceability for calibration of sound calibrators, artificial ears, 6cc couplers and torso simulators. It is performed according to IEC 61094-2 [2], it is realized by many NMIs and its methods have been checked in several key comparisons [3]. In the last one, the CCAUV.A-K3, realized in 2003 and conducted by Consultative Committee for Acoustics, Ultrasound and Vibration (CCAUV) with the participation of fifteen NMIs, Inmetro showed that already consolidated this method as result presented in the final report [4]. In this key comparison were calibrated LS2 microphones. They were calibrated in magnitude for the frequency range from 31.5 Hz to 31.5 kHz [4]. Figure 2 shows graphs of degree of equivalence per country for 250 Hz and 1000 Hz.

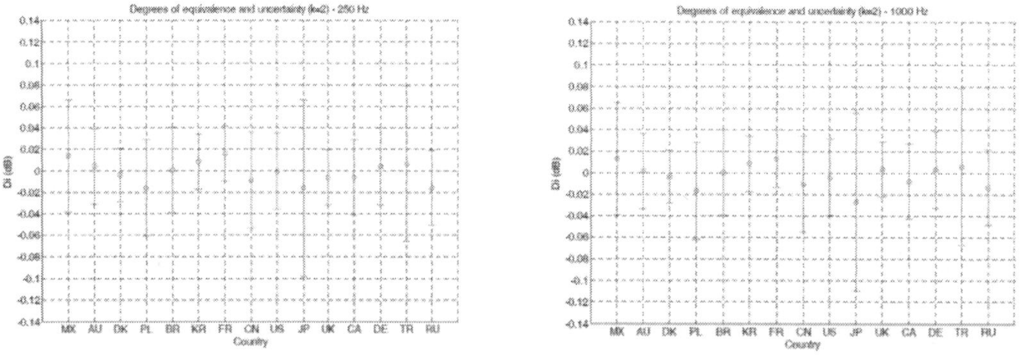

Figure 2. Graphs of degree of equivalence per country for 250 Hz and 1000 Hz [4].

Today Inmetro is participating in the key comparison CCAUV.A-K5 which began in 2011 also conducted by CCAUV with the participation of twelve NMIs [5]. In this key comparison are being calibrated LS1 microphones according to the new edition of IEC 61094-2 [6]. They are being calibrated in magnitude and phase for the frequency range from 2 Hz to 10 kHz [5].

4. Reciprocity calibration in free field

Reciprocity calibration in free field provides traceability for the calibration of sound level meters and work standard microphones, for the measurement of the frequency response of loudspeakers and any measurements of sound pressure level in free field. It is performed according to IEC 61094-3 [7], it is realized by few NMIs and its methods were checked only once in key comparisons [3], the CCAUV.A-K4, realized in 2007 also conducted by CCAUV with the participation of seven NMIs. In this key comparison Inmetro also showed that already consolidated this method as result presented in the final report [8]. In this key comparison were calibrated LS2 microphones. They were calibrated in magnitude for the frequency range from 1 kHz to 40 kHz. Figure 3 shows graphs of degree of equivalence per country for 2.5 kHz and 8 kHz.

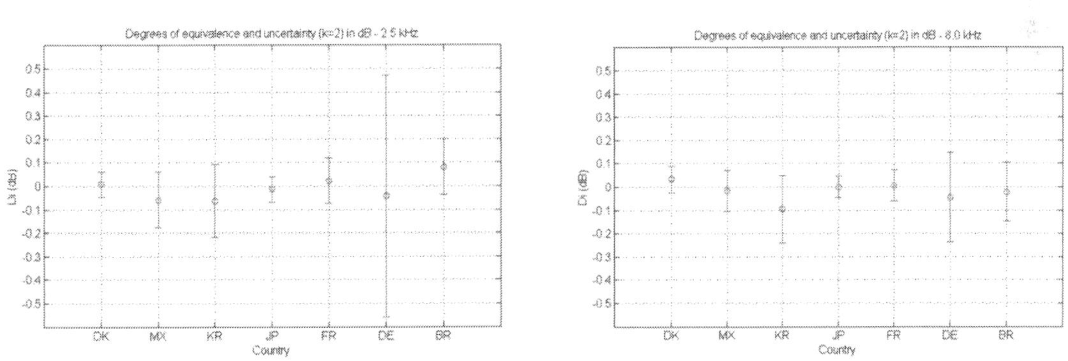

Figure 3. Graphs of degree of equivalence per country for 2.5 kHz and 8 kHz [8].

5. Reciprocity calibration in diffuse field

Reciprocity calibration in diffuse field provides traceability for the calibrations of sound level meters and work standard microphones and any measurements of sound pressure level in diffuse field. There isn't yet any international standard on this subject, any NMI declares this calibration to BIPM and any key comparison was realized to check this method [3]. Some NMIs, however, begun to develop research about this as DPLA [9], Denmark's NMI, and NPL [10], United Kingdom's NMI.

Follow these same lines Inmetro also begun to develop research about this. It was built a reverberation chamber of approximately 2 m^3 with of polished granite slabs mounted on a set of four steel springs [11]. In a first step was measured the acoustic transfer impedance. The procedures adopted proved adequate and the results pointed where the investigations should be concentrated. The procedures adopted and the results obtained will be presented at the international congress Internoise 2013 [12]. In a second step will be measured the electrical transfer impedance and will be calculated the sensitivity of LS1 microphones. Figure 4 shows the photo of the reverberant chamber before the measurements for determining the acoustic transfer impedance.

Figure 4. Photo of the reverberant chamber before the measurements for determining the acoustic transfer impedance.

References

[1] Milhomem T A B 2008 Technique for free field calibration of half inch microphone by reciprocity *Dissertation for the degree of Master of Science PEM / UFRJ / COPPE.* (In Portuguese)
[2] IEC 61094-2:1992 Measurement microphones – Part 2: Primary method for pressure calibrations of laboratory standard microphones by the reciprocity technique 1 ed.
[3] http://www.bipm.org. Access: jun. / 2013
[4] Henríques V and Rasmussem K 2006 Final report on the key comparison CCAUV.A-K3
[5] Barham R 2010 Technical protocol for the key comparison CCAUV.A-K5
[6] IEC 61094-2:2009 Measurement microphones – Part 2: Primary method for pressure calibrations of laboratory standard microphones by the reciprocity technique 2 ed.
[7] IEC 61094-3:1995 Measurement microphones – Part 2: Primary method for free-field calibrations of laboratory standard microphones by the reciprocity technique 1 ed.
[8] Barrera-Figueroa S, Nielsen L, Rasmussen K, Matzumoto A E P and Razo J N R 2010 Final report on the key comparison CCAUV.A-K4
[9] Barrera-Figueroa S and Rasmussen K 2008 A note on determination of the diffuse-field sensitivity of microphones using the reciprocity technique *Jasa* **124** 3 1505-12
[10] Jackett R 2012 Implementations of diffuse-field microphone calibration system *Internoise 2012*
[11] Müller S 2011 Final Report. (In Portuguese)
[12] http://www.internoise2013.com. Access: jun. / 2013

Compact fluorescent lamps, LED lamps and harmonic distortion

A M R Franco[1], R M Debatin[1], F C G Cotia[1], M V M Silva[1], R S Ribeiro[1] R R N Zampilis[1]
[1]Electrical Energy Metrology Laboratory, Electrical Metrology Division
National Institute of Metrology, Quality and Technology – Inmetro
Av. Nossa Senhora das Graças, 50 – Xerém – Duque de Caxias, Brazil

E-mail: amfranco@inmetro.gov.br

Abstract. The aim of this paper is to evaluate the harmonic distortion in the current waveform of Compact Fluorescent Lamps (CFL) and Lamps Lighting Emitting Diode (LED). For this, we analysed the power factor, voltage waveform, current waveform, total harmonic distortion (THD) and active power consumed.

1. Introduction

Compact fluorescent lamps were originally developed aiming the replacement of incandescent lamps. They have similar operating principle of the tube fluorescent lamps, but their dimensions are pretty reduced.

They have modern and compact design, high efficiency, long life (up to 20 times greater than that of ordinary incandescent lamp), light beams distribution diversity and a variation of products able to meet the most different applications: commercial, industrial or residential.

In Brazil, the use of this type of lamp is widely encouraged and intensified during the 2002 energy shortage.

Currently when thinking about innovation and energy efficiency in lighting people think in the Led lamps. Notes the desire of many people to make use of this technology, but there is still much lack of knowledge about its advantages and disadvantages. Led lamps as modernization of incandescent, halogen and fluorescent tube lamps are products increasingly found in the market.

Seduced by the idea of a long life, high luminous efficiency, no ultraviolet radiation, infrared and harmful metals such as mercury, more and more people want to use the LEDs as a light source.

Some care, however, must be observed when specifying or purchasing Led lamps for modernization of electrical installations. Currently, there are a variety of products on the market and we do not have in place any regulation or recommendation for the use of equipment that adopt this technology.

Concerning energy efficiency, it is evident the superiority of LED lamps when compared to incandescent and compact fluorescent lamps. However, when observing the effects of Compact Fluorescent lamps and LED lamps on the power quality of the system in which they are used, there is a certain commitment due to the emergence of unwanted distortion in the network.

These distortions are composed of multiple frequencies or sub-multiple of the fundamental frequency (60 Hz) and are known as harmonics.

2. Harmonic Distortion and Harmonics

Given a periodic signal with a frequency *f*, as the fundamental frequency, it is denominated the harmonic component of this signal, a periodic waveform with a frequency multiple (integer) of the fundamental frequency. [1]

Figure 1 illustrates this concept. The waveform of larger amplitude and lower frequency is the fundamental waveform and the waveform with smaller amplitude and higher frequency represents a harmonic component.

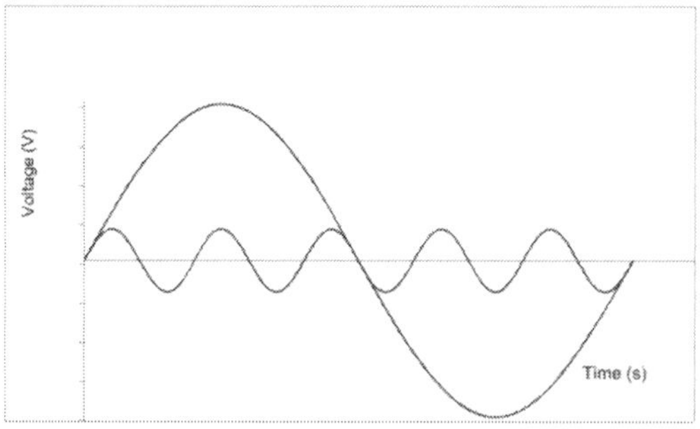

Figure 1. Periodic fundamental waveform and its harmonic component.

Figure 2 shows the signal resulting from the sum of the two waveforms. This signal represents the distortion caused by the presence of harmonics. Note that, the final waveform is no longer perfectly sinusoidal.

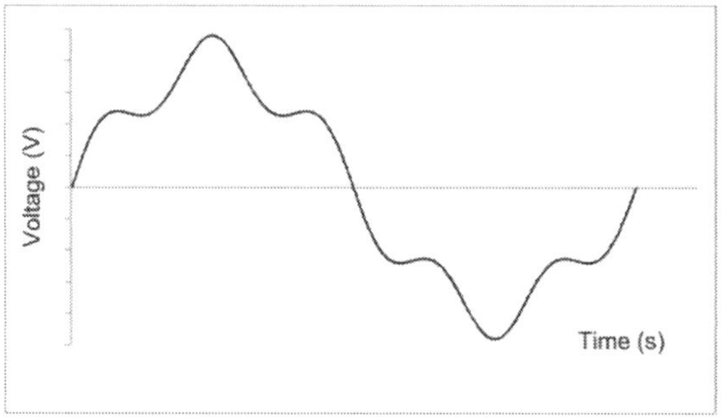

Figure 2. Periodic signal distorted by a harmonic component.

The harmonic waveforms occur continuously, should not be confused with signals that have few cycles or short durations.

To characterize the content of harmonics in practice, the distortion factor d or THD (Total Harmonic Distortion) is used, which is defined as the equation 2.1(for example, current (*I*)):

$$d = \left(\frac{\sum_{k=2}^{\infty} I_k^2}{I_1^2} \right)^{1/2} \tag{2.1}$$

i.e., the square root of the ratio of the quadratic sum of all harmonic divided by the rms value of the squared fundamental waveform.

3. Methodology

To measure power, two digital multimeters from Agilent 3458A [3] are used synchronously. Applying the method of digital sampling, a multimeter measures the voltage signal while the other multimeter measures the current signal on the voltage drop across a shunt [2]. The algorithm developed to analyse non-sinusoidal waveforms partially applies the discrete integration method and Discrete Fourier Transform (DFT) [1]. To perform the evaluations in the LED and CFL lamps were used the equipment shown in figure 3.

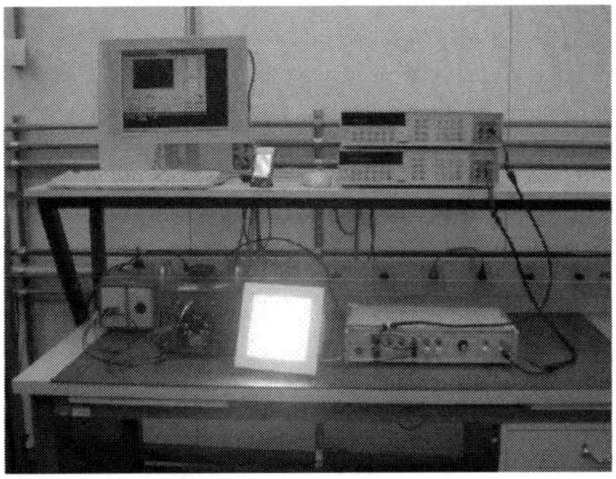

Figure 3. System used in the measurement of LED and CFL lamps.

The calculations of the voltage and current rms values and active power are made by applying the discrete integration method. For the reactive power (Q) are shown two of the most important definitions reported in the literature: QB according to the Budeanu definition and QF according to the Frize definition [4]. The accuracy of the developed methodology is ± 0.01%.

4. Exemplification

The lamps evaluated had active power similar characteristic.

Table 1 shows the results of the interest measurements. figures 4 to 7 show the behaviour of lamps 1 and 2.

Table 1. Measurements of LED lamps and CFL.

Quantity	LED Lamp 1	CFL Lamp 2
Voltage	220.07 V	127.99 V
Current	0.083 A	0.17 A
Active Power	17.30 W	16.98 W
Power factor	0.95	0.79
THD(current)	22.6 %	61.6%

Figures 4 and 6 show the program Harmonic Analyser control panel, with the results of the analysis of LED lamps and CFL.

For better visualization of the harmonics the panel presents also their spectra. As can be seen in figures 5 and 7, in these cases the odd harmonics of the currents are dominant. The table in the right of each screen shows the results of powers analysis.

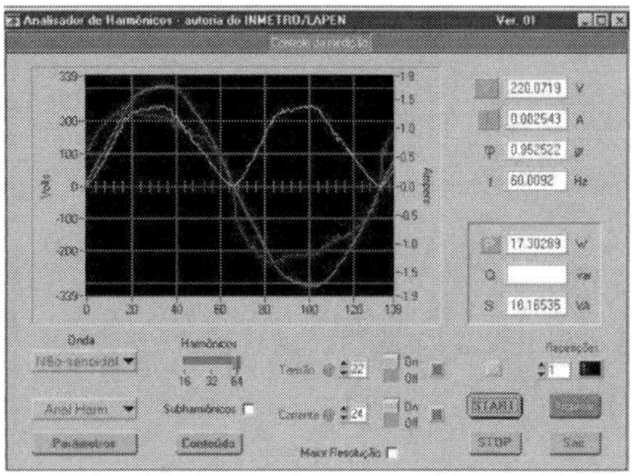

Figure 4. Lamp 1 Behavior

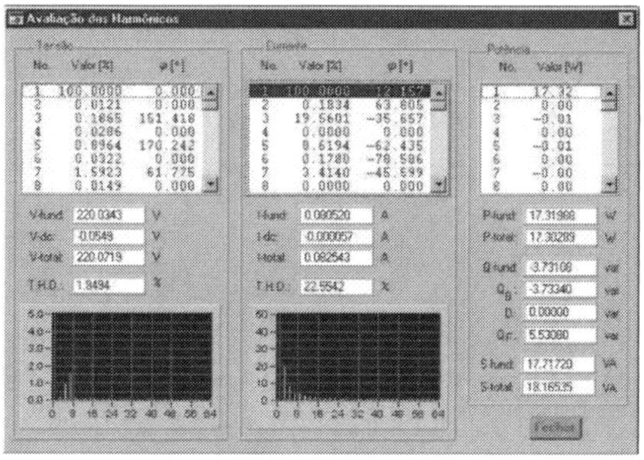

Figure 5. Lamp 1 Harmonic Components

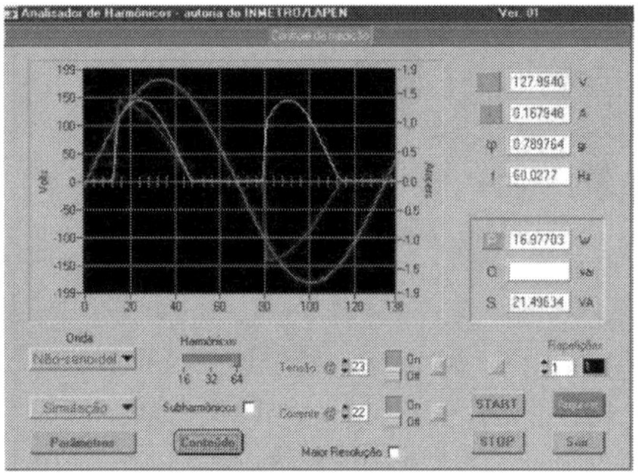

Figure 6. Lamp 2 Behavior

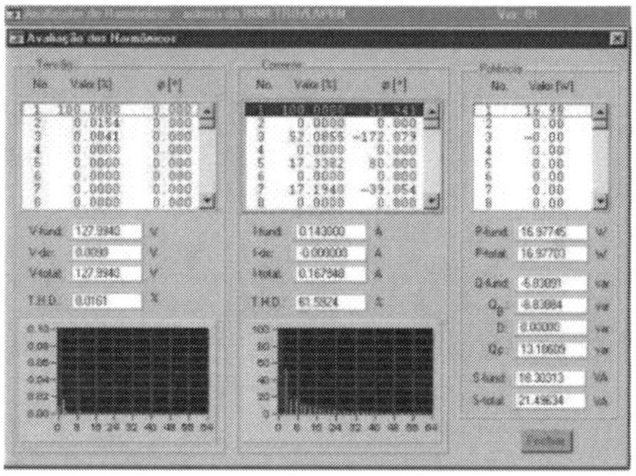

Figure 7. Lamp 2 Harmonic Components

5. Conclusion

This paper presented an evaluation of harmonic distortion caused by LED and CFL lamps and verified that these lamps cause significant harmonic distortion in the current circuit, since their waveforms have the highest harmonic waveform with respect to the voltage supplied.

It is concluded that, even with high harmonic distortion in LED and CFL lamps, is not observed the same behavior of compact fluorescent lamps in relation to the power factor [5].

6. References

[1] Oppenheim A V and Schafer W R 1989 *Discrete-Time Signal Processing*, pp 514-640, Englewood Cliffs (New Jersey 07632).

[2] Franco A M R, Debatin R M, Toth E 2008 *A Power Measurement System Under Non-Sinusoidal Loads*. 6th IMEKO TC4 Symposium and 13th Workshop on ADC Modelling and Testing (Florence).

[3] Agilent 1994 *HP3458A Multimeter. Operating, programming and Configuration Manual.*

[4] Arsenau R, et. al. 1996 *Practical Definitions for Powers in Systems with Nonsinusoidal Waveforms and Unbalanced Loads: A Discussion* (IEEE Trans. On Power Delivery) vol. 11, No. 1.

[5] Franco A M R, Tóth E, Debatin R M 2002, *Influência dos harmônicos na medição da potência*, V International Congress of Electrical Metrology (Rio de Janeiro).

Analysing the behaviour of LED module driver under variation of ambient temperature

A Bürger, O Sanchez
Laboratório de Equipamentos Elétricos e Ópticos, Instituto de Pesquisas Tecnológicas, São Paulo, Brazil

E-mail: osanchez@ipt.br

Abstract. Temperature variations affect the operation of electronic devices. In this article we analyze effects that temperature variation entails a LED lamp commonly used in Brazil. We simulated temperature variations through the use of an oven of 1 m³. Then obtained emission spectrum in the visible, CRI, Correlated Color Temperature, harmonic distortion on the current, supply current, and CIE Chromatic Coordinates x and y in two situations: first with the LED module and controller inside the oven and second with the controller outside. The temperature was monitored in different part of this set. Interesting correlations for specifiers and developers were found.

1. Introduction

The theoretical process of conversion of electric energy to light energy that occurs in a semiconductor diode is a well-known technology. Taking this into account the search for improvements in components and in materials leads to fast changes on the luminotechnical scenario, principally when we are talking about technologic applications.

Two different factors leads to this situation: from one side there is a demand to be filled and specifiers that needs products and resources to make possible the total generation of projects, always taking into account the needs of their customers and new conceptions of lightning technique. Most of the time these customers show their needs focusing in the quality of the product for the end user. By the other side, the financial impact of lightning services is taking place as an important variable, relevant in the budget of many public and private enterprises/services. In this context, there are situations that need particular metrics to measure their needs like energetic expenditure by illuminated area or the financial performance of a lightning service in the business or at the human productivity.

Figure 1 illustrates the process responsible for the conversion between electric energy and luminous energy when analyzing a classic model of physics. The semiconductor junction p-n maintain an electrical potential difference where electrons tends to go from the valence band in order to fulfill conduction band losing energy in the form of light.

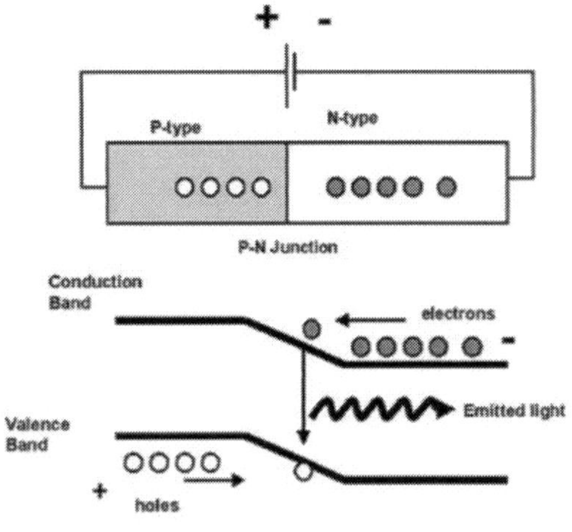

Figure 1. Conversion of electric energy to electric light occurring on a P-N junction. [1]

The quantum efficiency of this conversion depends on the crystal structure, its purity, the absence of imperfections in the crystal net and some other parameters. The temperature at the semiconductor junction influences the quantum efficiency inasmuch as disrupts the crystal net and changes the emission spectrum.

The emission spectrum is the key point to reproduce color of surfaces and stuff rightly. When we look at a stuff and see colors, what is actually happening is the surface of the stuff reflecting photons from a light source to our eyes, making possible its visualization. This way, if the incident light in the stuff changes, the apparent color of the object may also change. LED sources with different technologies, such as RGB LED and a PC LED is shown in figure 2. The stimulus in human vision can be calculated multiplying the energy distribution spectrum of a lamp by the reflectance spectrum of the stuff (in figure 2 we have a TCS09, a standardized lamp used to calculate the Color Reproduction Index using the 9 CIE standards – R9). Taking into account that visual stimuli is the main way that human receive information day-to-day it is relevant to understand the LED behavior when this product is being used in real situations, like lightning services.

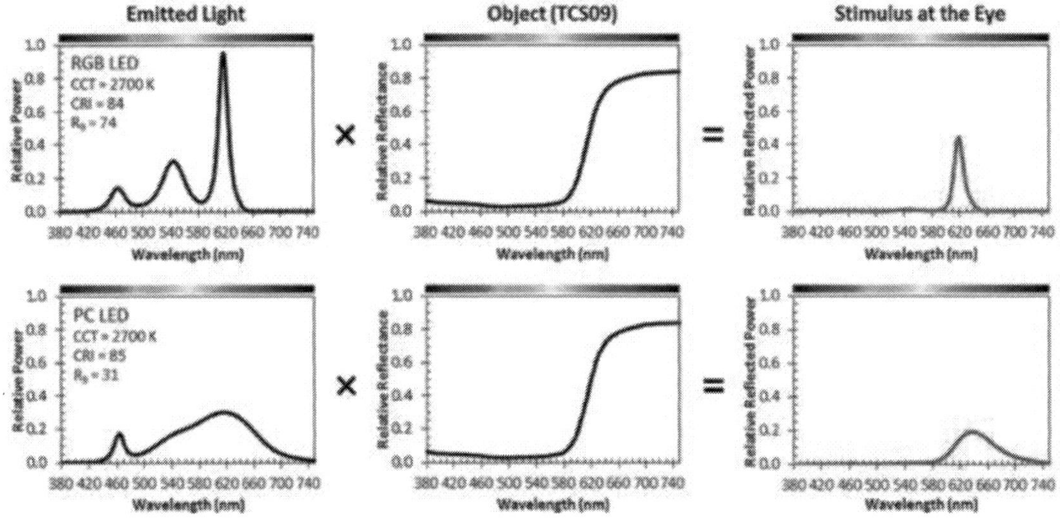

Figure 2. Color perception with two different LEDs models. [2]

Journal of Physics: Conference Series **575** (2015) 012010 doi:10.1088/1742-6596/575/1/012010

2. Motivation

The energy spectral distribution emitted by the LED determines the lightning efficiency of the radiation. Most white LEDs on the market shows the curve illustrated in figure 3 (continuous black line). The peak at 450nm, high intensity and thin, is due to the transition of electrons in the p-n junction whereas the second peak, at 550nm, less intense and more widespread is due to the florescence of the phosphorus inside the LED capsule. Still in figure 3 there is a green dotted curve that represents the standard photopic curve that represents the human eye sensibility to light. The combination of these two curves determines the human sensibility for colours.

A parameter that most affect on this behavior of White LED and determines its efficiency to convert electric energy into luminous energy is the ambient temperature. The ambient temperature interferes not only in the life cycle of LED and its driver, but also in the spectral emission, changing colorimetric characteristics of LED. Therefore study the behavior of the set LED and driver is important to obtain conditions that are the best to operate. So, it is fundamental to have a good knowledge about the luminaire that will be used in the set-up in a way to provide a specific light distribution and adequate heat dissipation of the heat emitted in the process of conversion of electric energy to luminous energy – about 5% of all energy is consumed and converted into heat.

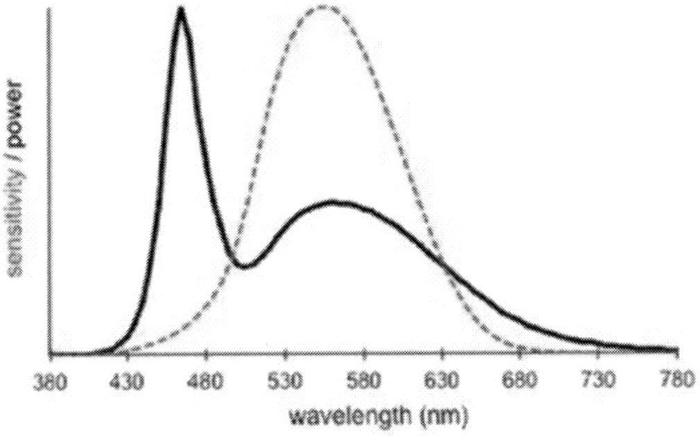

Figure 3. Energy distribution spectrum of the luminaire determines the efficacy of the lightning. [1]

So, the intent of this article is to identify changes in colorimetric parameters – like correlated color temperature, color rendering index and chromatic coordinates – when a white LED is in an ambient with variation of temperature.

3. Experimental

To study the colorimetric behavior of the white LED luminaire and be able to differentiate which parameters depends on the driver, and which depends on the driver and the LED module together. An luminaire that represent the ordinary product that are available in stores to general public was used.

The luminaire design is a metallic box with a ribbed diffuser and a heatsink. The driver of the LED module may be installed together or not with the luminaire box. The dimension of the luminair is 360mm x 240mm x 50mm, 220Vac, rated power of 50W – with driver it consumes 74W –, total flux of 4002lm, energetic efficiency of 54lm/W at ambient temperature of 25°C. Figure 4 is a general vision of the luminaire.

Figure 4. Overview of the test luminaire.

We simulated variation in ambient temperature and obtained colorimetric values for each situation. A Nova Ética oven of approximately $1m^3$ of useful volume, a power source California Instruments 1251RP, a Yokogawa wattmeter WT210 to monitor electric parameters, and in order to monitor temperature in 5 different spots of the luminaire a Yokogawa MV100 system integrated with 5 thermocouple was used. To obtain emission spectra from the luminaire, SpectraScan PhotoResearch 705 photometer was used. These spectra data were analyzed according to CIE protocol from 1995.

The experimental arrangement is illustrated in figure 5. The oven has only one hole on its top so an polytetrafluoroethylene (PTFE) target was placed in a way that PTFE target reflect the light directly from the luminaire to the PhotoResearch photometer.

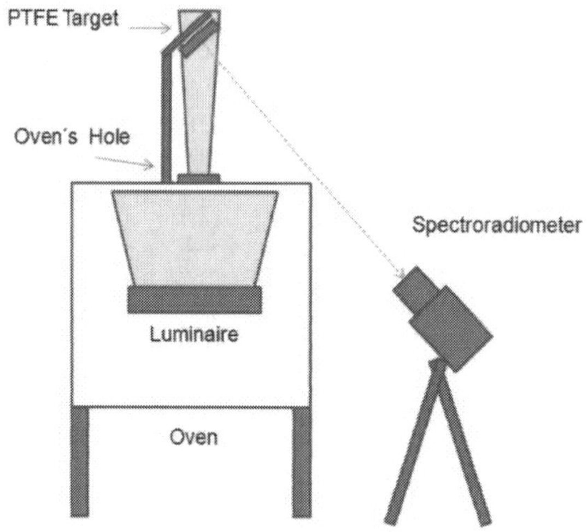

Figure 5. Experimental set-up to obtain LED emission spectrum. The luminaire inside the oven is on a settled temperature and aiming its light to PTFE target we can obtain the spectrum with the radiometer.

Therefore, the target was in a constant temperature during all the experiment. The data were obtained in a rise of temperature starting in 25°C and rising near to 75°C. Each measurement was performed after a stabilization period of 30 minutes. To avoid background radiation the room was kept dark.

The measurements were done in two steps: one with the set LED module and driver inside the oven and the second one with only the LED module inside and the driver at constant temperature. The first step was labeled as *M+C* while the second one was labeled as *M-C*.

The reflectance of PTFE target was taken into account in the measurement process, although it shows a linearity relation in visible range as we can see in figure 6.

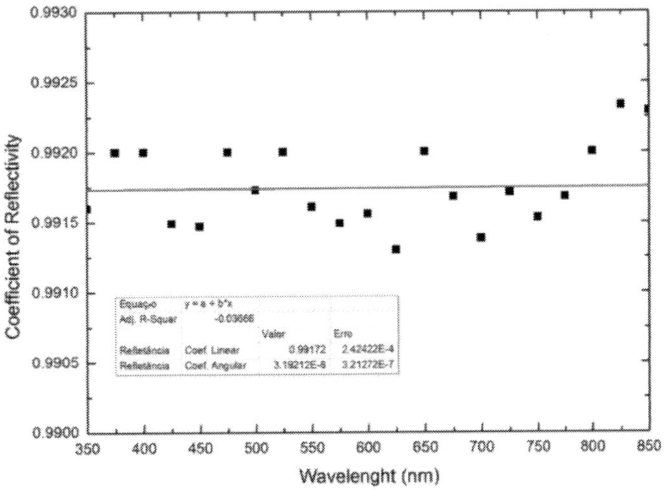

Figure 6. Coefficient of reflectivity of Avian PTFE target in the visible range.

4. Results and Discussion

The graphs shown in figure 7 to 15 expose the results of the measurements as a function of ambient temperature (oven temperature). The label M+C refers to the data obtained with the set driver and LED module inside the oven represented by red circular spots, while the label M-C refers to the situation that only the LED module was inside the oven while the driver was in constant temperature outside the oven. M-C is represented by black dots.

In figure 7 we can observe different behaviors of temperature in each spot of the set in test according to the variation of ambient temperature, inside the oven.

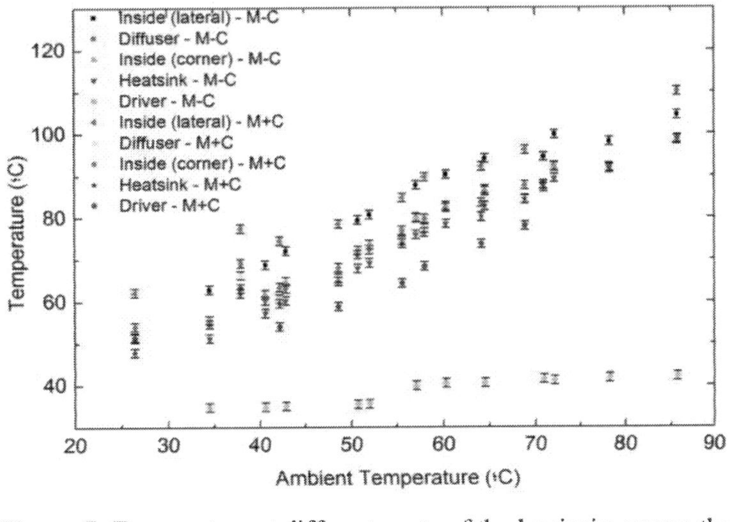

Figure 7. Temperature at different spots of the luminaire versus the ambient temperature inside the oven.

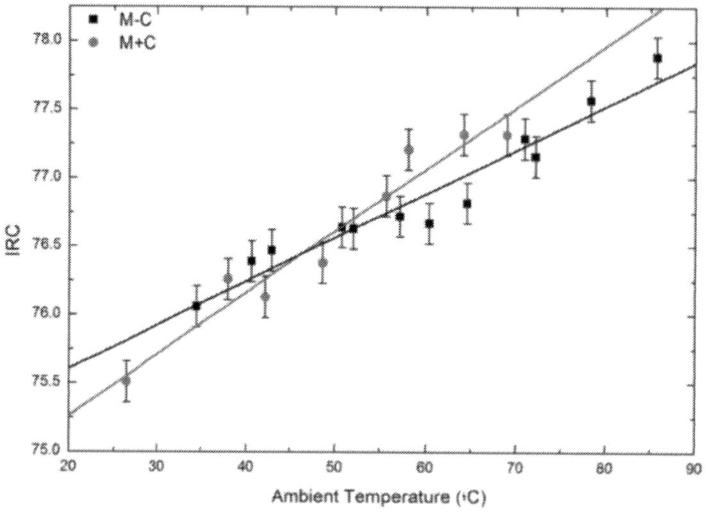

Figure 8. Effect of the variation of the temperature over the IRC.

In figure 9 and 10 we observe the effect of the elevation of the ambient temperature over the chromatic coordinate.

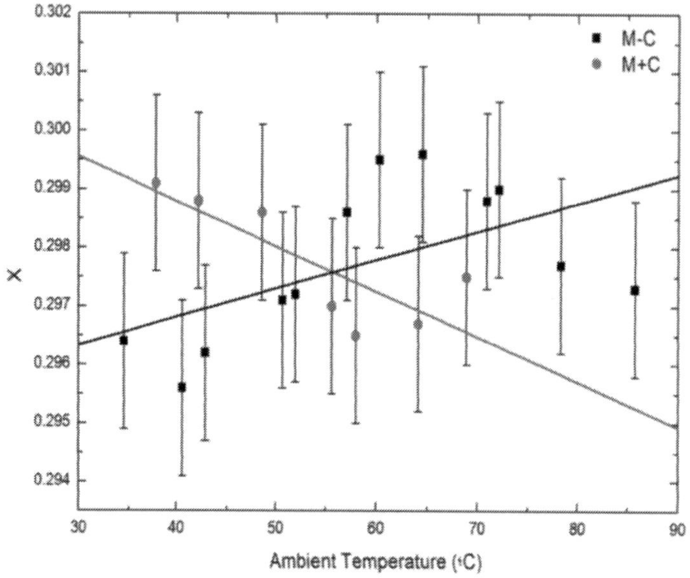

Figure 9. Effect of the variation of the ambient temperature over the chromatic coordinate X.

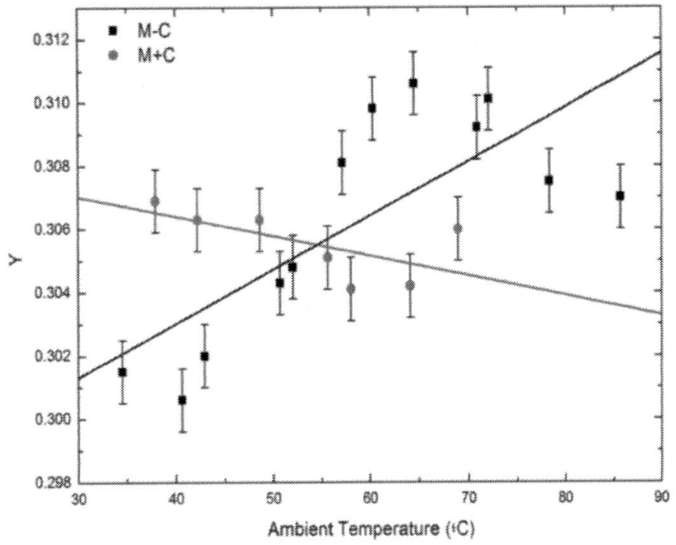

Figure 10. Effect of the variation of the ambient temperature over the chromatic coordinate Y.

In figure 11 the behaviour of the correlated color temperature is shown:

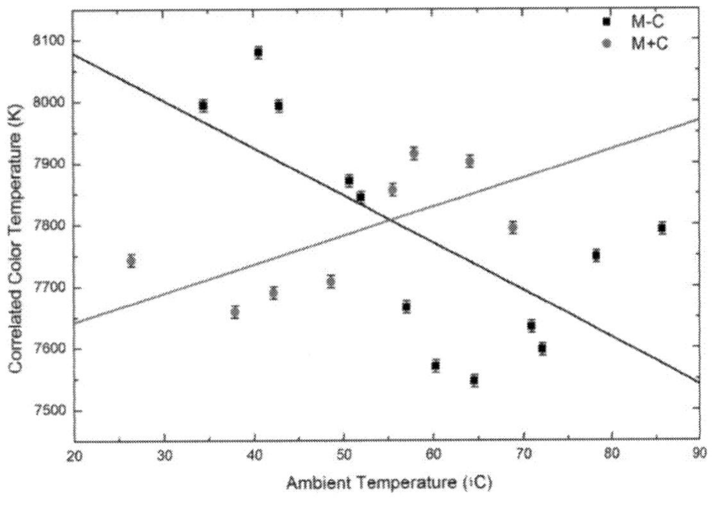

Figure 11. Effect of the variation of ambient temperature over the correlated color temperature.

In figure 12 we can note the relation between the ratio peak1/peak2 of a LED spectrum with the ambient temperature.

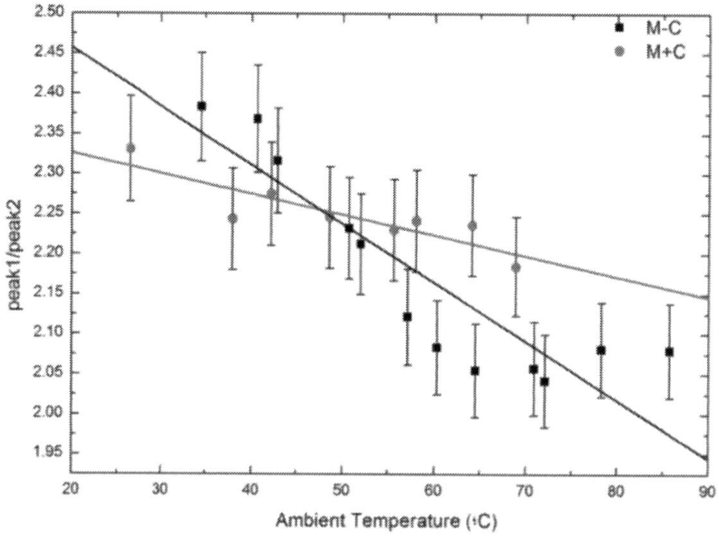

Figure 12. Effect of the variation of ambient temperature over the ratio peak1/peak2.

In figure 13 we have two spectra emitted by the luminaire. Both were obtained with the driver outside the oven. However the black dots represents the data obtained at 34.5°C and the red dots the data obtained at 85.7°C.

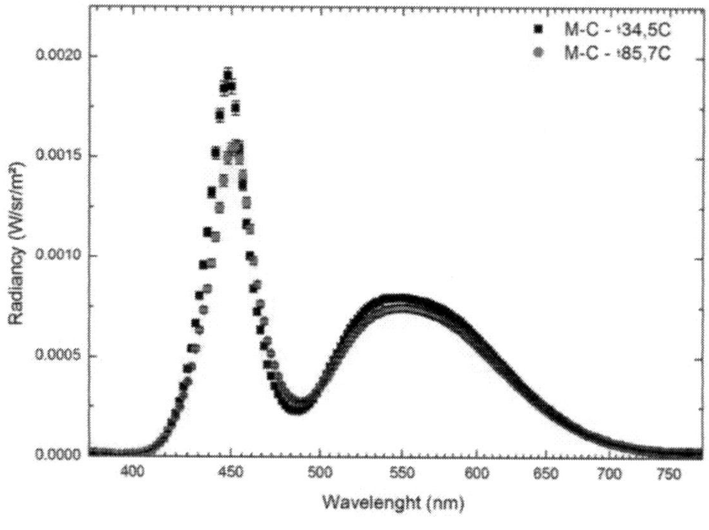

Figure 13. Emission spectrum of the luminaire during the M-C procedure under 34.5°C and 85.7°C.

In figure 14 we observe the effect of the increased temperature on total harmonic distortion.

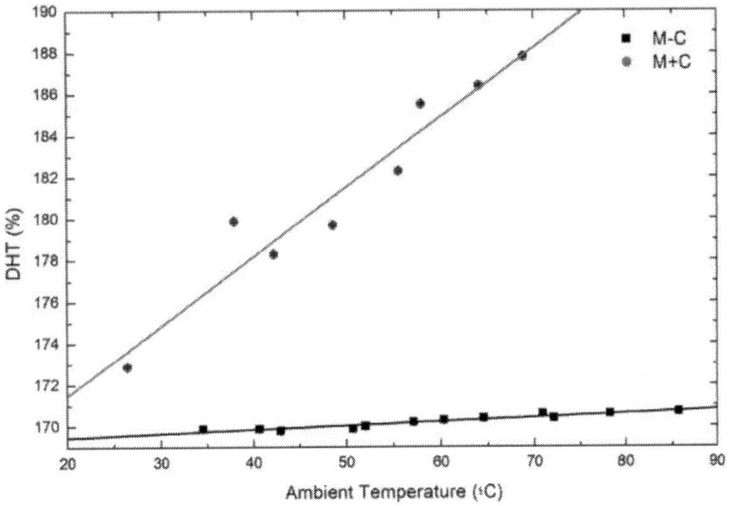

Figure 14. Effect of the variation of ambient temperature over the DHT.

In figure 15 we observe the effect of the ambient temperature on electric current.

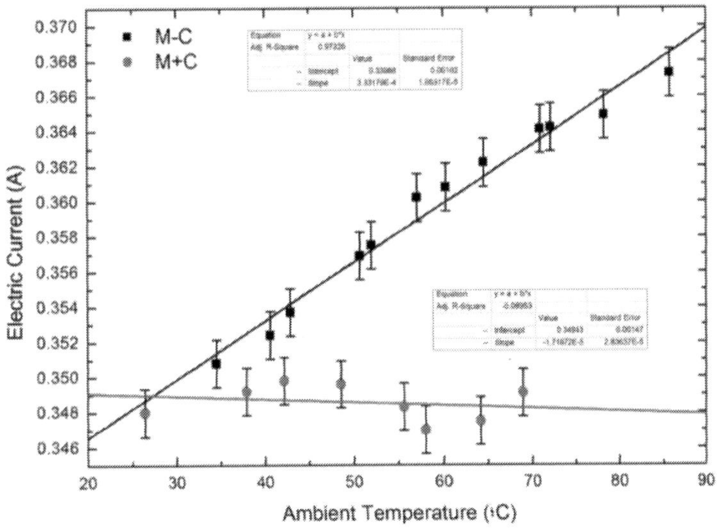

Figure 15. Effect of the variation of ambient temperature over the electric current.

From figure 7 we can notice the increase of the temperature in several points of the luminaire. For both situations we can note that the inner part of the luminaire shows the maximum value of temperature. An interesting fact since the LED module is installed inside this part the luminaire. At this compartment the temperature is about 30°C higher than in the other spots.

In figure 8 we can notice that for both situations, M+C and M-C, a rise in color rendering index occurs. However, analyzing the graph, for the set M-C, we notice a larger variation comparing to M+C, 0,045(4)/°C and 0,032(2)/°C, respectively. It indicated that the driver has a relatively sensibility for the variation of ambient temperature.

Talking about the chromatic coordinates, shown in figures 9 and 10 we notice that we have similar behaviors in both situations. When analyzing M+C we notice a tendency of decrease in X and Y, indicating the contribution of wavelengths corresponding to shades of purple. In M-C we notice an increase in values of X and Y, indicating the contribution of wavelengths corresponding to yellow.

The color correlated temperature, in figure 11, also shows two distinct behaviors for each case. When analyzing M+C, its linear fit shows a negative slope, whilst in M-C the opposite occurs. A curious fact is that M+C has a less variation than the M-C data. Maybe it happens because the driver compensates the effect of the temperature on the LED module and vice-versa.

Analyzing the spectra in figures 12 and 13 we can notice that the ratio between the radiance in peak 1 and in peak 2 decrease more sharply to M-C situation, indicating that the compensation occurs again. In figure 13 the variation in spectrum can be noted.

In figure 14 we can observe the behavior of total harmonic distortion – THD – under the variation of temperature. We note a wide increase of THD in the case of M+C, while a constant value represents the case M-C.

In the case of current, figure 15, the effect of the elevation of ambient temperature modifies mainly M-C data.

5. Conclusions

We can note a sharply influence of the variation of the temperature in distinct components that makes up a luminaire project. These components most of the time are sold as separated items and when these items are put together sometimes the compability between these items and the final product is not took into account.

This experiment may be improved by analyzing others colorimetric and electrical parameters that are part of the project of the luminaire. Also analyze different types of luminaires, such as those used for public lightning. The new parameters to study could be life cycle and the luminaire´s degradation suiting the boundary conditions to a Brazilian reality.

References

[1] National Research Council 2013 *Assesment of Advanced Solid State Lightning* 500 Fifth Street NW Washington The National Academies Press

[2] U.S. Department of Energy 2012 Building Technologies Program Solid-State Lightning Fact Sheet

Automated Evaluation of Dynamic Performance of Impulse Voltage Measurement Systems

L C Faria[1], E C Silva[2], M T Silva[3], C R H Barbosa[1], L C Azevedo[3]
[1] Posgraduate Program in Metrology, Pontifical Catholic University of Rio de Janeiro, Rua Marquês de São Vicente, 225, Gávea, 22453-900, Rio de Janeiro, Brazil
[2] Electrical Engineering Department, Pontifical Catholic University of Rio de Janeiro
[3] Centro de Pesquisas de Energia Elétrica, Av. Olinda 5800, Adrianópolis - Nova Iguaçu, 26053-121, Rio de Janeiro, Brazil

E-mail: hall@puc-rio.br

Abstract. This manuscript presents and describes an automated system for the evaluation of the dynamic performance of high-voltage measurement systems, according to the requirements of the standard IEC 60060-2/2010. The system was developed in LabVIEW and controls the acquisition, measurement and analysis of step response tests of measurement systems, automatically calculating the relevant amplitude and time parameters.

1. Introduction

The periodic performance verification of the measurement systems is one of the essential activities that ensure the technical quality of the results obtained in test laboratories of high voltage electrical equipment [1]. Such verification is required to prove and guarantee that the characteristics of the measurement systems are adequate for each particular test. The measurement systems used in high voltage and high current impulse tests are among the systems to which it is demanded more detailed performance verification.

Part of this verification refers to characterization of the dynamic performance, especially for voltage dividers and shunts, which are basic components of the systems. Such a characterization is based on the measurement and analysis of the voltage or current step response [2]. The step response of measurement systems is measured and processed in order to determine some characteristic parameters, according to known and standardized technical criteria. Such parameters delimit the application of the measurement systems in high voltage tests, related to the front time of lightning impulse voltage tests and current impulses.

In 2010, the International Electrotechnical Committee (IEC) revised the standard IEC 60060-2 [1], aiming at defining new requirements for test and calculation of the parameters that allow the evaluation of the dynamic performance of the measurement systems employed by high voltage test laboratories. Such a performance evaluation is performed in resistive, capacitive and hybrid dividers and in resistive shunts used in tests involving voltage/current impulses. With the advent of digital signal processing techniques, it was necessary to develop new software to control the acquisition and automatically analyze the step response parameters, as presented in this manuscript.

Next section describes the various components of the measurement chain. On following, the step response test is explained and the software STERAL 2.0 is presented, having been developed according to the new standard IEC 60060-2.

2. Measurement Chain

The evaluation of dynamic performance is necessary to obtain the limits of application of an impulse measurement system. The main components of the measurement chain are defined in more details in the next subsections.

2.1. System Under Test

In order to guarantee accuracy, traceability and reliability of the results, the measurement systems used by high voltage laboratories must be periodically calibrated. The calibration process can be summarized as a comparison between the system being calibrated and a standard system and, additionally, by the measurement and analysis of the dynamic performance of the measurement system, based on a step response evaluation.

2.2. Step Generator

The step response evaluation is performed by applying, to the system under test, a voltage step with 4 ns rise time and 90 V amplitude, produced by a step generator. Based on the response of the measurement system to the voltage step, it is possible to assess its measurement range. For lightning impulses, the front time ranges from 0.84 μs to 1.56 μs, and the measurement system dynamic performance must be compatible with such times. The measurement system includes the following components: divider/shunt, transmission system and digital recorder.

2.3. Digital Oscilloscope

The output signal of the system under test is acquired by a Tektronix digital oscilloscope, model TDS 5104B or TDS 7104, both having 4 channels and maximum sampling frequency of 4 GS/s. Both models have GPIB control interface and can be connected, through GPIB–USB adapter cable, to a microcomputer, what allows the oscilloscope to be automatically controlled by software developed specifically for this process. STERAL 2.0, as described below, controls the oscilloscope, acquires and automatically calculates the step response parameters, storing the data in a text file.

2.4. STERAL 2.0 Control System

The platform used for the development of STERAL 2.0 was LabVIEW 9. By using the adequate oscilloscope drivers provided by Tektronix, the program controls the acquisition of the step response signal, providing to the user all the controls available in the oscilloscope front panel.

3. Step Response Measurement

A voltage step is applied to the divider or measurement system and its response is measured and analyzed by an oscilloscope or digitizer. The step response test is performed based on standard IEC 60060-2, being presented in Figure 1 a typical signal [1]. Is is also automatically calculated the curve $T(t)$, the integral over time of the function $1-g(t)$, shown in Figure 2 [1].

Based on both curves $g(t)$ and $T(t)$ several parameters are measured:

- Reference Level (l_R): average amplitude of the acquired curve taken over the reference level epoch, i.e., between $0.5t_{min}$ and $2t_{max}$, where the interval $[t_{min}, t_{max}]$ defines the nominal epoch for the relevant time parameter (e.g. in the case of full lightning impulses, the relevant time parameter is the front time T_1, and the nominal epoch is typically [0.8 μs, 1.8 μs]). In practice, due to limitations of the oscilloscope, the reference level is calculated over the interval $[0.25t_{aq}, t_{aq}]$, where t_{aq} is the total acquisition time. The reference level l_R is used as a reference for the normalization of the step response curve in order to obtain $g(t)$, and identifies the stabilization level of the step response waveform.

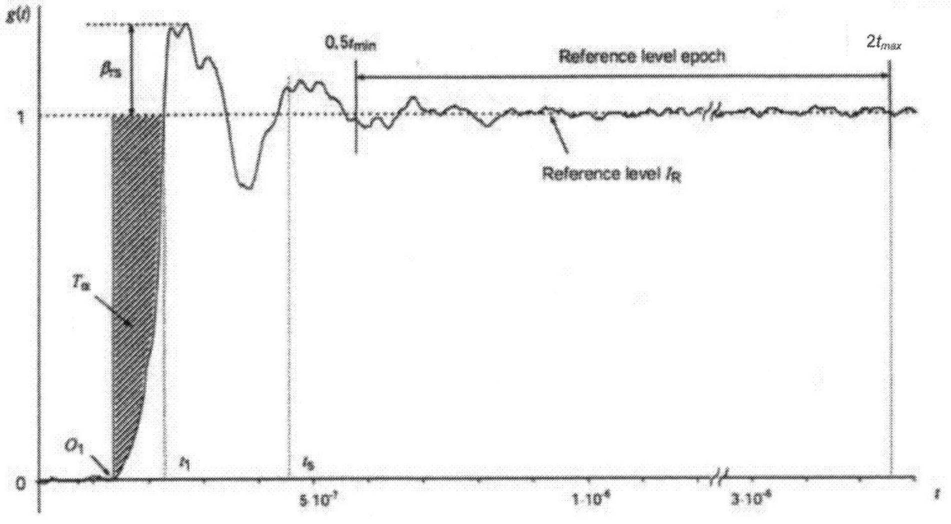

Figure 1. Typical Normalized Step Response Curve $g(t)$ [1]

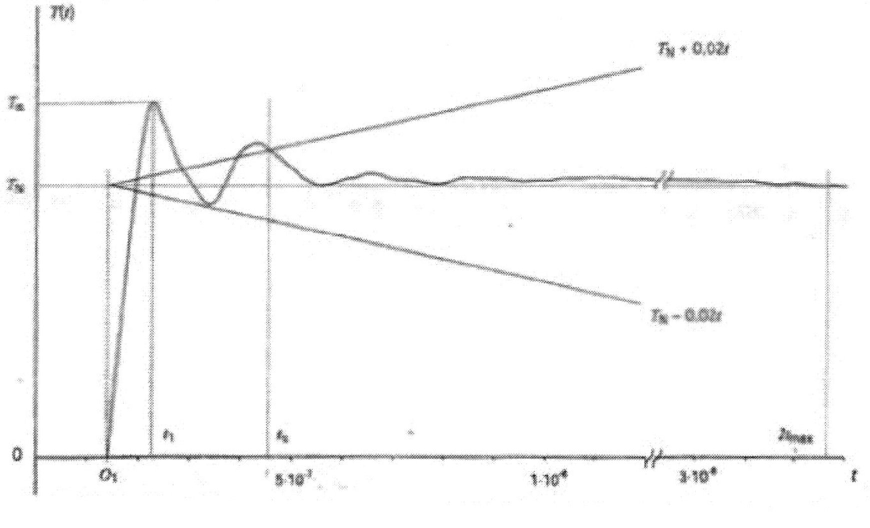

Figure 2. Typical Step Response Integral $T(t)$ [1]

- Virtual Origin of the Step Response (O_1): instant when the wave front initiates a monotonic rise above the amplitude of the noise at the zero level of the unit step response. All other time parameters are calculated using this instant as a time reference.
- Experimental Response Time (T_N): value of the step response integral $T(t)$ at $t = 2t_{max}$ (or $t = t_{aq}$)
- Partial Response Time (T_α): maximum value of $T(t)$ for $t \leq 2t_{max}$ (or $t \leq t_{aq}$)
- Residual Response Time ($T_R(t_i)$): T_N subtracted by $T(t_i)$, where $t_i < 2t_{max}$. In the case of STERAL 2.0, t_i is the instant when $g(t)$ first becomes higher than unity
- Overshoot of the Unit Step Response (β_{rs}): difference between the maximum value of the normalized waveform, g_{max}, and unity, as a percentage of unity.
- Settling Time (t_S): shortest time for which the residual response time $T_R(t)$ becomes and remains less than 2% of t.
- Initial Distortion Time (T_0): partial area between the line of zero and the virtual origin.

Journal of Physics: Conference Series **575** (2015) 012011 doi:10.1088/1742-6596/575/1/012011

4. STERAL 2.0

The software STERAL 2.0 comprises 2 modules: Acquisition and Analysis. The Acquisition module, shown in Figure 3, controls the oscilloscope and acquires the step response waveform, with a typical sampling frequency of 1 GHz and a length of 1000 samples. The software initially identifies the oscilloscope that is connected to the computer and, if it is not one of the compatible models, from the series DPO4000 and DPO7000, an error message is shown and the program is halted.

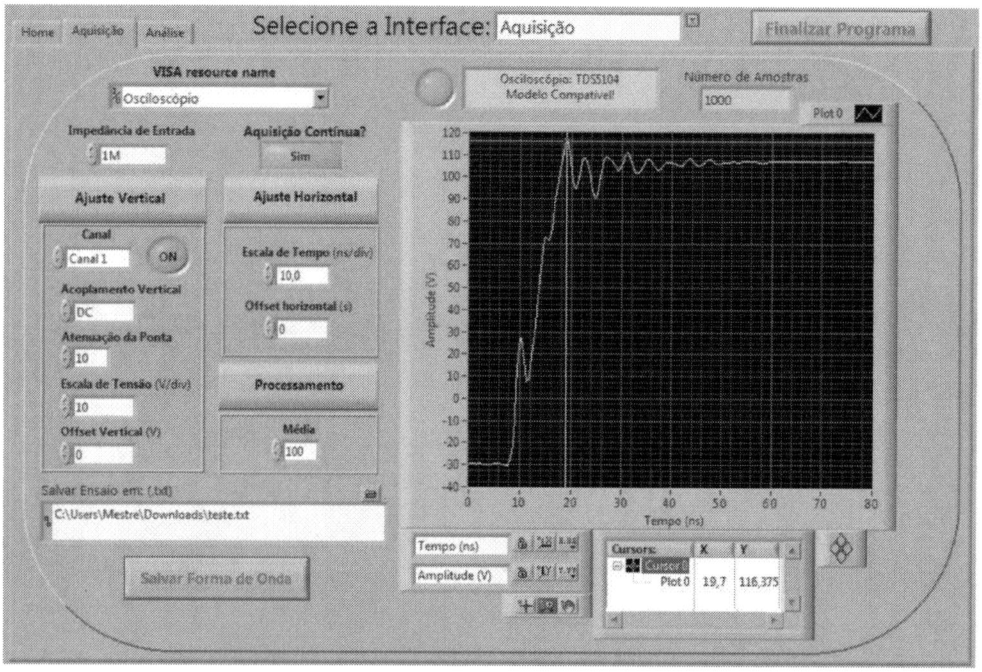

Figure 3. STERAL 2.0 Acquisition Module

On following, it is immediately started the reading of the oscilloscope selected channel, using the **Voltage Scale** and **Time Scale** values previously defined by the operator. Another important control in high voltage tests is the **Input Impedance**, which can be set to 1 MΩ or 50 Ω.

The interface also has controls for **horizontal offset** and **vertical offset**, so that the operator might adjust the waveform to the graph display and optimize the acquisition. It is also possible to include the attenuation factors introduced by dividers and other attenuators used by the measurement system.

The second module of STERAL 2.0, shown in Fig. 4, is the Analysis, presenting to the user an interface that allows the analysis of a previously acquired waveform. It is possible to analyze a series of tests with different voltage dividers, for example, without the need to follow a predefined sequence.

After the waveform to be analyzed is selected, the relevant parameters are automatically calculated and the waveforms related to the analysis are presented in 3 separate graph displays, called: $G(t)$, $g(t)$ and $T(t)$. $G(t)$ is the original step response waveform, being the reference level automatically estimated and indicated in the graph; $g(t)$ is the normalized step response, based on the reference level, being also shown the so-called adjustment line and the initial level; $T(t)$ is the integral over time of the normalized step response $g(t)$, used by the operator to verify the settling of the step response.

The initial level is automatically calculated, but can be manually adjusted should the operator desires to select a different value. It affects the determination of the virtual origin O_1. As described in the previous section, the parameters calculated by this module are: virtual origin (O_1), experimental response time (T_N), partial response time (T_α), residual response time (T_{re}), t_S, T_0 and Overshoot (%). Other important value presented are the Number of Samples and the acquisition interval dt (s).

As for the wave front adjustment line, the software allows two types: linear regression and maximum derivative, the selection being made by the operator based on his expertise. After the operator finalizes the adjustments, the parameters calculated can be saved in an Excel spreadsheet.

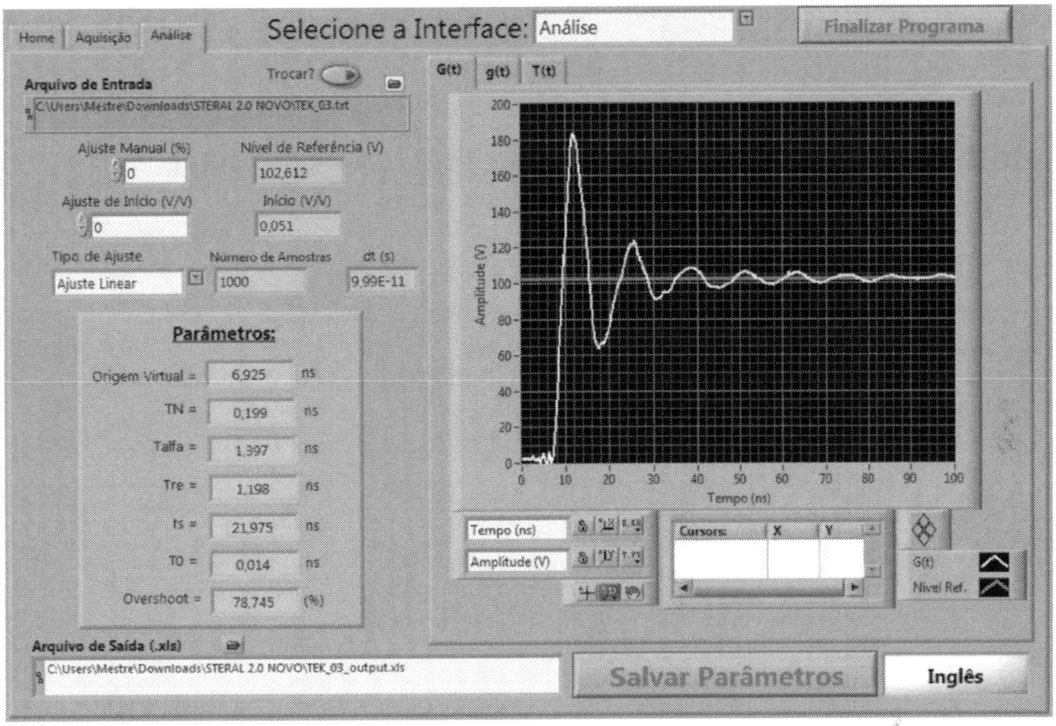

Figure 4. STERAL 2.0 Analysis Module

5. Conclusion

The implementation of the STERAL 2.0 software yields reliability in the evaluation of the dynamic performance of impulse measurement systems, aiming at guaranteeing the adequacy of voltage dividers used in high voltage impulse tests. Such a reliability is derived from the results that have a level of accuracy higher than the previously used software (STERAL 1.0), also presenting a more versatile operation, employing new technologies that allow better evaluation of the systems under test.

STERAL 2.0, having been developed following the new standard IEC 60060/2010, is updated and will be able to be used for several years without the need of further updates, aside from new versions of the IEC standards. One important feature of the new software, which provides a differential in relation to the prior version, is the portability for the tests, as STERAL 2.0 can be executed in a laptop computer connected to an oscilloscope, precluding the need for a large and complex experimental setup. Finally, STERAL 2.0 can also be used to evaluate the dynamic performance of measurement systems used with current impulse test according to standard IEC 62475/2010 [2].

6. References

[1] IEC 60060-2/2010
[2] IEC 62475/2010

Assessing Quality of Experience while comparing competing mobile broadband services from the user perspective

Ewerton L. Madruga, Rodrigo David, Rodolfo Sabóia de Souza, Romulo Dantas

INMETRO – Instituto Nacional de Metrologia e Qualidade e Tecnologia
Diretoria de Metrologia Científica (Dimci)
Divisão de Metrologia de Tecnologia da Informação e Comunicações (Dmtic)
Av. Nossa Senhora das Graças, 50 – Prédio 6
Duque de Caxias, RJ BRAZIL 25250- 020

{elmadruga, rpdavid, rssouza}@inmetro.gov.br, dantass.romulo@gmail.com

Abstract. The growth of mobile traffic is exploding globally, and users can already choose their best smartphone or tablet options from a handful of manufacturers based on specific criteria such as price and usability. It is much less clear when the user needs to pick from various mobile broadband service providers when choices are available. After all, how does one know what is the best provider for a given usage profile? This work uses drive tests to investigate the variation of radio frequency conditions and relate them to the quality of experience from the viewpoint of the user.

1. Introduction

Everything is going mobile. This evolution is driven by video, cloud-based services, the internet and machine-to-machine (M2M) connectivity. Access to the internet is a prerequisite and will drive further build-out of mobile networks. By 2017, an astonishing 85 percent of the world's population will be covered by 3G networks while 4G ramps up its global deployment [1].

This work focuses on studying different sources of information to develop objective tools users could leverage to find out about the mobile broadband service offered by different providers in a given geographical area. Therefore, it looks for relevant information made available by modems, tablets and smartphones with the purpose of letting users make a more educated guess of what operators could better provide service in the areas where people live, work or just entertain themselves.

2. Relevant technology components

When planning a new one or optimizing a network already in production, mobile broadband providers have to rely on a small set of radio frequency parameters to establish whether a given geographical region meets the target service level. And it is common practice to rely on parameters that measure signal strength, interference and link quality [3][4].

The first one is signal transmit power, which establishes how far out any information coming from the radio station will propagate. The second is signal interference, which depending on the cellular technology considered may be caused by the other cell phones and modems in the service area.

When considering HSDPA, other parameters like channel quality may be of importance. HSDPA requires user equipment to compute a metric known as Channel Quality Indicator (CQI), which represents the status of the radio frequency conditions in a given moment. It is important to note that the CQI is a metric computed at the user equipment and sent in the uplink to the radio station and user for the optimization of the packet scheduling policy chosen by the operator. As opposed to RSCP and Ec/Io, which are metrics associated with the entire cell, the CQI is a representation of the unique view from the perspective of the user. This indicator is also a parameter computed by user equipment and used also in LTE networks as an input for packet scheduling [2].

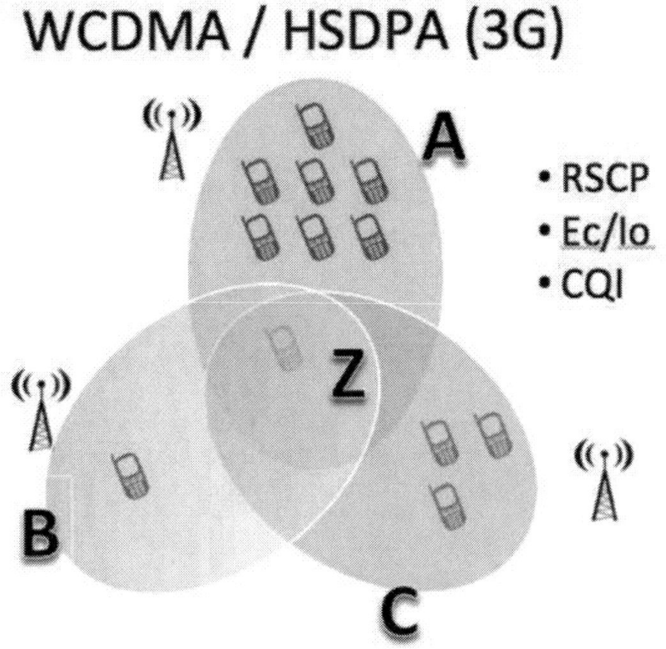

Figure 1. Multiple cellular radio stations

The diagram in Figure 1 shows a simplified example of a WCDMA/ HSDPA network with 3 distinct cells A, B and C, with an overlapping area Z. This diagram can be seen as a snapshot reflecting mobility and radio frequency conditions in a given moment in time.

All things being equal, and if the backhaul linking all three base stations to the core of the service provider have equivalent capacity, the best broadband experience tends to be the one for the user equipment in cell B. For this particular case, the Received Signal Code Power (RSCP), the usual signal strength metric for WCDMA-based networks, is expected to be ideal given the close range to the base station antenna.

Furthermore, this same user equipment is located into a cell with very few other devices: in this particular scenario, there is only the device in area Z. This device is thus less prone to suffer phone-based interference, which is a fact peculiar to WCDMA or any other technology where mobile phones and modems use the same frequency to communicate simultaneously. In WCDMA terminology, the status of signal and interference is frequently measured by the ratio of the received signal code power (RSCP) divided by the total received power (RSSI) in the channel bandwidth (Ec/Io) [3].

As mentioned above, HSDPA requires user equipment to compute a metric known as Channel Quality Indicator (CQI) which is a measurement of the state of radio frequency conditions on the downlink channel (from base station to user equipment) in a given moment. In a scale that ranges from zero to 30 (zero being the worst possible condition), the device computes this indicator based on the probability to transmit successfully with a high throughput to the base station it is associated with in a particular instant. After receiving this information, this measurement is used by the base station to make proper scheduling arrangements, assigning more resources to the devices with a better detected radio frequency condition.

3. Related Work

In the United Kingdom, the Office of Communications (Ofcom), the national regulatory agency, has been working with broadband service assessment for some time. As for mobility, a detailed study of nationwide mobile broadband service has been done by the agency, combining the subjective perception of service by the users with the objective evaluation of periodic automated tests [6].

These objective tests included software-based probes, static hardware probes and vehicle-based tests with smartphones collecting measurement data while moving in a geographical area. Ofcom has recently issued a request for input to its efforts on assessing the quality of mobile broadband, indicating it considers important to take the initial study further [6]. In the United States, the Federal Commission of Communications (FCC) is also observing closely how the industry evolves, with a focus on the state of market competition, as part of its annual reports to the American congress [7].

In Brazil, Anatel, the national telecommunications regulatory agency, and Inmetro, a government agency for standards and technology, teamed up with CGI.br, the Brazilian Internet governance body, to assess fixed broadband service [8]. This initial effort on assessing broadband at a national level has provided the necessary feedback for Anatel to issue new regulation with regard to managing quality in mobile broadband [9]. This regulation requires operators to constantly monitor download speeds, which must not be on average less than 60% of the advertised speed. Unfortunately, the rules do not require operators to make public the state of radio frequency parameters during drive tests, so that academic research could help users understand better in what conditions data was collected.

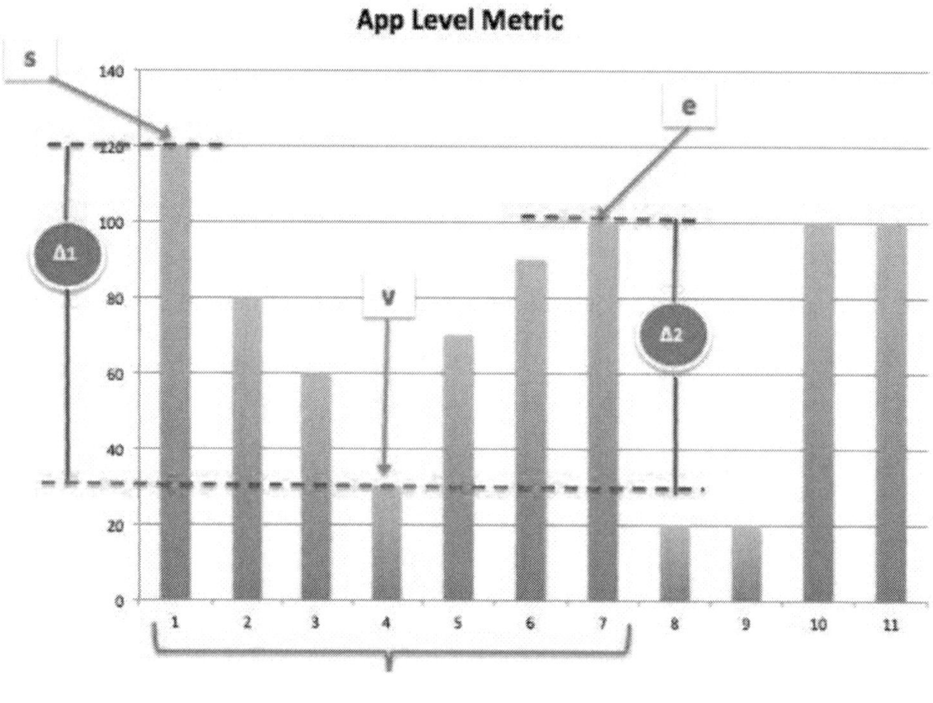

Figure 2. An application performance valley

4. Methodology

With the purpose to study how radio frequency parameters affect application level performance, all application measurement sessions were processed as to find a sequence of performance valleys. In this context, a valley of interest "v" is composed of a limited length sequence of samples, marked by the starting and ending sample, respectively denoted by the "s" and "e" in Figure 2.

A valley considered of interest has a maximum time window from the start to the end sample, lasting usually a few seconds. It is composed of two sections: a) the initial section where the application level metric decreases from the starting sample "s" all the way to the sample with the lowest performance value, denoted by "v" and b) the section with a gradual increase from sample "v" to sample "e". Also, a valley will only be accepted if the application performance at sample "s" differs from "v" at least by a given percentage ($\Delta 1$).

Furthermore, it is only accepted as a valid valley if performance at "e" is better than "v" by a given minimum percentage ($\Delta 2$).

5. Description of the Experiment

Located between the mountains and the sea, Rio de Janeiro is a challenging urban area to plan wireless services for. Specifically two areas in town were chosen due to its peculiarities. The first is the surrounding area of a lake in the southern section of Rio de Janeiro, known as "Lagoa."

The second area used for drive tests was the route that links two of the largest sports arenas in the city of Rio de Janeiro. In a stretch of an approximate 15-mile (22 km) round trip drive, this particular road crosses a residential area of middle class inhabitants that starts at the Maracanã stadium and ends at the Engenhão stadium. These are two of the most important venues of the upcoming sports events to be held in Rio de Janeiro, most importantly the Summer Olympics 2016.

The equipment used for the drive test data collection process was a SwissQual Diversity Benchmark, with a 3G spectrum analyzer [5] operating in the frequencies of 850 Mhz and 2.1 Ghz. The particular model used is equipped with 4 independent Intel Core 2 Duo 2 Ghz units with 2 Gb of RAM each, running Windows 7 Professional. Each unit is capable of starting and managing tests for one individual 3G service provider, for a total of four simultaneous possible tests. Each test unit is connected to a WCDMA/HSPA+ 850/2100 Mhz Sierra Wireless AirCard USB 308 modem. This particular model is of Category 14, and its maximum throughput is 21 Mbps.

FTP download was the application used as a reference for user experience of the perceived quality of service. The dedicated FTP server acting as the repository for file downloads is located in a datacenter. It is in a controlled environment with a 10 Gbps Internet link and dedicated computational resources enough to hold hundreds of simultaneous FTP sessions.

6. Results

The analysis of data collected during drive tests was broken down into a sequence of valleys of interest as defined above. By observing the defined reference values for RSCP, Ec/Io and CQI, it was possible to match cases where the performance drop coincided with a turn to the worse on any one of these RF parameters alone or in combination.

The reference values are set so that any occurrence of signal strength (RSCP) below -100 dBm is considered to be a possible cause for performance drops on FTP downloads, given a weak signal scenario.

Any occurrence of Ec/Io of -12 dB or smaller is considered an obstacle to good performance due to higher interference. And finally any occurrence of CQI equal to 14 or less is considered as a possible reason for the performance decrease.

Figure 6 shows a breakdown of possible causes for performance valleys for the drive tests we have done for each of the four major mobile broadband service providers. It holds an interesting finding that is associated with all four service providers studied in the drive tests. As shown by the dark blue bar in the graphs, a significant proportion of the throughput drops is not associated with any of the three radio frequency parameters considered.

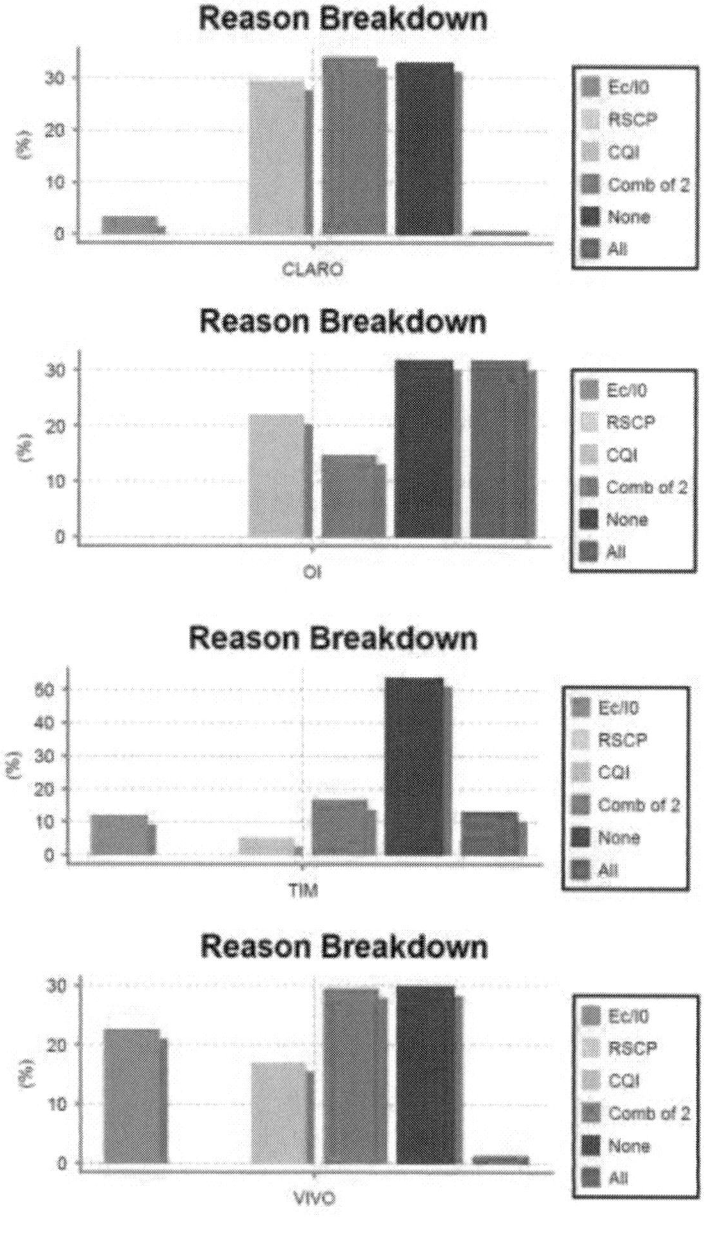

Figure 6. Possible reasons for performance valleys for 4 different providers

In the cases where none of the radio frequency parameters could be seen as possible causes for the speed decrease events, an excessive load on either end of the FTP test sessions could be a possible cause. But since all caution was used at the communication endpoints, either the backhaul connecting the radio station to the core network of the service provider cannot handle the aggregate traffic of all radio stations in the geographical area, or the core itself needs to adjust better to the traffic coming from the mobile broadband users.

7. Conclusion

In this work we have presented components of mobile broadband technology and introduced the radio frequency parameters that best measure parameters with important impact on user experience, that is, signal strength, interference and channel quality.

In this particular set of drive tests, all mobile service providers have shown application performance problems even when all radio frequency parameters are at adequate levels. This implies that even when service coverage is adequate, there is no UMTS cell oversubscribing and the user acknowledges a crystal clear wireless channel, our FTP download speeds suffered in performance at least 30% of the cases. This observation suggests limitation in the backhaul or in the network´s infrastructure in the areas where the drive tests occurred.

References

[1] Rysavy Research 2012 *Mobile Broadband Explosion*, 4G Americas.

[2] 3GPP 2013 *Evolved Universal Terrestrial Radio Access (E-UTRA) and Evolved Universal Terrestrial Radio Access Network (E-UTRAN): Overall description*, 3GPP TS 36.300 version 11.4.0 Release 11.

[3] Holma H. and Toskala A 2010 *WCDMA for UMTS – HSPA Evolution and LTE*, Fifth Edition, Wiley.

[4] Electronic Communications Committee (ECC) within the European Conference of Postal and Telecommunications Administrations (CEPT) 2007 *UMTS Coverage Measurements*.

[5] SwissQual 2010 *Diversity Benchmark User Manual*.

[6] Office of Communications - United Kingdom (Ofcom) 2013 *Measuring mobile voice and data quality of experience*.

[7] Federal Communications Commission - United States (FCC) 2011 *Measuring Broadband America: A Report on Consumer Wireline Broadband Performance in the U.S.*

[8] Madruga, E.L. et al. 2012 *Measuring nationwide residential broadband services*, IEEE International Conference on Innovations in Information Technology (IIT), Abu Dhabi.

[9] Agência Nacional de Telecomunicações 2011 *Quality Management of Mobile Service Provisioning*, Federal Regulation no. 575, RGQ-SMP (portuguese).

Kalman filtering applied to the calibration of a high-resolution analog-to-digital converter

Waldemar Guilherme Kürten Ihlenfeld [1], Gregory A. Kyriazis [1], Regis P. Landim [1]
[1] Instituto Nacional de Metrologia, Qualidade e Tecnologia - Inmetro

E-mail: wgkihlenfeld@gmail.com

Abstract. The paper proposes the application of the Kalman filter in dc calibrations of analog-to-digital converters with a Zener or a Josephson dc voltage reference standard. Repeated measurements over the entire full-scale range of an analog-to-digital converter with posterior Kalman filtering allow optimal estimation of its deviations in respect to a dc reference to be obtained, yielding improved assessment of converter's linearity.

1. Introduction

The Kalman filter is mainly used to tackle estimations problems from measurements of the state of nonstationary (and noisy) processes [1-3]. It has extensively been investigated since the 1960's and finds wide application in many different areas such as space crafting, determination of ballistic trajectories and orbiting, radar, and adaptive adjustment of antenna arrays among others. It plays an important role in the military and defense applications. This paper presents the application of the Kalman filter in metrology to estimate with enhanced denoising properties and to a very high quality, the deviations of the readings of an analog-to-digital converter (ADC) or digital voltmeter (DVM) by calibrating it against a dc voltage reference of quantum or classical base.

2. Definition of the problem

Figure 1 depicts schematically the calibration problem of an ADC corrupted by noise in which only the initial state $\mathbf{x}(0)$ of the reference (at the instant zero) is known with a variance σ^2_0 (i.e., assumed Gaussian distributed, $N(x(0), \sigma^2_0)$). The ADC readings $\mathbf{y}(n)$ may eventually be biased (i.e., with an offset) and to vary randomly with a variance σ^2_v. Although this problem seems of trivial nature, normally solved by averaging readings up to a certain number to avoid the effect of flicker noise, a Wiener filter is, by concept, predestined to deliver best estimations under stationary conditions. However, aiming at tackling in the future more complex and nonstationary problems related to the traceability of alternating (ac) quantities, we propose the use of the Kalman filter as a general treatment. Only the dc calibration of an ADC is here of interest.

The treatment of the Kalman filter can be found in extensive literature (see references). Figure 2 shows the Kalman filter in the form of a block diagram with feedbacks, as originally devised from control theory. In this case $\mathbf{x}(n)$ denotes the state vector of a dynamic system at time n described by a transition state matrix $\mathbf{A}$ based on previous (n-1) states. Best linear estimates of the state of the system at time n given the observations or measurements $\mathbf{y}(k)$ for index k = 1, 2,..., n are denoted with vector $\hat{\mathbf{x}}(n|n)$.

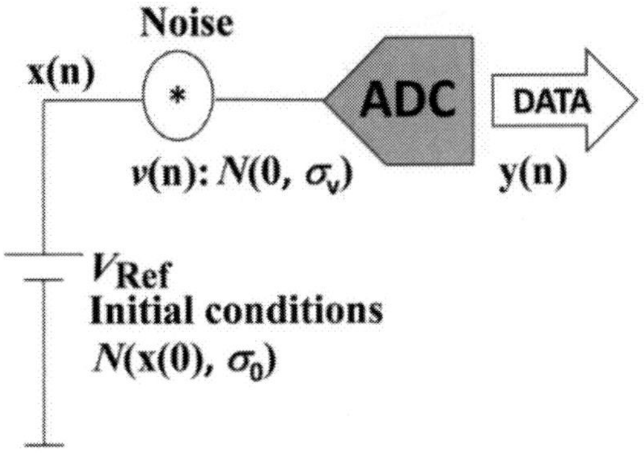

Figure 1. Direct calibration of an ADC with a reference at state **x**(n) from measurements (or sampled values) **y**(n) at instants n corrupted by noise.

Similarly, $\hat{\mathbf{x}}(\mathrm{n}|\mathrm{n}-1)$ stands for the best estimate at time n given the (n-1) prior observations (*i.e.*, y(k) for k = 1 to (n-1)). Noise vectors of the dynamic system and of measurements are denoted by *w*(n) and *ν*(n) respectively. The measurement vector **y**(n) is related to **x**(n) by an observation matrix **C**(n) with additive noise *ν*(n) as shown in the insert in figure 2. Measurements at time n, *i.e.*, **y**(n) are compared with measurement estimates $\hat{\mathbf{y}}(\mathrm{n}|\mathrm{n}-1)$ employing the past (n-1) observations, and the deviations are corrected with a Kalman gain **K**(n), which is continuously updated in a feedback loop (or tracking loop) so that the error vector **e**(n|n) of estimates of **x**(n) (*i.e.*, $\hat{\mathbf{x}}(\mathrm{n}|\mathrm{n})$) with respect to **x**(n) tends to a minimum. The z^{-1} factor stands for the delay operator of one sample, as is usually done in digital signal processing. Feedback loops of the type shown in figure 2 are central in electronics and akin to the known feed-forward feedback, which gives the utmost performance in respect to the antagonic requirements of accuracy and speed.

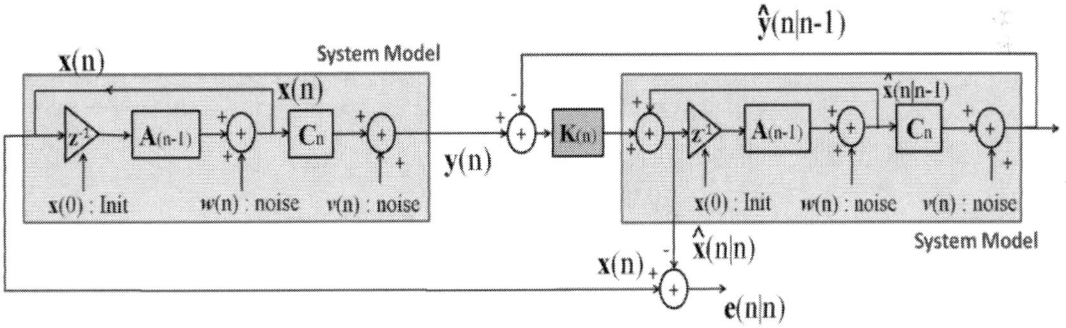

Figure 2. General block diagram of the discrete Kalman filter.

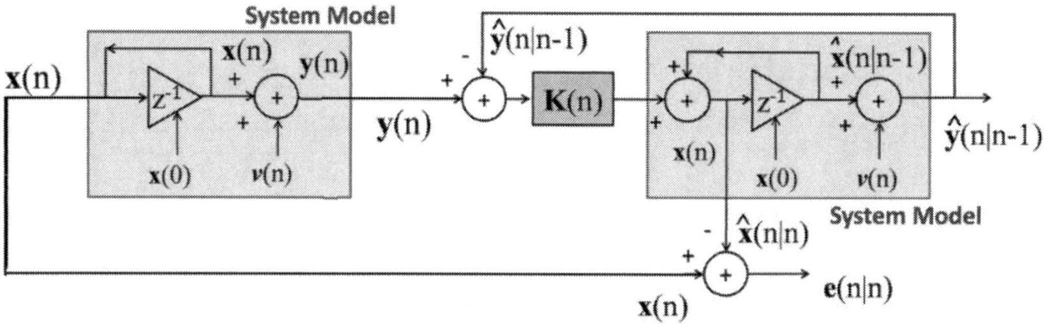

Figure 3. The Kalman filter in the form of a block diagram for the ADC calibration problem as posed in figure 1 and according to figure 2.

3. Kalman filtering

From figure 2, estimates at time n employing all n samples are expressed by

$$\hat{\mathbf{x}}(\mathrm{n}|\mathrm{n}) = \hat{\mathbf{x}}(\mathrm{n}|\mathrm{n}-1) + \mathbf{K}(\mathrm{n})[\mathbf{y}(\mathrm{n}) - \mathbf{C}(\mathrm{n})\hat{\mathbf{x}}(\mathrm{n}|\mathrm{n}-1)] \quad (1)$$

Similarly, estimates at time n employing the (n-1) samples are recursively dependent on previous estimates as

$$\hat{\mathbf{x}}(\mathrm{n}|\mathrm{n}-1) = \mathbf{A}(\mathrm{n}-1)\hat{\mathbf{x}}(\mathrm{n}-1|\mathrm{n}-1) \quad (2)$$

The Kalman gain **K**(n) is dependent on the noise covariances and on the Hermitian transpose of the observation matrix **C**(n), i.e., $\mathbf{C}^{H}(n)$ [2-3].

For the particular case of a dc calibration with a "constant" dc voltage reference (i.e., x(n) = x(n-1)) in which $w(n) = 0$ and only the initial conditions are previously known (as stated in figure 1), it follows that

$$\hat{\mathbf{x}}(0|0) = \mathrm{E}\{\mathbf{x}(0)\} = \mathrm{x}(0) \quad (3)$$

and the initial covariance

$$\mathbf{P}(0|0) = \mathrm{E}\{\mathbf{x}(0)\mathbf{x}^{\mathbf{H}}(0)\} = \sigma_0^2 \quad (4)$$

where E{·} stands for the expectancy of a random quantity. Further, according to figure 3, other matrices simplify to **A** = 1, **C** = 1, and **x**(n|n) to a scalar quantity, whose Kalman estimates for n ≥ 1 are:

$$\hat{\mathrm{x}}(\mathrm{n}) = \hat{\mathrm{x}}(\mathrm{n}-1) + \frac{\sigma_0^2}{\mathrm{n}\sigma_0^2 + \sigma_v^2}[\mathrm{y}(\mathrm{n}) - \hat{\mathrm{x}}(\mathrm{n}-1)] \quad (5)$$

In calibrations, the ADC deviations at time n, *i.e.*, δ(n) from the reference (or nominal) are of interest and are expressed as

$$\delta(\mathrm{n}) = \hat{\mathrm{x}}(\mathrm{n}) - \mathrm{x}(0) \quad (6)$$

To illustrate the application of the Kalman filter, suppose that a 28-bit integrating ADC with a noise spectral density of 200 nV/(Hz)$^{1/2}$ has to be calibrated against a 10 V dc-reference with a known "uncertainty" σ_0 of 1 μV (root-mean-square). Acquiring equally spaced samples y(n) with an aperture time of nearly 1 s at a 1 s sampling rate leads to a σ_v of approximately 141 nV. Figure 4 displays the ADC δ(n) deviations for simulated observations y(n) with added Gaussian distributed noise. Application of equation (5) demonstrates the enhanced noise suppression of the Kalman filter as $\delta(n) \rightarrow 0$ for large n. Although the ADC readings spread considerably, the estimated values $\hat{x}(n)$ barely vary. Some small systematic deviations though persist, which are dependent on the start conditions.

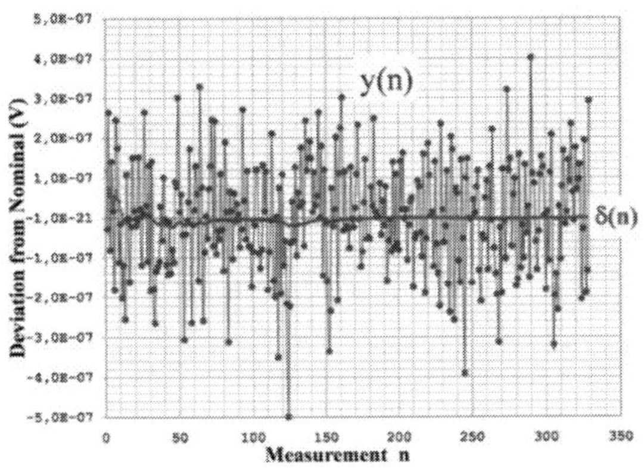

Figure 4. Simulated Kalman filtering for an ADC calibration at its full-scale of 10 V.

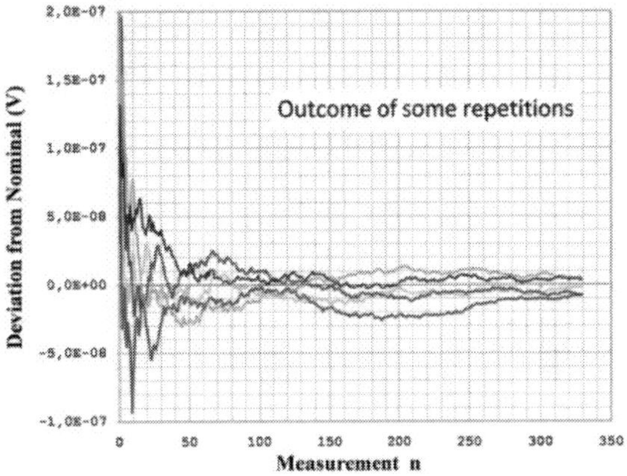

Figure 5. Simulated outcome of five repetitions of the Kalman filter on five different data records of 330 values each.

Figure 5 shows the outcome of five independent repetitions of the Kalman filter on five different records. For n > 300 and the case in question, the δ deviations are smaller than 10 nV or 1 into 10^9 parts of full-scale (10 V).

This procedure can be repeated for many x(0) start values within the whole full-scale range of the ADC to access its linearity.

Figure 6 shows an experimental result of the Kalman filtering for an 8 ½ digit 3458A digital multimeter (DMM) at 10 V. Zener x(0) start value was obtained from a conventional Josephson Voltage Standard calibration performed right before the DMM calibration. In this case, the voltage offset was not compensated and an estimation of its value can be made from figure 6 (around 1 μV).

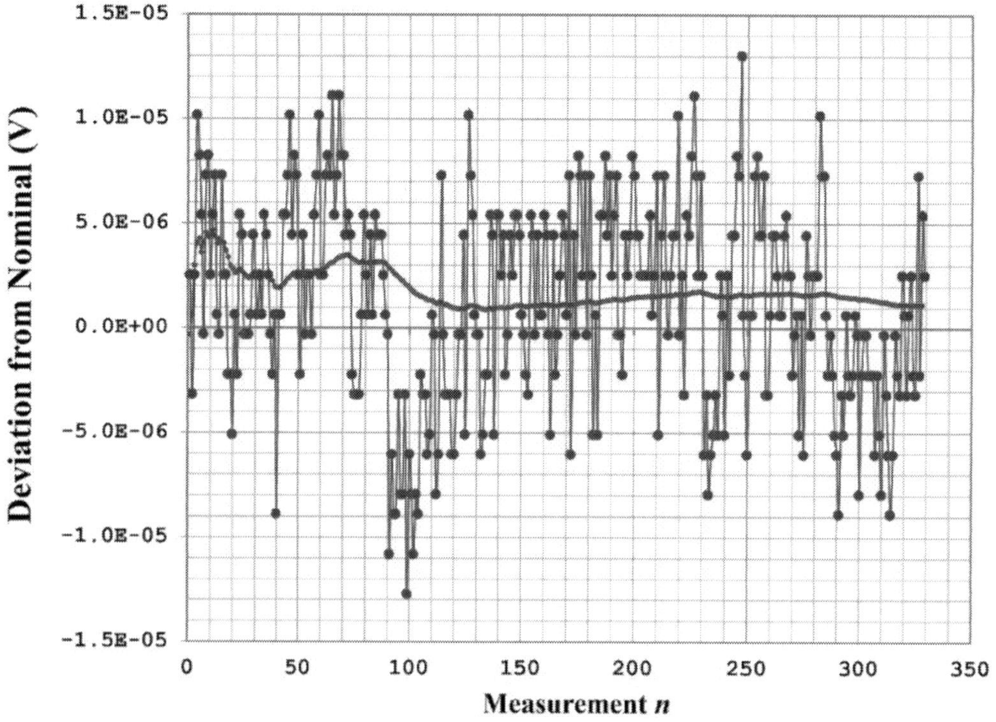

Figure 6. Experimental Kalman filtering for an 8 ½ digit 3458A digital multimeter (DMM) calibration at its full-scale of 10 V using a Zener reference.

4. Uncertainty of the Kalman estimates

The Kalman filter has strong ties to Bayesian inference, where prior information is used for estimation purposes. An approach which takes into account only the moments of the posterior probability distribution to determine the dynamic state of a system [4] demonstrates that the optimal solution for this problem is the Kalman's filtering algorithm.

For the steady model considered in this paper, information about the state at time n is described by a Gaussian distribution whose mean and variance are respectively

$$\hat{x}(n) = \frac{\sigma_v^2 \hat{x}(n-1) + \sigma_{n-1}^2 y(n)}{\sigma_v^2 + \sigma_{n-1}^2} \tag{7}$$

$$\sigma_n^2 = \frac{\sigma_0^2 \, \sigma_v^2}{n\sigma_0^2 + \sigma_v^2} \tag{8}$$

Equation (7) is equivalent to equation (5). The moments of the distribution will approach limits as n gets large or approaches infinity, *i.e.*, σ^2_n approaches σ^2_v/n and $\hat{x}(n)$ the average of the observations, becoming increasingly insensitive to additional observations y.

It is interesting to compare these findings with results obtained from the Bayesian framework used in traditional ′stationary′ metrology. In this case, information about the state at time n is described by a Gaussian distribution whose mean is [5]

$$\hat{x}(n) = \frac{\sigma_v^2 \hat{x}(0) + \sigma_0^2 \sum_1^n y(n)}{n\sigma_0^2 + \sigma_v^2} \tag{9}$$

The variance is also given by equation (8).

Equations (5), (7) and (9) are all equivalent leading to the same numerical result.

Figure 7 shows the differences of the mean of observations from 1 to n and the Kalman estimates at time n. For large n and neglecting 1/f - noise effects, these differences get close to zero on the limit as expected. Although these differences appear negligible in this example, the Kalman filter delivers the optimum estimates when nonstationarity is rampant.

Figure 8 shows a simulated outcome of Kalman estimates with the evaluated Kalman uncertainties as given in equation (8).

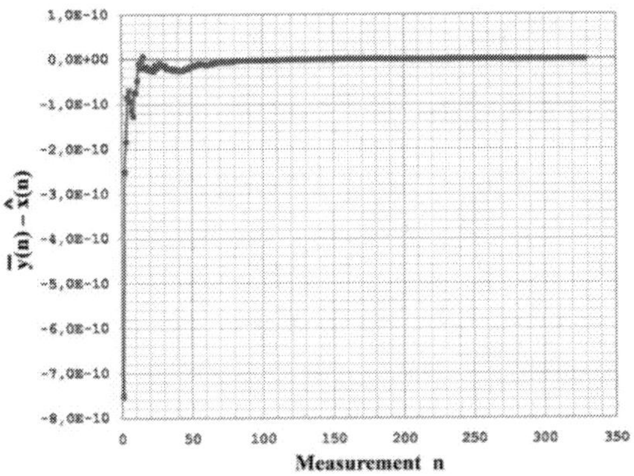

Figure 7. Deviations of the mean of measurements in respect to Kalman estimates in dependence of n.

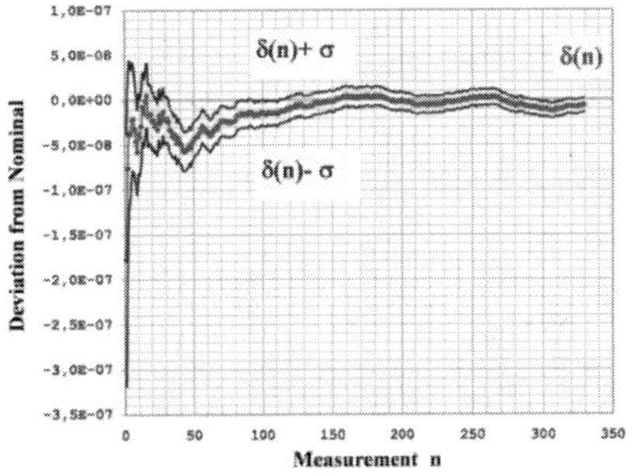

Figure 8. A simulated outcome of Kalman estimates with the evaluated Kalman uncertainties as given in equation (8).

Acknowledgements
This work was conducted in the frame of the project Pronametro of the Instituto Nacional de Metrologia, Qualidade e Tecnologia – INMETRO.

References
[1] Kalman R E 1960 Trans ASME *J. Basic Eng* Ser 82D 35.
[2] Haykin S, *Adaptive Filter Theory*, Prentice Hall, Englewood Cliffs, NJ, 1986.
[3] Hayes M H, *Statistical digital signal processing and modeling*, John Wiley & Sons, NY, 1996.
[4] Kyriazis G A, Martins M A F and Kalid R A 2012 *Measurement* 45 1558-1563.
[5] Lira I and Kyriazis G A 1999 *Metrologia* **36** 163-169.

Journal of Physics: Conference Series **575** (2015) 012014 doi:10.1088/1742-6596/575/1/012014

Inmetro's Quantum Hall System verification on the basis of BIPM calibration results

H R Carvalho, R E M Briones, J R B Silva
Instituto Nacional de Metrologia, Qualidade e Tecnologia – Inmetro,
Av. Nossa Senhora das Graças 50, Xerém - Duque de Caxias - RJ - Brazil

E-mail: hrcarvalho@inmetro.gov.br

Abstract. This article presents the analysis generated by the primary resistance measurement system based on the Quantum Hall Effect, currently being implemented in the Quantum Electrical Metrology Laboratory (Lameq) of the National Institute of Metrology, Quality and Technology (Inmetro). It has as reference the calibration results emitted by the *Bureau International des Poids et Mesures* (BIPM) for the reference resistors maintained by the Electrical Standardization Metrology Laboratory (Lampe), also of Inmetro, thus offering a preliminary parameter of the compatibility status between the primary systems of these two institutions.

1. Introduction

Since the installation of its primary resistance measurement system based on the Quantum Hall Effect, in April, 2008, Inmetro has been improving its abilities in handling the various components that integrates it, especially related to conditioning of Hall samples and the Cryogenic Current Comparator operation. Our main goals regarding this system are: the achievement of the international recognition of the electrical resistance primary standardization based on the Quantum Hall System (QHS) at Inmetro; and the insertion of this part of the scope into the Calibration and Measurement Capabilities (CMC) of the Mutual Recognition Arrangement of the International Committee for Weights and Measures (CIPM) – nominated CIPM MRA [1].

In order to reach such objectives, according to the CIPM MRA, a National Metrology Institute (NMI) must take part on international interlaboratory comparison programs (named key comparisons), regional comparison (named regional key comparisons) or other regional or bilateral comparison modalities (nominated supplementary comparisons), basic directive which allows technical competence analysis of the participants [2].

The line of conduct traced by Lameq for the international recognition of its Quantum Hall System consists in, as a preliminary step, performing an unofficial interlaboratory comparison; after this participating of the BIPM.EM-K12 (Quantum Hall resistance standards and their scaling to other resistance values) key comparison.

Until now, two unofficial interlaboratory comparisons were performed:

a) Comparison of the results generated by the quantum system with Inmetro classic resistance measurement system (maintained by Lampe);

b) Comparison of results generated by the quantum system with Inmetro standards certificates results issued by BIPM, which is the subject of this article.

2. Equipment and metodology

2.1. Inmetro Quantum Hall System Equipment

Lameq QHS is composed of:

- A cryostat manufactured by Cryogenic Ltd. which contains the quantum Hall device (a semiconductor sample composed of a GaAs/GaAlAs heterostructure) which is the electrical resistance primary standard.
- An auxiliary reservoir where ^{3}He gas is stored.
- A cryogenic current comparator (CCC), assembled by National Physical Laboratory – NPL in a partnership with Cryogenic.
- The electronics pieces for data acquisition and control.

The cryostat enables reaching an approximate temperature of 0.3 K in the quantum Hall device and has a superconductive coil which allows the application of a magnetic field of up to 16 T in this device.

The CCC enables the calibration of Lameq reference measurement standards (two 100 Ω standard resistors) through its QHS. These reference standards are compared to other resistors, also through the CCC on a scale, thus composing Lameq traceability chain.

2.2. Setting up the calibration historic of the 100 Ω reference standard

In the period of 2011 and 2012, the two 100 Ω nominal value standard resistors were calibrated several times, against the quantum Hall device using the CCC. One of these resistance standards is the 100 Ω nominal value manufactured by IET Labs, model SR-102/DC, which gets the code "7D" at Lameq. In this way it was built the historical evolution of this reference standard to be used in the verification described in this article.

Each calibration during this period is the result of the mean value of various ratio measurement between this resistor and the quantum standard. The mean of these values is used to obtain the resistor value and it is registered as a measurement point in its historic. Thus it is defined the straight line that best fits these historic values through the linear regression based on the method of least squares. This way, it is obtained the resistor estimate value and the respective uncertainty for any time.

2.3. Traveling standard resistor calibration using the 100 Ω reference standard standard

The 100 Ω standard (resistor 7D), with its value traceable to Lameq quantum standard, was then used with the CCC for the 10 kΩ resistor calibration (whose Lameq code is R1) which, periodically, is also calibrated by BIPM. The value obtained in this calibration may be compared with the historic estimate of this resistor plot by means of BIPM calibration values.

3. Results

3.1. Historical evolution of 7D

The graph in figure 1 shows each calibration of 7D having the quantum Hall device as reference on the $i = 2$ plateau (squares) and on the $i = 4$ plateau (circles), thus building, the historical evolution of this resistor.

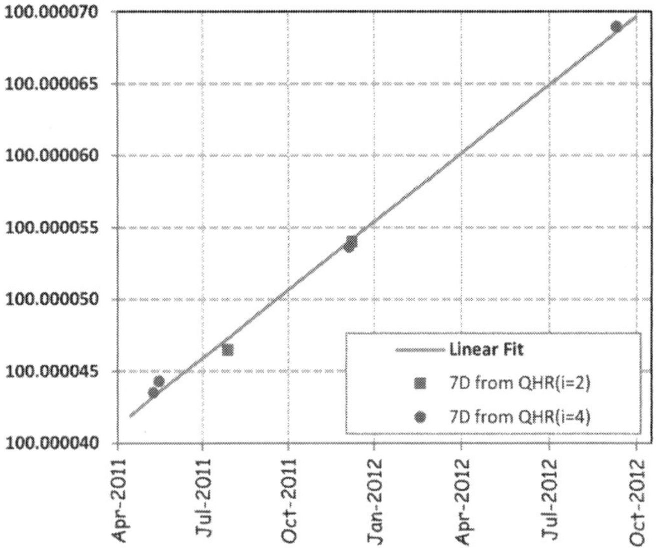

Figure 1. 7D historical evolution graph.

The linear fit (straight line in figure 1) becomes the best estimation of the 7D value for any time and is calculated by:

$$R_{7D}(\Delta t) = a \cdot \Delta t + b \, , \tag{1}$$

where a and b are the angular and linear coefficients, $\Delta t = t - t_0$, t is the date (horizontal axis) and t_0 (reference) is the beginning of 2000.

Hence, 7D resistor was characterized and may be used during R1 calibration.

3.2. BIPM calibrations of R1

Figure 2 shows all eight calibrations of R1 made at BIPM (diamonds) and a straight line that represent the best fit of these data. Each point is associated with a bar of uncertainty for k = 2 (a coverage probability of 95.45 %, assuming a Gaussian distribution). The mean value of this expanded uncertainty is:

$$\overline{U_{BIPM}} = \frac{1}{8}\sum_{i=1}^{8} U_{BIPM\_i} = 0.033 \ \mu\Omega/\Omega \, . \tag{2}$$

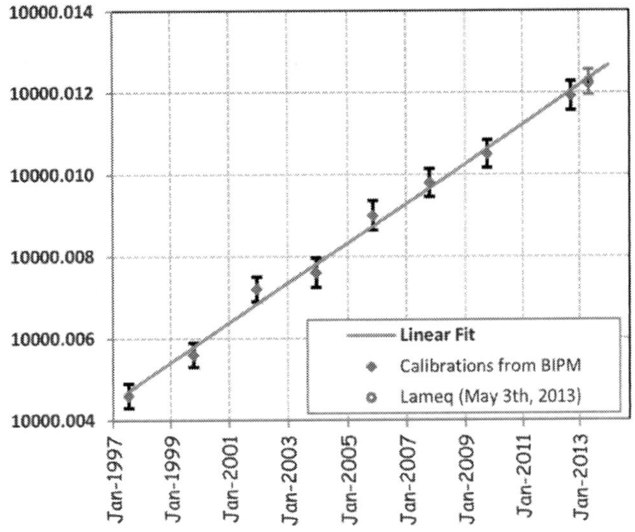

Figure 2. R1 historical evolution graph.

The value of R1 resistor at t = *May 3th, 2013* (to compare with Lameq calibration made at the same date), predicted by the linear least squares fitted straight line to BIPM calibrations using equation (1) is:

$$R_{R1\_BIPM}(t) = 10000.01234\ \Omega \tag{3}$$

with an expanded combined uncertainty:

$$U_C\left(R_{R1\_BIPM}(t)\right) = 0.034\ \mu\Omega/\Omega \tag{4}$$

estimated by combination of the least squares linear regression uncertainty [3, 4] and the uncertainty of the last BIPM measurement (September, 2012).

3.3. Lameq calibrations of R1

Resistor 7D was used as a standard for Lameq R1 resistor calibration.

In figure 2 it is also marked this calibration value found on May 3th, 2013 by using Lameq system (circle) with its uncertainty (coverage probability is 95.45 % and k = 2.51).

The value based on Lameq calibration at the same date, traceable to Lameq quantum Hall device, is:

$$R_{R1\_Lameq}(t) = 10000.01225\ \Omega \tag{5}$$

with associated relative expanded uncertainty:

$$U\left(R_{R1\_Lameq}(t)\right) = 0.031\ \mu\Omega/\Omega, \tag{6}$$

This uncertainty is close to each one of BIPM calibrations in the equation (2).

3.4. Comparison between Lameq and BIPM calibrations

The result was compared with the estimated value from BIPM periodic calibrations.

The normalized error is:

$$En = \frac{|R_{R1_{Lameq}} - R_{R1_{BIPM}}|}{\sqrt{\left(U_{ABS_{BIPM}}\right)^2 + \left(U_{ABS_{Lameq}}\right)^2}} = 0.13 \quad (7)$$

where "ABS" means the absolute expanded uncertainty.

The relative difference between both values is:

$$\frac{|R_{R1\_Lameq} - R_{R1\_BIPM}|}{R_{R1\_BIPM}} = 9.0 \times 10^{-9} \quad (8)$$

which is a difference less than the mean of deviations of each BIPM calibrations points related to the correspondent point on the fit line.

4. Conclusion

A comparison was made between the predicted 10 kΩ resistor value for periodic calibrations performed by BIPM and the value found for this same resistor by calibration performed in Lameq traced to the quantum Hall system of this laboratory.

The difference between the two values was found to be 9.0 parts in 10^9 indicating that both the quantum Hall device and Lameq CCC have good technical and metrological performance and that could be soon approved as the new Brazilian standard of electrical resistance.

Acknowledgement

The authors gratefully acknowledge Inmetro Lampe staff, for the loan of their standard resistor (R1), which has been necessary to execute this research.

References

[1] Comité International des Poids et Mesures 1999 *Mutual recognition of national measurement standards and of calibration and measurement certificates issued by national metrology institutes* (Paris, French: BIPM)
[2] Comité International des Poids et Mesures 2012 *Guide to the implementation of the CIPM MRA - CIPM MRA-G-01* (Paris, French: BIPM)
[3] Bussab W O and Morettin P A 2002 *Estatística Básica* (São Paulo, SP, Brazil: Saraiva)
[4] Working Group 1 of the Joint Committee for Guides in Metrology JCGM/WG 1 2008 *Guide to the expression of uncertainty in measurement* (Paris, French: BIPM)

Harmonic Distortion as an Influence Quantity on Reactive Static Electrical Energy Meters

M T Vasconcellos, J H M Luna, J C Mateus, S A Portugal

Researchers, National Institute of Metrology, Quality and Technology - Inmetro

E-mail: mtvasconcellos@inmetro.gov.br

Abstract. This article presents test results from static electrical energy meters, indicating their susceptibility to the total harmonic distortion (THD) present at the mains line. This particular characteristic impacts directly in measuring reactive energy and power factor for billing purposes.

1. Introduction

Historically, generation, transmission, and distribution of electrical energy is held in a fixed frequency of 60Hz in Brazil. However, the extensive use of non-linear loads changed continuously over time, turning purely sinusoidal waveforms into complex nonsinusoidal signals.

Although the definition of active electrical energy for measurement purposes has long been established – even in terms of its equations and mathematical formulation [1] – the same doesn't occur for reactive electrical energy. There is no international agreement on the concepts of power factor and reactive electrical energy in nonsinusoidal situations for measurement and billing purposes [1].

The Brazilian Electrical Energy Agency (ANEEL) defines active power, reactive power and power factor as (in a free translation):

XXXI – active electrical energy: that one which can be converted in another form of energy, expressed in, kilowatt hour (kWh);

XXXII – reactive electrical energy: that one which flows through diverse both electric and magnetic fields in an alternate current system, without producing work, expressed in reactive kilovolt-ampere hour (kvarh);

...

XXXV – power factor: quotient between active electrical energy and the square root of the sum of the squares of the active electrical energy and reactive electrical energy, consumed at the same period of time.

Active electrical energy can be seen as the integral of average power through time, yielding simple mathematical formulas [1] suitable for both sinusoidal and nonsinusoidal situations.

However, it is not possible to derive simple formulas using the above reactive electrical energy definition in nonsinusoidal situations. The above definition does not arise from the concepts of voltage and current neither contemplates the scenario of modern electrical systems, where total harmonic distortion (THD) levels makes the waveforms substantially nonsinusoidal.

The power factor calculation is also influenced by the reactive energy definition, yielding a quantity strongly dependent on harmonics in the mains line.

An attempt to characterize the electrical powers and energies that arise in non-sinusoidal environment is described in IEEE Standard 1459-2010 [1], which was taken as a basis for the present study.

In order to improve the regulation regarding power factor and billing for excess reactive power, ANEEL called the public hearing 065/2012 [3]. Among its key proposals is that, even in environments with non-sinusoidal waveforms, one should perform the calculation of the power factor for billing purposes exclusively on nominal frequency of the power grid.

In practical terms, the measurement of electrical energy for power factor calculation should be performed at 60 Hz for both, active and reactive energy. This would require electricity meters extremely selective in frequency, capable of perform the measurement of energy only at 60 Hz.

This study aims to evaluate how electronic electricity meters, with models approved by Brazilian Institute of Metrology, Quality and Technology (Inmetro) [4], would behave in an environment with non-sinusoidal waveforms [5-8]. To do so, shall be measured reactive power measurement errors in various situations of harmonic distortion in waveforms of voltage and current, trying to characterize them as influence quantities in the measurement of electricity.

In order to obtain experimental data, various electronic electricity meters were subjected to a series of situations with forms of non-sinusoidal wave. The meters were submitted to two test plans in order to characterize:

A. A most unfavorable condition for reactive power measurement in an harmonic environment;
B. The behavior of the tested static meters in the condition identified in (A), searching for correlations between errors and THD.

It was found that the meters are particularly susceptible to the harmonic content of the power grid, provided specific harmonic voltage and current of the same order occurring simultaneously, showing the correlation between the harmonic distortion and errors of energy measurement, characterizing the former as influence quantity of the second.

2. Power in the IEEE 1459-2010 Standard Definitions

This sections aims to state a brief description of IEEE 1459-2013 Standard Definitions to the reader not familiar with its concepts.

Aware of the challenge of standardize the concepts of electric power at both sinusoidal and nonsinusoidal environments, for symmetrical and nonsymmetrical loads, IEEE took efforts creating an workgroup to propose modern and broadly accepted concepts of electrical power, reflecting the state of art in electrical energy measuring understanding.

The IEEE 1459-2010 Standard Definitions are based on apparent power calculations through voltage and current root mean square (RMS) values. Here one analyses the single phase case, which can be easily extended to three phase systems.

Assume that voltage and current continuous time signals, $v(t)$ and $i(t)$, respectively. Their RMS values can be calculated through their integrals over time along an integer number ($k \in \mathbb{N}$) of time periods (T):

$$V^2 = \frac{1}{kT}\int_{\tau}^{\tau+kT} v(t)^2 dt \qquad (1)$$

$$I^2 = \frac{1}{kT}\int_{\tau}^{\tau+kT} i(t)^2 dt \qquad (2)$$

Apparent power S (VA) is calculated through the product between voltage and current RMS values (1) and (2):

$$S = VI \tag{3}$$

First harmonic waveforms, both for voltage and current, as well as the sum of the remaining harmonics, are defined as:

$$v_1(t) = \sqrt{2}V_1 \sin(\omega t - \alpha_1) \tag{4}$$

$$i_1(t) = \sqrt{2}I_1 \sin(\omega t - \beta_1) \tag{5}$$

$$v_H(t) = V_0 + \sqrt{2}\sum_{h\neq 1} V_h \sin(h\omega t - \alpha_h) \tag{6}$$

$$i_H(t) = I_0 + \sqrt{2}\sum_{h\neq 1} I_h \sin(h\omega t - \beta_h) \tag{7}$$

where V_0 and I_0 are average (DC) values of voltage and current, V_1 and I_1 are voltage and current fundamental harmonic RMS values, V_h and I_h are voltage and current high order harmonic RMS values of h order, α_h and β_h are phase angles for voltage and current h order harmonics, $v_1(t)$ and $i_1(t)$ are voltage and current fundamental waveforms, $v_h(t)$ e $i_h(t)$ are h order harmonics voltage and current waveforms.

Full voltage and current waveforms (8) and (9) can be posed as the sum of fundamental waveforms (4) and (5) with the respective harmonic waveforms (6) and (7).

$$v(t) = v_1(t) + v_H(t) \tag{8}$$

$$i(t) = i_1(t) + i_H(t) \tag{9}$$

RMS values of both fundamental and harmonic voltages (10) and currents (11) are intertwined by:

$$V_H^2 = V_0^2 + \sum_{h\neq 1} V_h^2 = V^2 - V_1^2 \tag{10}$$

$$I_H^2 = I_0^2 + \sum_{h\neq 1} I_h^2 = I^2 - I_1^2 \tag{11}$$

One can notice that, although voltage and current average values V_0 e I_0 aren't harmonics, their values are included in V_H and I_H, considering that in alternate waveform systems, they are Always low amplitude values.

Active power (12) is calculated as the instantaneous power ($p(t) = v(t)i(t)$) average value:

$$P = \frac{1}{kT}\int_{\tau}^{\tau+kT} v(t)i(t)dt \tag{12}$$

Fundamental active power P_1 (W) and harmonic active power P_H (W) are defined as:

$$P_1 = \frac{1}{kT}\int_{\tau}^{\tau+kT} v_1(t)i_1(t)dt = V_1 I_1 \cos\theta_1 \tag{13}$$

$$P_H = V_0 I_0 + \sum_{h\neq 1} V_h I_h \cos\theta_h = P - P_1 \tag{14}$$

In the same trend, fundamental reactive power Q_1 (VAR) is defined as:

$$Q_1 = \frac{\omega}{kT}\int_{\tau}^{\tau+kT} i_1(t)\left[\int_{\tau}^{\tau+kT} v_1(t)dt\right]dt = V_1 I_1 \sin\theta_1 \tag{15}$$

where ω is the angular frequency and θ_1 is the fundamental harmonic phasor angle between voltage and current.

Fundamental apparent power S_1 (VA) is defined as the product between voltage and current fundamental RMS values:

$$S_1 = V_1 I_1 \quad (16)$$

$$S_1^2 = P_1^2 + Q_1^2 \quad (17)$$

Apparent power S (VA) is split into fundamental apparent power (S_1) and nonfundamental apparent power (S_N):

$$S^2 = (VI)^2 = (V_1^2 + V_H^2)(I_1^2 + I_H^2) \quad (18)$$

$$S^2 = (V_1 I_1)^2 + (V_1 I_H)^2 + (V_H I_1)^2 + (V_H I_H)^2 \quad (19)$$

$$S^2 = S_1^2 + S_N^2 \quad (20)$$

Nonfundamental apparent power S_N (VA) can be defined as the difference between apparent power (18) and fundamental apparent power (17):

$$S_N = \sqrt{S^2 - S_1^2} \quad (21)$$

Nonfundamental apparent power can be described also as a three term quadratic sum: current distortion power D_I, voltage distortion power D_V and harmonic apparent power S_H:

$$S_N^2 = D_I^2 + D_V^2 + S_H^2 \quad (22)$$

Current distortion power D_I (VAR) is stated as:

$$D_I = V_1 I_H \quad (23)$$

Voltage distortion power D_V (VAR) is stated as:

$$D_V = V_H I_1 \quad (24)$$

Harmonic distortion power S_H (VA) can be stated as composse by harmonic active power P_H(W) and harmonic distortion power D_H (VAR):

$$S_H = V_H I_H = \sqrt{P_H^2 + D_H^2} \quad (25)$$

Harmonic distortion power, P_H, defined by (14), is determined by the product of voltage and current harmonics of order. Harmonic distortion power is composed by the product of voltage and current harmonics of different orders.

$$D_H = \sqrt{S_H^2 - P_H^2} \quad (26)$$

Nonactive power N (VAR) is defined as the difference between apparent power and active power taken as orthogonal vectors:

$$N = \sqrt{S^2 - P^2} \tag{27}$$

The fundamental power factor can be defined by taking into account only fundamental active and fundamental apparent power:

$$PF_1 = \cos\theta_1 = \frac{P_1}{S_1} \tag{28}$$

Power factor is defined by taking into account active and apparent power:

$$PF = \frac{P}{S} \tag{29}$$

According to [1], one can define the following power zoo (Table 1):

Table 1 – Power Zoo according to IEEE 1459-2010

Quantity or indicator	Combined	Fundamental Powers	Nonfundamental Powers
Apparent	S (VA)	S_1 (VA)	S_N e S_H (VA)
Active	P (W)	P_1 (W)	P_H (W)
Nonactive	N (VAR)	Q_1 (VAR)	D_I, D_V e D_H (VAR)
Line utilization	$PF = P/S$	$PF_1 = P_1/S_1$	
Harmonic pollution			S_N/S_1

One can notice that in [1] the concept of Budeanus's reactive power Q_B (30) was ruled out, because it represents an immeasurable and nonphysical indicator :

$$Q_B = \sum_{h\neq 1} V_h I_h \cos\theta_h \tag{30}$$

In a nutshell, one can summarize quantities and indicators presented previously through Figure 1, where each rectangle represents one species in the power zoo. Each power is defined as a quadratic sum of the internal rectangles, as defined in (17), (20), (22) and (25).

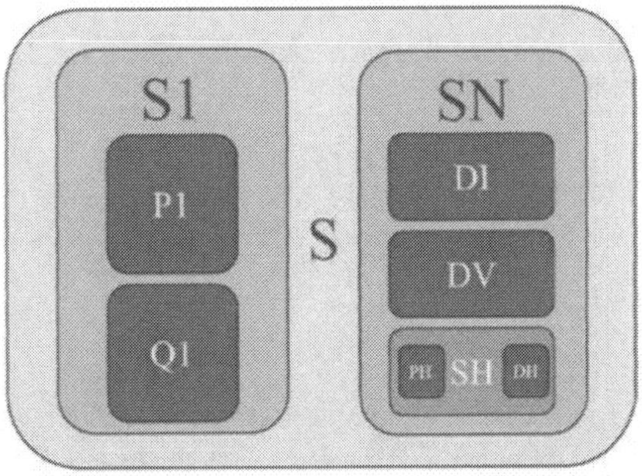

Figure 1 – Pictorial Representation of Powers in IEEE 1459-2010

Journal of Physics: Conference Series **575** (2015) 012015 doi:10.1088/1742-6596/575/1/012015

3. Harmonic Distortion

In this section one presents a brief description of harmonic distortion concepts into order to unify the reader's understanding. One considers that a waveform of voltage $v(t)$ or current $i(t)$ is represented by the generic signal $x(t)$.

According to [2], the total harmonic distortion is defined by (1):

$$\mathrm{THD}_X = \frac{X_H}{X_1} = \sqrt{\left(\frac{X}{X_1}\right)^2 - 1} \qquad (1)$$

where X represents the rms value of the quantity in question, V or I, calculated from the integration of the squared waveforms over an integer ($k \in N$) of periods (T) of the fundamental harmonic (2):

$$X^2 = \frac{1}{kT}\int_{\tau}^{\tau+kT} x(t)^2 dt \qquad (2)$$

The voltage signals or current can be decomposed as:

$$x(t) = x_1(t) + x_H(t) \qquad (3)$$

where $x_1(t)$ is the fundamental waveforms of voltage or current, given by:

$$x_1(t) = \sqrt{2}X_1 \sin(\omega t - \alpha_{x1}) \qquad (4)$$

where X_1 is the RMS value of (4), ω is the fundamental angular frequency and α_{x1} is the phase angle for the current or voltage.

Likewise, one may define the waveforms of voltages and currents by the sum of its harmonic components plus each of their respective average values:

$$x_H(t) = X_0 + \sqrt{2}\sum_{h\neq 1} X_h \sin(h\omega t - \alpha_{xh}) \qquad (5)$$

where h is the order of harmonics, α_{xh}, the phase angle for the current or voltage for the h-th order harmonics, X_0, the $v(t)$ or $i(t)$ mean values, defined by:

$$X_0 = \frac{1}{kT}\int_{\tau}^{\tau+kT} x(t) dt \qquad (6)$$

Finally, X_h in (5) is defined as the RMS value of the h-th harmonic of voltage or current.

4. Methodology

This study consists in two sets of tests: set A with two static electrical energy meters and set B, with six meters, all of them approved by Inmetro [4]. Table 2 depicts the static meters tested:

Table 2 – Static meters under test

Meter	Meter type	Rated Voltages (V)	Rated Current (A)	Accuracy Class	Kh (Wh/pulse)	Test Set
M2	Direct	120	15	1%	2	A
M3	Indirect	120 / 220	2.5	0.2%	0,2	A/B
M7	Indirect	120 / 220	5	0.2%	1,8	A/B
M10	Indirect	120 / 240	2.5	1%	0,2	A/B
M11	Indirect	120 / 240	2.5	0.5%	1,8	A/B
M12	Indirect	120 / 220	2.5	0.2%	1,8	A/B
M16	Direct	120 / 240	30	0.5%	3,6	A/B

The test sets area:

A. Identification of worst scenario for reactive energy metering in an harmonic environment;
B. Correlation between THD and measurement error.

The two sets of tests are described below.

4.1 Test set A – Scenario identification

The aim of this set of tests is to identify the worst scenario for reactive energy measurement in a concise schedule of harmonics. Each meter was subject to eight three-phase tests, each one with a different set of harmonics. Voltage was set at 120V, current was held at rated values when possible, with a maximum of 8A, and power factor of 0,707. At each test, none or three harmonics of magnitude 17% from fundamental were injected at both voltage and current. This yields a THD of approximately 30% in each phase.

Table 3 – Test Set A

Test	Voltage Harmonics			Current Harmonics			Energy Measured
1	-	-	-	-	-	-	Active
2	2	3	4	2	3	4	Active
3	3	5	7	2	4	6	Active
4	-	-	-	-	-	-	Reactive
5	2	3	4	2	3	4	Reactiva
6	3	5	7	2	4	6	Reactive
7	3	5	7	2	3	4	Reative
8	2	3	4	3	5	7	Reactive

Tests 2 and 5 presents overlapping harmonics at both current and voltage while measuring, respectively, active and reactive energy. Tests 1 and 4 presents no harmonics at all. Tests 3 and 6 presents no overlapping harmonics at both current and voltage.

4.2 Test set B – Correlation between THD and error

Each tested static meter was subjected to three-phase energy with harmonic distortion, in an electric energy meter bench Nansen brand, Precision Lab model, with internals single-phase energy patterns Radian RD-20, each one with 0,01% of accuracy.

Simultaneously, data from voltage and current was acquired through an oscilloscope Tektronix, model TDS-5104B, with voltage and current pointers. Data acquired was processes in Scilab in order to calculate the power zoo proposed by [1].

The electrical energy meter test bench can generate harmonics from 2nd to 15th order, with maximum THD of 30% per phase. The digital meters analyzed were subject to a test plan, consisting of eighteen different measurement accuracy tests, with increasing harmonic amplitudes. Voltage and current harmonics are coincident in order, subjecting the meter to a more severe environment, according to [10].

Journal of Physics: Conference Series **575** (2015) 012015 doi:10.1088/1742-6596/575/1/012015

At all three phases, harmonics of 2nd, 3rd and 4th orders were injected, along with fundamental voltage and current set at rated values. Table 4 summarizes the harmonic profile of the test plan, with harmonic amplitudes rated as percentage from nominal values.

Table 4 – Test Set B

Test	**Voltage Harmonic Amplitude (%)**			**THDv (%)**	**Current Harmonic Amplitude (%)**			**THDi (%)**
	2	**3**	**4**		**2**	**3**	**4**	
0	0	0	0	0.00	0	0	0	0.00
1	1	1	1	1.73	1	1	1	1.73
2	2	2	2	3.46	2	2	2	3.46
3	3	3	3	5.20	3	3	3	5.20
4	4	4	4	6.93	4	4	4	6.93
5	5	5	5	8.66	5	5	5	8.66
6	6	6	6	10.39	6	6	6	10.39
7	7	7	7	12.12	7	7	7	12.12
8	8	8	8	13.86	8	8	8	13.86
9	9	9	9	15.59	9	9	9	15.59
10	10	10	10	17.32	10	10	10	17.32
11	11	11	11	19.05	11	11	11	19.05
12	12	12	12	20.78	12	12	12	20.78
13	13	13	13	22.52	13	13	13	22.52
14	14	14	14	24.25	14	14	14	24.25
15	15	15	15	25.98	15	15	15	25.98
16	16	16	16	27.71	16	16	16	27.71
17	17	17	17	29.44	17	17	17	29.44

Table 2 shows some rated characteristics from the digital energy meters tested in the study, where Kh is the calibration constant, that relates the pulse rate to the energy measured. Its unity is Watt.hour per pulse emitted.

The digital electric energy meters in Table 2 have had their measurement errors registered, for each one of the eighteen tests, gradually increasing THD_V and THD_I (since their value is the same).

One should notice that the test plan in Table 1 is only one scenario among several others. Although 30% THD_V is improbable for a real power system, 30% THD_I is very likely in several situations. Test conditions 6 and 7, for example, with THD between 10% and 12%, are very probable both for voltage and current at delivery.

Although the harmonics injected as not usual, they were chosen as a logical order sequence. More tests are being planned with several harmonics combinations.

Measurements have been repeated several times, discarding random errors as main interference source.

5. Results

Results were obtained from test sets A and B. Test set A results are showed in Tables V and VI for meters M2 and M3, respectively. From them, it is possible to conclude that test 5, measuring reactive energy with overlapping harmonics at both voltage and current is the worst case for harmonic influence. Therefore, test set B was run with condition 5 for meters M3 to M16.

Table 5 – Errors for test set A meter M2

Test	Energy Measured	Error (%)	THD (%)
1	Active	-0.06	0.00
2	Active	0.14	29.44
3	Active	-0.12	29.44
4	Reactive	0.11	0.00
5	Reactive	-4.96	29.44
6	Reactive	0.08	29.44
7	Reactive	-0.85	29.44
8	Reactive	-0.94	29.44

Table 6– Errors for test set A meter M3

Test	Energy Measured	Error (%)	THD (%)
1	Active	0.07	0.00
2	Active	0.09	29.44
3	Active	0.24	29.44
4	Reactive	0.08	0.00
5	Reactive	-6.53	29.44
6	Reactive	0.06	29.44
7	Reactive	-1.23	29.44
8	Reactive	-1.24	29.44

After running tests with the method presented in the previous section, measurement errors on reactive electrical energy have been obtained, for each meter in Table 2, at each test point in Table 1, as seen in Tables 5 and 6.

Figure 2 shows measurement errors against THD for meters M3, M7 and M12, of accuracy class D. Horizontal lines represent the boundaries of class D accuracy, 0.2%. One should notice that, starting from a relatively low THD (between 5% and 7%), all meters present measurement errors greater than their accuracy class.

Figure 3 is similar to Fig. 2, but related to digital meters M11 and M16, accuracy class C. One should notice that the errors of the meter M16 are outside its accuracy class from 4% THD up. Errors of M11 meter are outside its accuracy class from 11% THD, approximately.

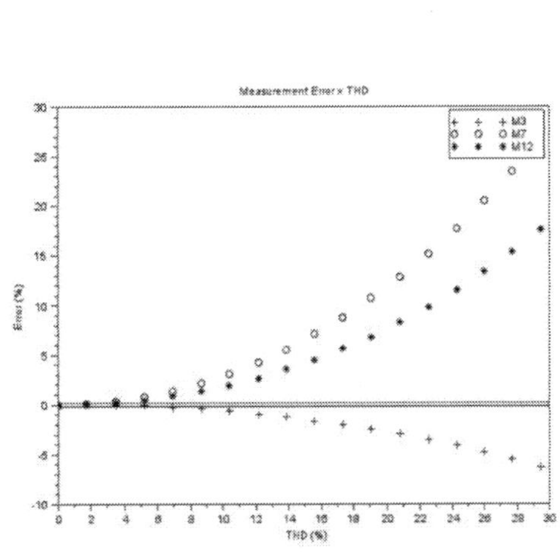

Figure 2 - Measurement errors for class D meters

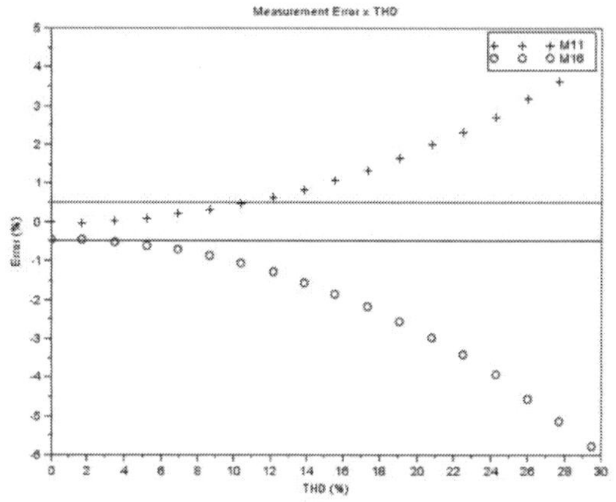

Figure 3 - Measurement errors for Class C meters

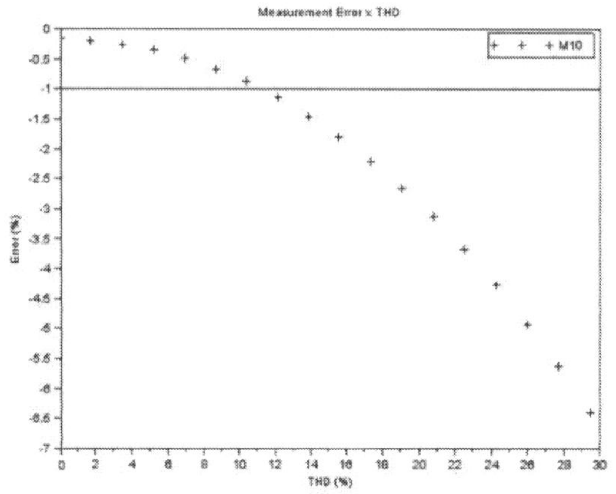

Figure 4 - Measurement Errors for class B meter

Similar behaviour is shown by meter M10, accuracy class B, with measurement errors greater than 1% just above 11% THD, as seen in Fig. 4.

Supposing a linear relationship between THD and measurement errors, Table 7 shows correlation coefficients [11] for each meter tested.

One should notice, in Table 7, the high values of correlation coefficients between THD and measurement errors. One can conclude that, considering the meters under test and the test plan executed, measurement errors are directly influenced by THD values, so one can consider THD as a direct influence quantity in digital reactive energy meters.

Despite 30% THD values are very improbable (at least in voltage), the extent of the curves in Fig.1, 2 and 3 clearly depicts the strong positive correlation between THD and reactive power measuring error.

Table 7 – Errors for test set A meter M2

Meter	Correlation Coefficient
M3	-0,9673
M7	0,9608
M10	-0,9573
M11	0,9595
M12	0,9599
M16	-0,9585

6. Conclusions

Analyzing results obtained from the tests performed using previously stated methods, one can pinpoint the following remarks:

The presence of overlapping harmonics at both voltage and current waveforms represent the worst environment for energy measurement, among the eight tests in set A.

The digital reactive electrical energy meters tested are susceptible to THD as influence quantity in reactive electrical energy measurement;

The correlation coefficients (R) [11] between THD and measurement errors are significantly high ($|R| \geq 0,95$) for all meter tested. As the THD is the only quantity altered along the tests, this confirms the meter susceptibility on the harmonic content present in the electric grid;

A unified, clear and consistent definition of both reactive power and energy would be helpful in establishing criteria for billing reactive power and low power factor, independent of harmonics content.

Acknowledgments

The authors are grateful to Inmetro for supporting this research through Pronametro.

References

[1] IEEE, Standard Definitions for the Measurement of Electric Power Quantities Under Sinusoidal, Nonsinusoidal, Balanced, or Unbalanced Conditions 1459-2010, IEEE Power & Energy Society, New York, 2010.
[2] ANEEL, General Condicions for Electrical Energy Supply (in Portuguese), Resolution 414/2010.
[3] ANEEL Technical Note 0083/2012-SRD/ANEEL", June 12th, 2012.
[4] INMETRO, Portaria Inmetro 431, November 5th, 2012, Inmetro, Brasil
[5] INMETRO, Portaria Inmetro 587, December 4th, 2011, Inmetro, Brasil.
[6] J. Driesen, T. Craenenbroeck, D. Dommelen, The Registration of Harmonic Power by Analog and Digital Power Meters, IEEE Trans. on Instr. and Meas., vol. 47, no. 1, Fev. 1998.
[7] P. S. Filipski, P. W. Labaj, Evaluation of Reactive Power Meters in the Presence of High Harmonic Distortion, IEEE Trans. on Power Delivery. Vol. 7, No. 4, Outubro 1992.
[8] L. R. Lisita, D. Pinheiro Neto, P. C. M. Machado, J. W. L Nerys, M. G. S. Figueiredo, Avaliação de Desempenho de Medidor Trifásico de Energia Elétrica Tipo Eletrônico Operando com Cargas Não-Lineares, IEEE/PES – Trans. and Dist. Latin America, São Paulo, 2010.
[9] M. R. Suhett, J. E. R. Alves Jr., S. C. G. Oliveira, E. H. Watanabe, Medição de Potência Reativa em Sistemas com Formas de Onda Não-Senoidais, VIII CBQEE, Blumenau/SC, 2009.
[10] M. T. Vasconcellos, J. H. M. Luna, S. A. Portugal, Medição de potência elétrica em ambiente não senoidal segundo a norma IEEE 1459-2010, X CBQEE, Araxá/MG, 2013. Accepted paper.
[11] E. W. Frees, Data Analysis Using Regression Models, Prentice Hall, 1996.
[12] IEC 62053-23 - Electricity metering equipment (a.c.) - Particular requirements - Part 23: Static meters for reactive energy (classes 2 and 3, IEC, 2003.
[13] IEC 62053-24 (Draft) - Electricity metering equipment (a.c.) - Particular requirements - Part 24: Static meters for fundamental component reactive energy (classes 0,5 S, 1S and 1), IEC, 2010

Bilateral comparison of radiated emission measurements

M S Dias[1], M F Barbosa[1], A G F Junior[1], M L P Filho[1], L C Silva[2], A S Andrade[2]

[1] Institute for Technological Research of the State of São Paulo, São Paulo, SP, 05508-901, Brazil
[2] National Institute For Space Research, São José dos Campos, SP, 12227-010, Brazil

E-mail: mdias@ipt.br

Abstract. The objective of this study is to perform a proficiency testing by bilateral comparison of radiated emission measurements between accredited laboratories. The results show that both laboratories are in conformity with the criterion of acceptance proposed by a comparison, which is the uniformity between the measured values according to normalized error ratio.

1. Introduction

The electrical and optical laboratory of Institute for Technological Research, IPT, and the integration and testing laboratory of National Institute For Space Research, INPE, started in April 2013 a bilateral measuring electromagnetic interference radiated. The comparison was carried out in accordance with CISPR 22:2006 [1] providing an important result to participants as required by the standard ISO/IEC 17025:2005 [2] – "Item 5.9 Assuring the quality of test and calibration results; b) participation in interlaboratory comparison." – The comparison of the results was done according to ISO 13528:2005 [3] (Statistical methods for use in proficiency testing by interlaboratory comparisons.)

The comparison allowed each participating laboratory to evaluate their working routines, the performance of their technicians in using different equipment, from a bias of measured results with respect to the reference value.

2. Methodology

Bilateral comparison for measuring electromagnetic field radiated during the second quarter of 2013. The laboratories involved measurements performed using a standard noise source, as shown in figure 1, which was used in the vertical and horizontal position. In the figure the noise emitter is positioned in vertical polarization. For horizontal polarization turn 90 degree the noise emitter leaving the button red (on/off) upper face.

Figure 1. Noise emitter.

The setup of the radiated emissions was installed in accordance with CISPR 22, the results were sent to the coordinating laboratory, IPT. The reference value was considered to be the value obtained by INPE, measurement antenna INPE is traced to the ETS-Lindgren laboratory that is accredited American network A2LA (American Association for Laboratory Accreditation) and the receiver is tracked RBC (Brazilian Calibration Network).

The frequency bands have been agreed in advance by both laboratories as per table 1.

Table 1. Frequency bands.

Band	Frequency (MHz)
01	40.0 – 60.0
02	540.0- 560.0
03	940.0- 960.0

The reference value for comparison was obtained by averaging the values of each track performed by the reference laboratory. Evaluation of measures was performed using the ISO 13528 [3] standard error, we constructed a diagram to identify laboratories with standard errors and the reference value for each test. The normalized error (En) is defined by equation (1). The evaluation criteria of the standard error were made by its value module, so that |En| ≤ 1 the result is satisfactory and |En| > 1 the result is questionable.

$$E_N = \frac{M_{LAB} - M_{REF}}{(U_{LAB}^2 + U_{REF}^2)^{\frac{1}{2}}} \tag{1}$$

Where:

- M_{LAB} is the measure of laboratory IPT;
- M_{REF} is the measure of laboratory INPE;
- U_{LAB} is the uncertainty of laboratory IPT;
- U_{REF} is the uncertainty of laboratory INPE.

2.1. Measurement protocol of radiated emission

The measurements were carried out using standard noise emitter in two positions, vertical and horizontal transmission, positioned in the center of the table to 3.0 m from the antenna and maximum values recorded.

To perform the comparison, optimizing test time, the band was divided into three frequency bands with a width of 20 MHz each, repeating each measurement three times at the beginning, middle and end of the frequency established by the CISPR 22 (30.0 MHz to 1.0 GHz). As shown in Table 1.

The quasi-peak detector was used to measure the electromagnetic waves (dBuV/m).

3. Results and Comments

Table 2 presents field measurements of the participant (M_{LAB}) and reference (M_{REF}) laboratory, taking into account the uncertainty of each laboratory (U_{LAB} = 4.6) and (U_{REF} = 6.4) and the standard error module (En). The values obtained indicate standard error result in accordance with criteria of this comparison.

Table 2. Results.

Noise emitter polarization	Band	M_{LAB} (dBμV/m)	M_{REF} (dBμV/m)	|En|
Vertical	01	44.73	45.53	0.10
	02	73.93	73.63	0.04
	03	67.60	65.40	0.24
Horizontal	01	39.27	39.43	0.02
	02	71.73	67.30	0.56
	03	67.17	63.53	0.46

Figure 2 shows the setup used for testing radiated emissions showing the position of the antenna, turntable and noise emitter at a certain moment of the test.

Figure 2. Setup of radiated emission test.

Figure 3 shows the result of a complete measurement in according of the standard CISPR 22 with the noise emitter positioned vertically and the reception antenna polarized horizontally.

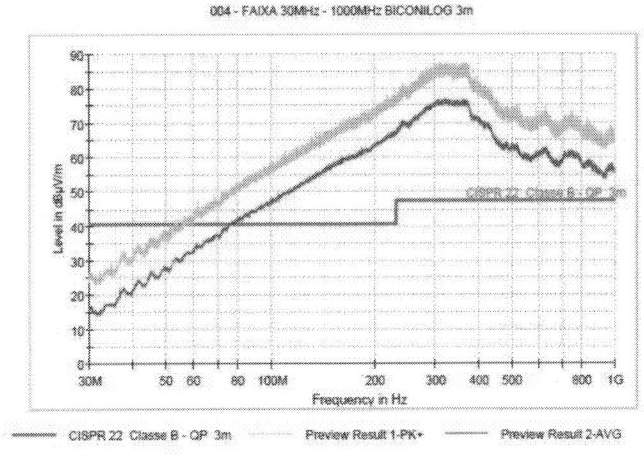

Figure 3. Measurement results in the full range of CISPR22.

4. Conclusion

The realization of bilateral comparison on measures of electromagnetic field radiated provided sufficient information, from calculations of standard errors, the two laboratories are in accordance with the criteria established by this comparison, showing that the tests between these laboratories are consistent with each other.

References

[1] CISPR22:2006 Information technology equipment – Radio disturbance characteristics – Limits and methods of measurement;
[2] ISO/IEC 17025:2005 General requirements for the competence of testing and calibration laboratories;
[3] ISO 13528:2005 Statistical methods for use in proficiency testing by interlaboratory comparisons.

Comparison of different methods for liquid level adjustment in tank prover calibration

D A Garcia, E C Farias, P C Gabriel, M H Aquino, R S E Gomes and V Y Aibe

Inmetro, Fluid Dynamics Metrology, Rio de Janeiro, Brazil

E-mail: dinam@inmetro.gov.br

Abstract: The adjustment of the liquid level during the calibration of tank provers with fixed volume is normally done by overfill but it can be done in different ways. In this article four level adjustment techniques are compared: plate, pipette, ruler and overfill adjustment. The adjustment methods using plate and pipette presented good agreement with the tank's nominal volume and lower uncertainty among the tested methods.

1. Introduction

Tank provers are essential instruments used as transfer standards for liquid volume quantity. There are two common types of tank provers: fixed volume and graduated volume type, as presented in figure 1.

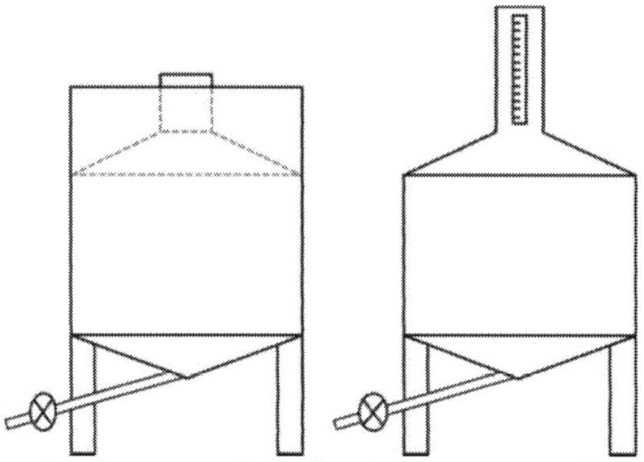

Figure 1. Common types of tank provers (left: fixed volume type; right: graduated volume type).

The calibration of a fixed volume tank prover is done generally by overfilling the tank with liquid and the meniscus volume on the edge of the tank's neck depends on the liquid surface tension and other external influences. The tank's volume can be determined either by the gravimetric or volume transfer method. The calibration results are very dependent on the meniscus level adjustment (figure 2).This adjustment can be done by many methods, depending on the tank prover installation facilities and the calibration uncertainty requirements.

Figure 2. Liquid meniscus in a fixed volume tank's neck.

2. Objective

This study aims the comparison of different methods commonly used for the liquid level adjustment during the calibration of fixed volume tank provers. The adjustment was done using the following methods: plate, pipette, ruler and overfill levelling. The calibration results are compared in terms of cost-benefit ratio (uncertainty achieved versus convenience of execution).

3. Methodology

In this study a fixed volume tank prover with $50 dm^3$ was used. The calibration of the tank prover was done by the gravimetric method, using a balance with 150kg capacity and 1g of resolution. The balance was calibrated using F1 class calibration weights. The tank was calibrated with filtered water and the water density was determined by means of a digital densimeter. The calibration facilities are presented in figure 3.

The tank prover was levelled over a platform, through its adjustable feet. Once filled, the liquid temperature was measured and the liquid level was adjusted in the tank. The liquid in the tank was then transferred to a container by means of a plastic pipe. The liquid transfer was done by opening smoothly the valve in the tank's bottom part. The assembly (container + liquid + pipe) was then moved to the balance platform to measure its mass. After the weighing, the container was emptied and the empty assembly (container + pipe) was also weighed.

Hereafter, the four methods used for the liquid level adjustment during the calibration are described.

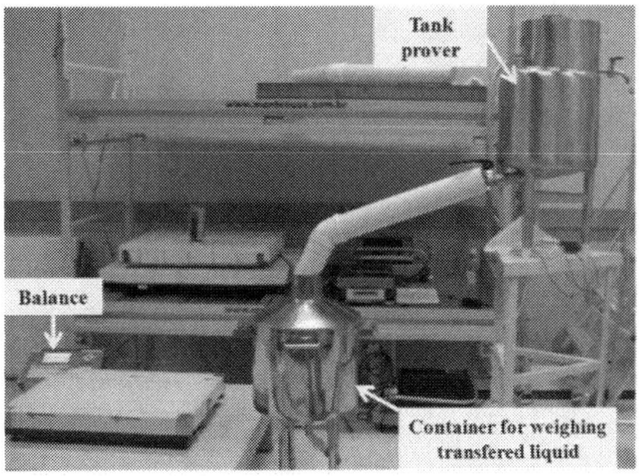

Figure 3. Calibration facilities.

3.1. Plate adjustment

In this method, the meniscus was adjusted by means of an acrylic plate, as presented in figures 4 and 5.

Figure 4. Meniscus adjustment using an acrylic plate

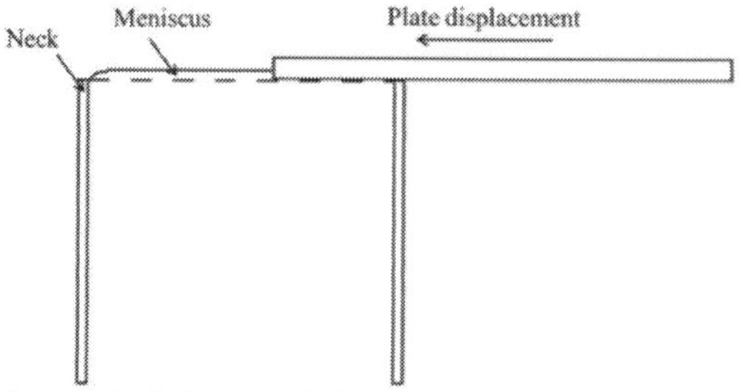

Figure 5. Scheme of the meniscus adjustment using an acrylic plate

The tank was filled and then the neck was overfilled until the formation of a meniscus. The meniscus was adjusted by moving slowly the acrylic plate, tangential to the neck's surface. An acrylic plate was chosen as it allows the view of air bubbles entrapped in the meniscus.

3.2. *Pipette adjustment*

This method consists of adjust the liquid level in the tank's neckedge by means of a pipette, as presented in figure 6 and 7. The tank was filled and the liquid level was then adjusted with the aid of a pipette, until it was visually at the neck's edgelevel (no meniscus is formed).

Figure 6. Meniscus adjustment using a pipette.

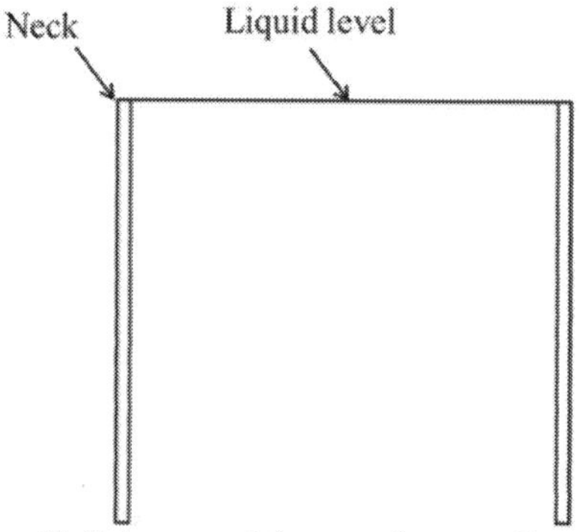

Figure 7. Scheme of the meniscus adjustment using a pipette.

3.3. *Ruler adjustment*

In this method, a metallic ruler was used to adjust the liquid meniscus, as presented in the figure 8 and 9.

Figure 8. Meniscus adjustment using a ruler.

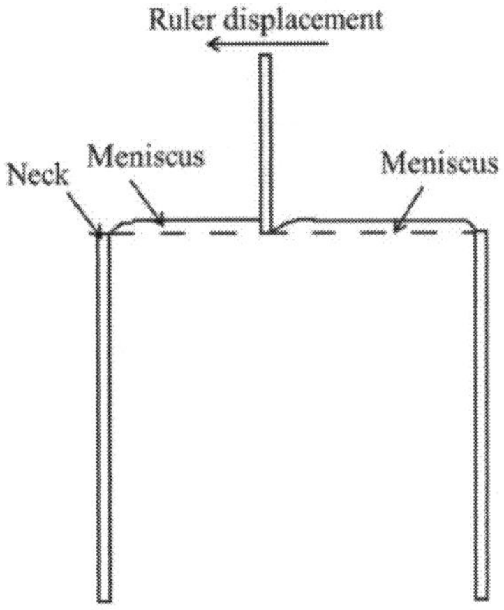

Figure 9. Scheme of the meniscus adjustment using a ruler.

The tank was filled and then the neck was overfilled until the formation of a meniscus. The meniscus was adjusted by moving the ruler, in the vertical position, tangential to the neck's surface.

3.4. *Overfill adjustment*

In this method, the meniscus was adjusted by pouring water in the tank's neck with the aid of a beaker, as presented in the figure 10. The liquid was poured from the beaker to the tank until the meniscus rupture.

Figure 10. Meniscus adjustment by pouring water with a beaker.

4. Mathematical model

The mathematical model used for determine the tank volume through the gravimetric method is presented below:

$$V_L(T_r) = \frac{M_{a1} - M_{a2} + M_E}{(\rho_L(T_L + \delta T_L) - \rho_A)} \cdot \left(1 - \frac{\rho_{ab}}{\rho_b}\right) \cdot (1 - \alpha_V \cdot (T_d - T_r)) + \delta M + \delta V \quad \text{Eq.1}$$

Where:

- $V_L(T_r)$ is the tank prover volume at its reference temperature T_r= 20°C;
- $M_{a1}$is the apparent mass of the full container, after the transference of liquid;
- $M_{a2}$is the apparent mass of the empty container, before the transference of liquid;
- M_E is the apparent mass of the evaporated liquid;

- $\rho_L(T_L)$ is the liquid density in the tank prover, at the temperature T_L;
- ρ_A is the air density during the apparent water mass measurement;
- ρ_{ab} is the air density during the balance calibration;
- ρ_b is the density of the standard weight used in the balance calibration;
- α_V is the thermal expansion volumetric coefficient of the tank prover;
- T_d is the tank prover temperature;
- δM is the error due to meniscus volume variation;
- δV is the random error due to repeatability of volume measurement.

Note: the apparent mass is the mass indicated by the balance display corrected by its calibration certificate data.

5. Results

The results obtained for each method are presented in the table 1 and figure 11, in terms of the measurand and its expanded measurement uncertainty.

Table 1. Tank calibration results.

Meniscus Adjustment Method	Calculated Volume (dm^3)	Expanded Uncertainty (dm^3)	Coverage Factor $k_{95\%}$	Relative Expanded Uncertainty (%)
Plate	50.0000	0.0032	2.000	0.0064
Pipette	50.0009	0.0032	2.010	0.0063
Ruler	50.0110	0.0033	2.000	0.0065
Pour	50.0197	0.0044	2.000	0.0088

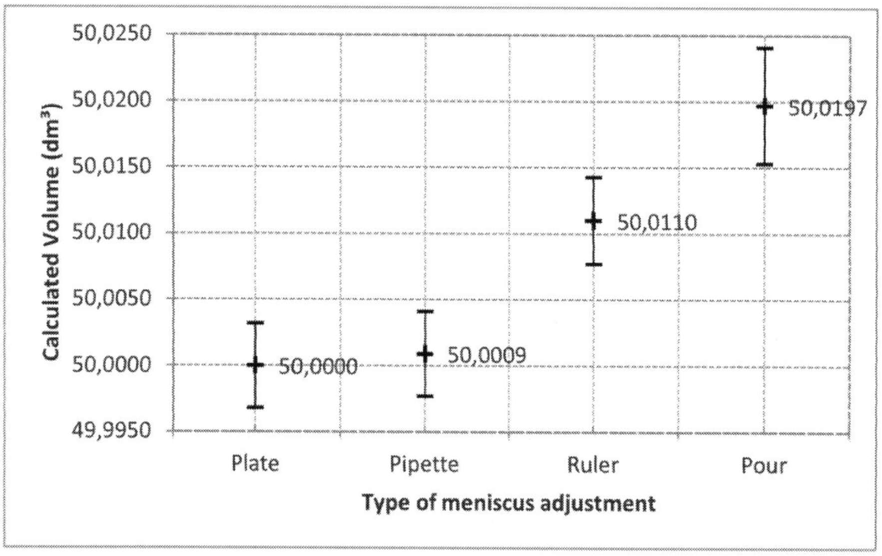

Figure 11. Results for the various meniscus adjustment methods

6. Conclusions

The tank's calibration results presented in the previous section show that the lower uncertainty levels were achieved using both the pipette and plate methods. Also, the calibration results were strongly affected by the calibration method due to contained volume variation.

For all tested methods, the contribution of the random error due to repeatability of the volume *(δV)* was one of the main sources of uncertainty. The ruler and overfill methods presented also an additional uncertainty source, the random error due to meniscus volume variation (δM), which has increased the uncertainty for these methods.

According the results, the plate method is the most recommended method for an interlaboratory comparison because it is a method of simple and fast execution and it is independent on the visual criteria used by the calibration personnel for the liquid volume adjustment. Besides, the execution of the plate method was easier than the pipette method (time of execution, easiness and convenience), with the best cost-benefit ratio.

Laboratory experience prove that the overfill method is most convenient method to be executed for fixed volume tank calibration. In cases where the uncertainty requirements are narrow, other calibration method can be chosen.

The tests carried out in this study denote also the dependency of the calibration results on the calibration method. Therefore, special care should be given upon the calibration method versus the utilization method of the tank. Both methods should be the same or the results will not be compatible.

7. References

BEAN, V. E.; ESPINA, P. I.; WRIGHT, J. D.; HOUSER, J. F.;SHECKELS, S.D.; JOHNSON, A. N. March 24, 2006. *NIST Calibration Services for Liquid Volume.*NIST Special Publication 250-72.

Colorimetric characterization of led luminaires

C L M Costa[1], R R Vieira[1], R C Pereira[1], P V M Silva[1], I A A Oliveira[1], A S Sardinha[1], D D Viana[1], A H Barbosa[1], L P Souza[1] and A D Alvarenga[1]

[1]Optical Metrology Division, National Institute of Metrology, Quality and Technology- Inmetro, Rio de Janeiro, Brazil

E-mail: clcosta@inmetro.gov.br

Abstract. The Optical Metrology Division of Inmetro – National Institute of Metrology, Quality and Technology has recently started the colorimetric characterization of lamps by implementing Correlated Color Temperature (CCT) and Color Rendering Index (CRI) measurements of incandescent lamps, followed by the CFL, and LED lamps and luminaires. Here we present the results for the verification of the color characterization of samples of SSL luminaires for public as well as indoor illumination that are sold in Brazil.

1. Introduction

Over the past few years producers and consumers have become more demanding about lighting quality, requiring lighting products that are not only of good quality but also present alleged protocols for lighting and electrical characteristics, as well as the report of conformity to normative documents issued by qualified experts. There is a clear shift from the incandescent lamps to these more economic models (LED lamps), what is in accordance with last year's Brazilian government passed legislation on removal from the market of all incandescent bulbs by 2016. That will increase the consumption of LED illumination products.

Here we present the analysis for the verification of the color quality performance of SSL luminaires for public as well as indoor illumination that are sold in the Brazilian market [1].

2. Methods

2.1. The SSL luminaires measurements

In the measurement system used for SSL luminaires, the spectral and angular distributions of LED lamps are measured with a goniophotometer with rotating mirror arm used in association with the spectroradiometer as illustrated in figure 1.

Spectral power distribution (SPD) measurements of a reflectance white tile are carried out in the standard 0°:45° illumination observation geometry. This arrangement results in a simple structure of the equipments and reliable measurements in a continuous range of observation angles. The CCT and CRI calculations take into account the measurement geometry and the procedures according to IES LM-79-08 [2].

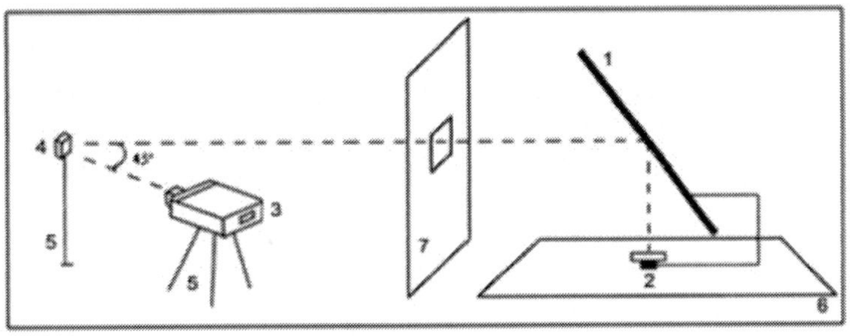

Figure 1. Schematic arrangement for SSL luminaires measurements

(1) Goniophotometer mirror, (2) SSL luminaire, (3) Spectroradiometer, (4) White tile, (5) Holders. (6) Well of goniophotometer, (7) Adjustable baffle wall.

2.2. Colorimetry measurements of LED luminaires

The ambient conditions, temperature and relative humidity, are maintained at (25 ± 1) °C and (50 ± 5) % respectively. The stabilization of the luminaire system is monitored by following up the uniformity of luminous flux, which occurs when all light emitters (LEDs) are in equilibrium. This occurs in 1 hour approximately. The SSL luminaire is positioned on a goniophotometer LMT-GO DS-2000 at a distance of 16.5 m of the reference tile. An ensemble of walls with adjustable apertures is installed between them to prevent stray light. The distance of the tile to the spectroradiometer is 0.8 m. The measuring directions are the combination of two parameters: the angle ϕ representing the position of the luminaire around its vertical axis and the angle θ representing the angle measured in a vertical plane in manner that $\theta = 0°$ to the downward direction of vertical axis and $\theta = 180°$ to the upward vertical axis. There are two sets of SPD measurements, as determined by the procedure IES LM-79-08: the first is set the angle ϕ at 0° and vary the angle θ at intervals of 10° up to the intensity limit for its acquisition, and the second consists of setting the angle ϕ at 90° and varying the angle θ in intervals of 10° up to the intensity limit as illustrated in figure 2.

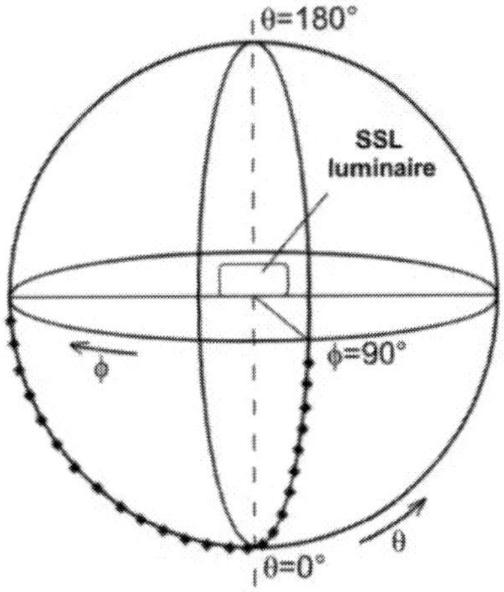

Figure 2. Schematic arrangement for goniophotometry of SSL luminaires measurements according [2]

2.3. Colorimetric calculations

Both the CCT and the CRI are obtained by measurements of the spectral power distribution of the luminaire performed by the spectroradiometer. The Robertson method is applied to obtain the CCT. We follow Gardner [3] to derive analytical expression for uncertainty in u and v chromaticity

coordinates and an uncertainty in CCT could be achieved. The CRI is calculated according to CIE 13.3 – 1995 – Method of Measuring and Specifying Colour Rendering Properties of Light Sources.
The CRI uncertainties are obtained by applying the ISO GUM [4] to derive analytical expressions for the uncertainties following the works of Gardner. In our work, partial derivatives of the calculation equations for color quantities are obtained in analytical form, what is not straightforward, due to the correlations among the chromaticity coordinates. The CCT and CRI calculations for SSL luminaires take into account the measurements geometry and procedures in IES LM-79-08.

3. Results

CCT and CRI were measured for a set of four SSL products, identified as samples J, K, L and M, as presented in table 1. The sample J, K and L are being designed and produced by national manufacturers that performs tests and essays to develop and improve their product. The sample M is sold in the Brazilian market and is produced by an international manufacturer. According to IES LM-79-08, the angle accuracy is not very critical for SSL luminaire measurement. That was experimentally verified using a SSL luminaire named sample M in table 1.

Table 1. Results

SSL Luminaires				
Sample J 72 LEDs 65 W	**Quantities**	**Nominal**	**0° Plane measure**	**90° Plane measure**
	CCT (K)	4500	4669	4665
	U_{CCT} (K)	275	83	83
	Ra	>70	77	78
	U_{Ra}	-	4	4
Sample K 90 LEDs 125 W	**Quantities**	**Nominal**	**0° Plane measure**	**90° Plane measure**
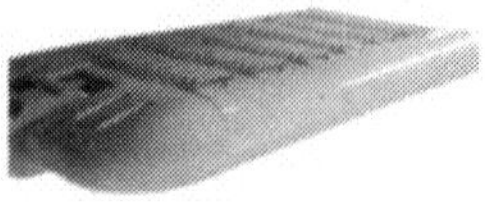	CCT (K)	5000	4958	4903
	U_{CCT} (K)	300	138	83
	Ra	-	65	78
	U_{Ra}	-	4	4
Sample L 96 LEDs 100 W	**Quantities**	**Nominal**	**0° Plane measure**	**90° Plane measure**
	CCT (K)	-	5244	5292
	U_{CCT} (K)	-	151	146
	Ra	-	73	64
	U_{Ra}	-	4	4
Sample M 49 LEDs 43 W	**Quantities**	**Nominal**	**0° Plane measure**	**90° Plane measure**
	CCT (K)	4000	4357	4305
	U_{CCT} (K)	-	74	63
	Ra	-	66	65
	U_{Ra}	-	4	4

Polar diagram in figure 3 shows the luminous intensity distribution for the SSL luminaire sample M in the vertical planes corresponding to $\phi = 0°$ (red line) and $\phi = 90°$ (green line). Values outside the frame correspond to the θ.

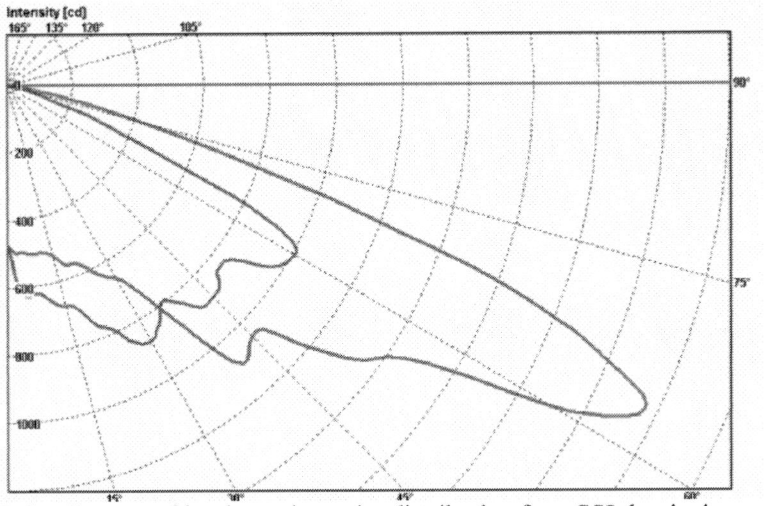

Figure 3. Polar diagram of luminous intensity distribution for a SSL luminaire sample M

Figures 4 and 5 shows the SPD measured for sample M acquired at intervals of 10° from 0° to 70° for the plane 0° and the 0° to 60° for the plane 90°, and those were the limiting angles due to intensity dropping. Another set of SPD measurements for the 0° and 90° plane were performed at steps of 1° around the angle where the intensity peaks occur. The maximum deviation for 0° plane was found to be around 5 K, and this difference is negligible considering, for example, the calculated 74 K uncertainty value that comes mainly from the colorimetric coordinates at this angle and observation plane.

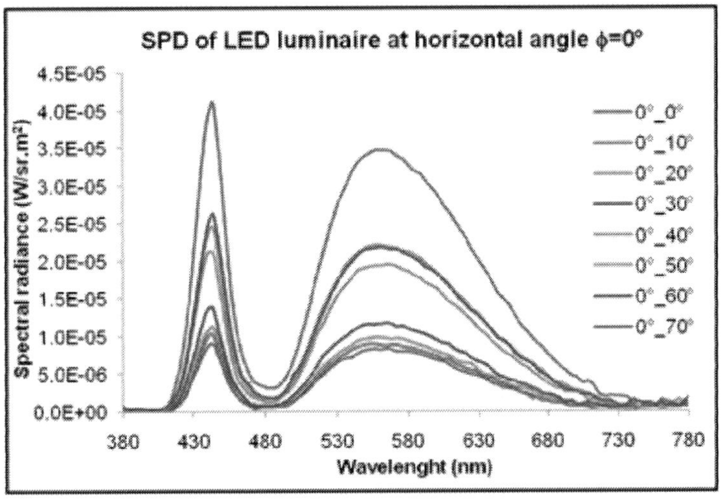

Figure 4. Spectral power distributions of sample M at horizontal angle $\phi = 0°$

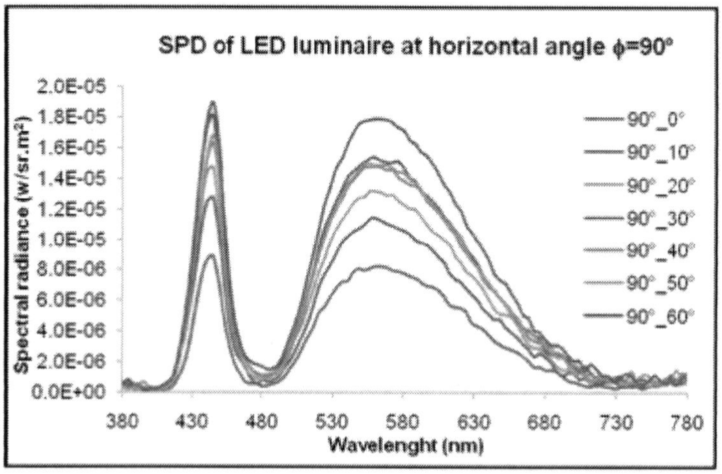

Figure 5. Spectral power distributions of sample M at horizontal angle $\phi = 90°$

For the 90° plane the maximum deviation was found to be around 12 K. At the angle where the more intense peak occurs, the maximum uncertainty from colorimetric coordinates also occurs, and consequently the great value for CCT and CRI uncertainties, and thus we usually do not present the CCT and CRI uncertainties for all the measured angles because they are smaller. Uncertainties are reported in Table 1 for the peak intensity of the measured luminaires. Other studies were carried out in our laboratory in order to examine the influence of the temperature of a typical single LED in the values of the CCT. These studies were carried out employing a thermal controller while measuring different SPD at each temperature. The temperature in the body of a SSL luminaire measured in the laboratory during the measurement is estimated to be around 50 °C using a copper constantan thermocouple probe. A luminaire using LEDs will present an expected shift in correlated color temperature [5] due to higher LED temperature during the luminaire operation, and this explains the differences between nominal and measured CCT. Sample M presents a CCT value around 4300 K, however through calculation and Duv table the CCT nominal category should be ranked 4500 K.

4. Uncertainty Budget

Table 2 shows the uncertainty budget considering a SSL luminaire sample M as an example. It takes into account the components associated with the measurements and the calculations as well. The last component is associated with geometric factors.

Table 2. Uncertainty budget for a SSL luminaire sample M

Components	Type	Uncertainty(K)
Colorimetric coordinates	A	38
Standard lamp certificate	B	2.3
Repeatability	A	0.2
Stability	A	0.3
Angle variation	A	2.5
Expanded uncertainty (k=2)		74

5. Conclusion

In these work we present the measurements methods to provide oversight and guidance on the evaluation of SSL luminaires. This will encourage and accelerate the adoption of high performance

SSL lighting by leading end-user manufacturer collaboration in the areas of performance, evaluation, application, and standardization. Solid-state lighting systems are a viable solution for all those who wish performance, economic and sustainability products. Metrology laboratories are the technology base to support government labeling programs and in a future these programs are important to ensure products of high quality.

Acknowledgements
We acknowledge CNPq, Inmetro and Faperj the financial support.

References

[1] Costa C L, Vieira R R, Alvarenga A P D and Grieneisen H P H *2011 Calibration of Standard Lamps in Color Temperature* 6° Congresso Brasileiro de Metrologia Conference proceedings published on line: http://limcserver.dee.ufcg.edu.br/metrologia_2011/viconbr/85821.pdf
[2] IES 2008 *IES LM-79-08 Electrical and Photometric Measurements of Solid-State Lighting Products*
[3] Gardner J L 2000 *Uncertainty estimation in colour measurement.* Colour Res & Appl Volume 25 Issue 5 pages 349–55 October 2000
[4] ISO GUM 2008 *Guia para a expressão de incerteza de medição* (Duque de Caxias, RJ) Translated from Evaluation of measurement data – Guide to the expression of uncertainty in measurement – GUM 2008, BIPM.
[5] Ohno Y 2004 *Colour Rendering and Luminous Efficacy of White LED Spectra* Fourth International Conference on Solid State Lighting Proc of SPIE Vol 5530

Assembly of a Faraday modulator for polarimetric measurements

C R Silva, M B Guedes, N C E Pereira and A D Alvarenga

Laboratory of Colorimetry and Spectrophotometry (Lacoe), Optical Metrology Division (Diopt), National Institute of Metrology, Quality and Technology (Inmetro), Av. N. Sr.ª das Graças 50, Xerém, Duque de Caxias, RJ, Brazil, ZIP.25250-020

E-mail: claudio.ribeiro@live.com

Abstract. We present the construction of a Faraday modulator for use in a polarimeter set up for quartz control plates optical rotation calibration. The choice of a glass type and relevant parameters for modulation are shown as well as initial test results.

1. Introduction

Industrial polarimeters and saccharimeters are calibrated by using quartz control plates as standards to ensure traceability. It was built at the Optical Metrology Division (Diopt) – Inmetro a high resolution polarimeter for calibration of quartz control plates by measuring its optical rotation by the polarimetric method [1]. The actual method using a rotating analyzer is detailed at [1, 2]. In this paper we show a Faraday modulator assembly, which will be part of a new set up of the polarimeter [3] with the purpose of comparing the results obtained by two different methods.

2. Lock-in modulation

A Faraday modulator consists of a solenoid containing a glass rod inside and an AC power supply wich output is modulated by a lock-in amplifier. The alternated magnetic field inside the solenoid causes the polarization plane of the He-Ne laser radiation to be rotated and modulated as it passes through the glass rod. The radiation is detected at the silicon photodetector connected to the lock-in. The lock-in recovers the modulated signal at the chosen frequency and thus very weak signals can be discriminated against large background noise.

2.1. Signal noise

The radiation at the photodetector generates an electrical signal as shown below:

$$V_{\text{signal}}(t) = V_0 + V_{\text{noise}} \tag{1}$$

where V_0 is the signal of interest. It is observed that the noise frequency is proportional to $1/f$ where f is the signal frequency. In very low measurements this noise is accepted as a common characteristic called flicker noise (that can be lessened) to the electronic systems caused by several factors. Incident photons in a photodetector is a random quantity that generates other noise source, shot or quantum noise, as known. In the equation (1), V_{noise} noise shot and Johnson noise are taken into account among others, and are the photodetector offset. In very low signal measurements, this total noise has great influence in the result of the measurement.

2.2. Operation

Basically, lock-in amplifier works in the following way: a sinusoidal signal is generated with the following form:

$$V(t) = V_0\cos(\omega t) \tag{2}$$

where V_0 is amplitude (voltage) of the signal and other sinusoidal signal, called reference signal (generated by the lock-in itself) at the same frequency $\omega = 2\pi f$, but with different angles θ (detuned) which reference signal is described as below:

$$V_R(t) = V_{0R}\cos(\omega t + \theta) \tag{3}$$

where V_{0R} is amplitude (voltage) of the sinusoidal reference signal. The product $V(t)V_R(t)$ of these two signals is:

$$V(t)V_R(t) = \frac{V_0V_{0R}}{2}\cos\theta + \frac{V_0V_{0R}}{2}\cos(2\omega t + \theta) \tag{4}$$

The averaged value of $V(t)V_R(t)$ gives the resulting low noise exit signal.

2.3. Modulation method with Faraday Effect applied in the polarimeter

In the Faraday Effect the radiation linear polarization axis rotates as it passes through a medium immersed in a magnetic field by an angle that is proportional to the Verdet constant of the material. In the modulator an AC current generates an alternating magnetic field. Verdet constant is expressed in International System of Quantities as [*rad/T·m*] and frequently encountered expressed in [*min/G·cm*]. For assembling the modulator we started from an existing hollow solenoid and we did a study in order to specify the glass rod type and size and later we made the characterization of the solenoid plus glass assembly.

2.4. Modulator characterization study

The most important characteristic for the glass rod inside the solenoid is the generation the largest rotation of the linear polarization axis with the smallest attainable magnetic field, in order to keep the solenoid body temperature as low as possible. The best choice was a SF4 glass with Verdet constant of 0.061 *min/G·cm*. The Verdet constant is defined as the rotation per unit path per unit field strength.

$$B(G) = 3.39654 + 348.4039I(A) \tag{8}$$

And so, a current (I) = 0.15 A generates a magnetic field (B) = 55.66 G or 0.006 T approximately. And for the proportionality relationship among field, Verdet constant and length of the glass:

$$\theta = VBl \tag{9}$$

where θ is the polarization rotation angle in degrees and V is the Verdet constant value, B is the magnetic induction and l is the length of the medium. The generated field results approximately 0.29° (0.005 *rad*) or ≈ 17 *min*, enough for the modulating signal necessary for the lock-in amplifier operation. The study showed that a glass rod of 5 *cm* length positioned in the center of the solenoid would provide the ideal rotation for radiation modulation. The glass Verdet constant was recalculated again after measurements of its optical rotation in the polarimeter resulting a value of ≈ 0.062 *min/G·cm* in agreement with the manufacturer's declared value.

Another accomplished study was the temperature increasing due to the current applied in the solenoid as a function of time and for a given current, with sensors positioned in the solenoid body and in the air at 1 *cm* away from it. We concluded that a current close to 0.15 A causes a very small temperature increase in the air around the Faraday modulator and that will not affect the temperature of the quartz control plates because they are isolated inside a thermalization chamber [1].

2.5. Test of the modulation system assembly

In the solenoid, an AC current generates an oscillating magnetic field through the glass rod, which causes the rotation of the radiation polarization axis to oscillate at the same applied AC current frequency. The solenoid is positioned between two linear polarizers and the laser radiation passes through them and is detected by the photodetector and so the input modulating signal for the lock-in amplifier is obtained. To provide the current for the coil, the lock-in amplifier generates a sinusoidal reference signal adjusted in its internal oscillator that feeds in an external amplifier that drives the solenoid.

Figure 1 presents the Faraday modulator experimental test set-up detailed. (1) He-Ne laser; (2) linear polarizer; (3) neutral density filter; (4) chopper (to compare the solenoid modulation – it will not use in final set-up); (5) solenoid; (6) power amplifier; (7) rotating linear polarizer (analyzer); (8) lock-in amplifier.

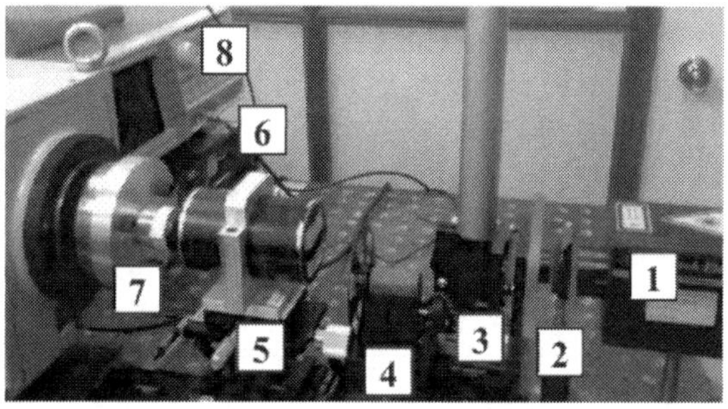

Figure 1. Faraday modulator experimental set-up.

With the assembled system we defined the lock-in reference signal voltage (0.500 ± 0.001) *V*, the frequencies (commons 50 *Hz* and 60 *Hz* harmonics free) which we used (177.05 ± 0.05) *Hz*, (137.05 ± 0.05) *Hz*, and (380.05 ± 0.05) *Hz*, and thus with this parameters the power amplifier provided the expected and necessary current to the system modulation. Table 1 shows a resulting gain in accuracy measurement when using the lock-in amplifier instead of just the voltmeter.

Table 1. Voltmeter – lock-in comparative.

Measurement	Voltmeter (*V*) ($\times 10^{-5}$)	Lock-in (*V*) ($\times 10^{-5}$)	Accuracy increase %
1	2.9	2.583	12.8
2	2.4	2.135	13.0
3	2.1	1.853	13.6
4	1.7	1.509	13.4
5	1.6	1.379	12.5
6	1.5	1.370	11.5
7	1.5	1.314	14.5
8	7.5	6.439	16.7
9	5.6	4.874	14.5

3. Acquisition system and first results

The lock-in amplifier was connected in a PC with GPIB interface, and a LabView lock-in control system programm was developed to set the parameters, acquire and record the data measurements. With the assembled Faraday modulator the first evaluation measurements were realized. It was studied the signal obtained in the configuration where polarizer and analyzer had their polarization axis at 90 degrees, a situation of minimum irradiance reaching the detector. The modulator was able to recover approximately 88% of the signal without noise.

4. Conclusions

Preliminary tests of a Faraday modulator built in the laboratory for polarimetric measurements presented satisfactory results as a considerable attenuation in the noise at low radiation level was reached. In next steps the modulator will be assembled inside the polarimetry system described in [1] to compare both methodologies and realize a complete metrological characterization of all system.

Aknowlwdgements

A D Alvarenga thanks CNPq and Faperj, and N C E Pereira thanks CNPq support. C R Silva thanks Diopt/Inmetro for supporting this research.

References

[1] Alvarenga A D, Pereira N C E, Tarelho L V G, França R S and Belaidi H 2009 *Calibration of Quartz Control Plates by High Resolution Polarimetry.* OIML Bulletin V.11, 5 (2010). Proc. V Brazilian Congress of Metrology, Salvador, Brazil, *http://repositorios.inmetro.gov.br/bitstream/10926/306/1/Alvarenga_AD_Metrologia2009.pdf*

[2] Alvarenga A D, Pereira N C E, Gomes B S and Grieneisen H P H 2011 *Evaluation of measurement uncertainties for polarimetric calibration of quartz control plates.* Proc. VII Brazilian Congress of Metrology, Natal, Brazil, *http://limcserver.dee.ufcg.edu.br/metrologia_2011/viconbr/85882.pdf*

[3] Schulz M, Fricke A, Stock K, Alvarenga A D and Belaidi H. High accuracy calibration of quartz control plates. Proc. IMEKO XVIII World Congress 2006, Rio de Janeiro, Brazil, *http://www.imeko.org/publications/wc-2006/PWC-2006-TC8-022u.pdf*

New infrared radiation temperature measurement system applied to the Laser Flash Method

D M Camarano[1], F F Alves[2], I Finelli[2], F L Migliorini[1], O Miranda[1] and P A Grossi[1]

[1] Centro de Desenvolvimento da Tecnologia Nuclear (CDTN)/ Comissão Nacional de Energia Nuclear (CNEN), Av. Presidente Antônio Carlos 6627, Campus da UFMG – Pampulha, Caixa Postal 941, CEP 30.161-970, Belo Horizonte, MG, Brasil

[2] Universidade Federal de Minas Gerais, Departamento de Engenharia Elétrica, Campus Pampulha, CEP 31.270-901, Belo Horizonte, MG, Brasil

E-mail: dmc@cdtn.br

Abstract. The laser flash method is based on measuring the temperature transient. The effect of the thermal inertia of the temperature measurement system is not negligible and cannot be ignored due to the fast dynamics of the measured thermal transient. This paper presents the new infrared radiation temperature measurement system applied to the thermal diffusivity measurements based on the laser flash method. The developed system consists in a detector cooled in liquid nitrogen, collimating lenses and a mathematical modeling developed in LabView. As a result, the generated electrical signal is converted into temperature. A detailed description of this new temperature measurement system is presented, as well as its mathematical model, computational implementation and calibration. For validate the new system, reference samples of homogeneous materials (Pyroceram 9606 and Inconel 600) were tested and the results of thermal diffusivities were compared with its reference certificates. The thermal diffusivity values measured by using the new system are in agreement with reference values, with less than 2 % of relative deviation.

1. Introduction

The determination of thermal diffusivity by laser flash method [1-4] is based on the variation of temperature on the opposite side of a thin cylindrical sample resulting from a short pulse of radiant energy received on its front side. This thermogram is measured and the thermal diffusivity is calculated as a function of sample thickness and the time required to reach half of the maximum rate of temperature on the opposite side. The Laboratório de Medição de Propriedades termofísicas (LMPT) of the Centro de Desenvolvimento da Tecnologia Nuclear (CDTN/CNEN) is expanding the measurement capabilities of its metrological platform devoted to the determination of the main thermophysical properties of materials. Then, the development of new temperature measurement system with a response time of nanoseconds is a priority to improve the quality of the thermal diffusivity measurement of materials. The new temperature measurement system was designed and built to improve experimental apparatus used to determined thermal diffusivity. The mathematical modelling of this system is also presented on this paper as well as its computational implementation and calibration. Reference samples of homogeneous materials (Pyroceram 9606 and Inconel 600) were used during the validation process and the results of thermal diffusivities were compared with its reference certificates.

2. Experimental apparatus and procedure

2.1. Physical design and construction

The design of the temperature measurement system has been divided into two phases. The first phase is related to the design and construction of the system. The HgCdTe detector is the photoconductive element which undergoes a change in resistance proportional to the incident infrared radiation. It is mounted in the metal dewar with ZnSe window offering optimum performance in the 8 to 12 µm. It was used the J15D series HgCdTe detector and the preamplifier model PA-101 supplied by Teledyne Judson Technologies. The ZnSe lens are the optical system used to collect the infrared radiation emitted by the sample. The system is equipped with an LED pilot light for accurate alignment of the sensor with the object to be measured. The signal is digitized using a 16 bits A/D converter of an NI PCI-6052 data acquisition device. The second phase is related with the signal processing using LabVIEW. Table 1 summarizes technical data of HgCdTe and the preamplifier used in the system. Figure 1 shows an overview of the new temperature measurement system.

Table 1. Summary of technical data of detector and preamplifier.

Item	Technical data
HgCdTe Detector	
Active area size:	4 mm^2
Cutoff wavelength (20 %)	> 12 µm
Peak wavelength	(11 ± 1)µm
Peak detectivity @ 10 kHz	min 1 x 10^{10} cm $Hz^{1/2}$·W^{-1} typical 1.5 x 10^{10} cm $Hz^{1/2}$·W^{-1}
Responsivity @peak	100 V/W
Time constant	0.5 µs
PA 101 Preamplifier	
Bandwith	t1 stage 10 Hz to 1 MHz 2nd stage 10 Hz to 200 kHz
Gain	1 st stage 100x 2nd stage 10x
Input impedance	10 kΩ through 100 µF capacitor
Input noise	1.5 nV Hz
Maximum Output level (high impedande load)	10 V p-p

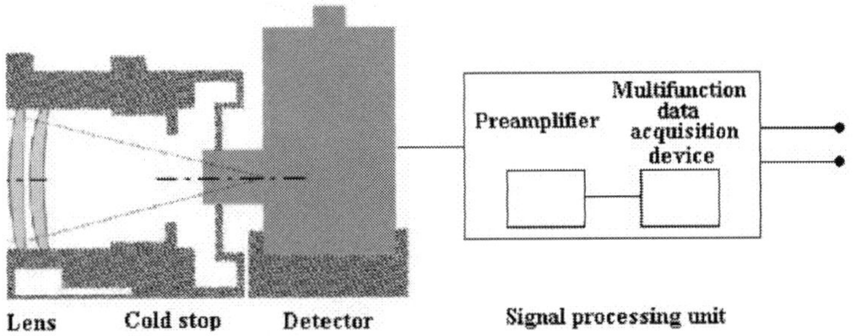

Figure 1. Structure of the new temperature measurement system.

2.2. Mathematical model

The electrical signal corresponding to the radiation emitted from a blackbody, S_{bb} (T_{bb}) is described by [5]:

$$S_{bb}(T_{bb}) = g \cdot \frac{R^* \cdot A_d \cdot \tau_0}{4 \cdot F^2 + 1} \cdot \int_{\lambda_1}^{\lambda_2} M(T_{bb}, \lambda) \cdot s(\lambda) \cdot d\lambda. \quad (1)$$

Where g is the amplification of the electronic block, R^* is the peak detector spectral sensitivity given in $V \cdot W^{-1}$, A_d is the detector area in cm^2, F represents the ratio of focal length of the lens and diameter of the objective, λ_1 and λ_2 are the limits of the spectral band of the detector in nm, M (T, λ) is the spectral exitance at the temperature T and wavelength λ ($W \cdot m^{-2} \cdot \mu m^{-1}$), τ_0 is the transmittance of the lens used and s (λ) is the detector relative spectral detectivity function. The measuring signal, S_r (T) related to the temperature of any object is be expressed by [5,6]:

$$S_r = \varepsilon\tau S_{obj}(T) + (1-\varepsilon)\tau S_{refl}(T_{refl}) + (1-\tau)S_{env}(T_{env}) - (1-\tau)S_{opt}(T_{opt}) - S_{det}(T_{det}) \quad (2)$$

Where S_{obj} (T), ε, τ is the electrical signal corresponding to the radiation emitted by the object, the effective emissivity of the real target and the effective transmissivity of the atmosphere, respectively. S_{refl} (T_{refl}) is the electrical signal corresponding to the radiation reflected from the surrounding objects, S_{env} (T_{env}) is the signal corresponding to the radiation emitted (or absorbed) by the environment between the object and the detector. S_{opt} (T_{opt}) is the electrical signal corresponding to the radiation emitted by the optical components and S_{det} (T_{det}) is the electrical signal caused by the by the detector itself. The temperature T_{out} of the object is calculated based on the value determined from the corrected signal S_r (T) converted to temperature using the calibration data (figure 2).

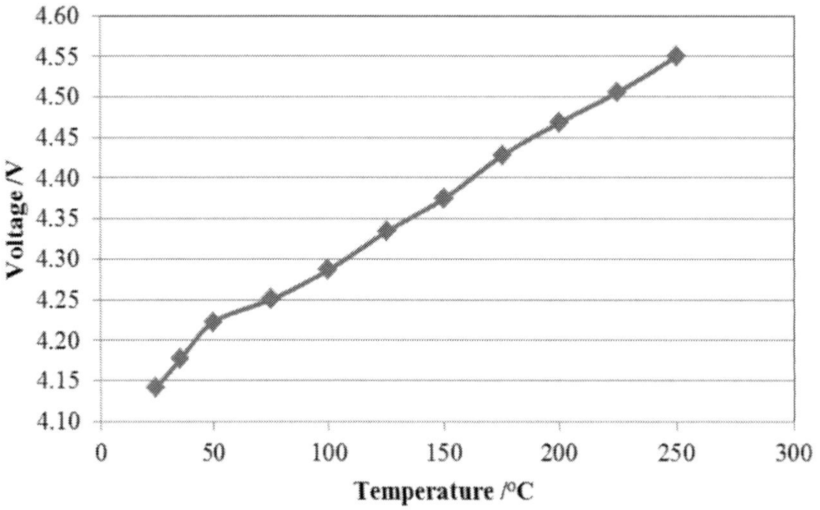

Figure 2. Calibration curve.

The mathematical model developed was implemented using the LabVIEW platform. The signal data are acquired and processed and the corrections are performed according to the equations described above. The corrected electrical signal is converted to temperature signal, considering the calibration of the system (figure 3). The program generates a screen containing the signal graph in temperature over time (figure 4).

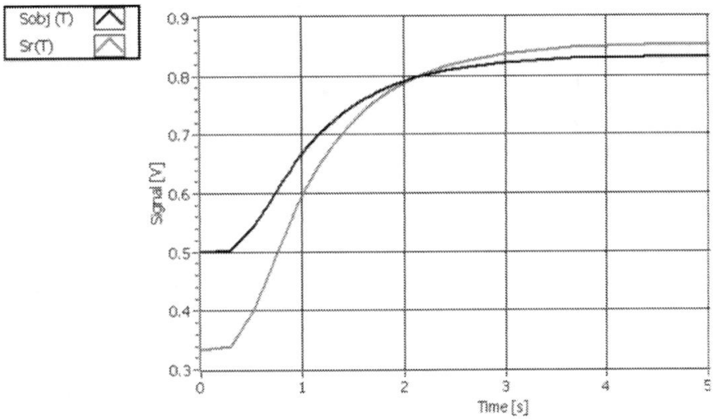

Figure 3. Typical signals obtained from laser flash measurements: S_{obj} and S_r.

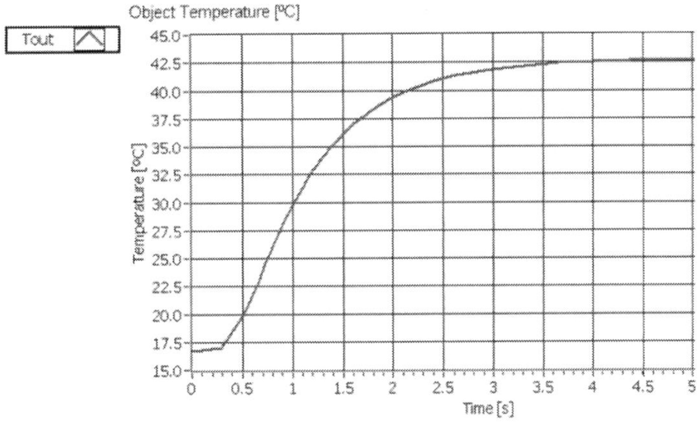

Figure 4. Typical temperature signal obtained from laser flash measurements.

2.3. Thermal diffusivity measurement
To verify the validity and desired features of the new temperature measurement system, it was used to measure the thermal diffusivity of two reference samples (Pure Inconel 600 and Pyroceram 9606) using the laser flash method [1-4].The system was coupled to the apparatus [4] developed by the staff of the Laboratório de Medição de Propriedades Termofísicas of Centro de Desenvolvimento da Tecnologia Nuclear (figure 5). This apparatus is composed of sample holding device, heating furnace, laser CO_2 (of 100 W total power) working at 10.6 μm wavelength, vacuum pump, measuring control and data processing system. The thermal radiance signal is digitized using a 16 bits A/D converter of an NI PCI-6052 data acquisition device and a LabView programming is used to acquire data. A thin carbon layer was applied over the front surfaces of the samples (8 mm in diameter and about 2.5 mm thickness). The rear surfaces of the samples were also coated to ensure an uniform emissivity.

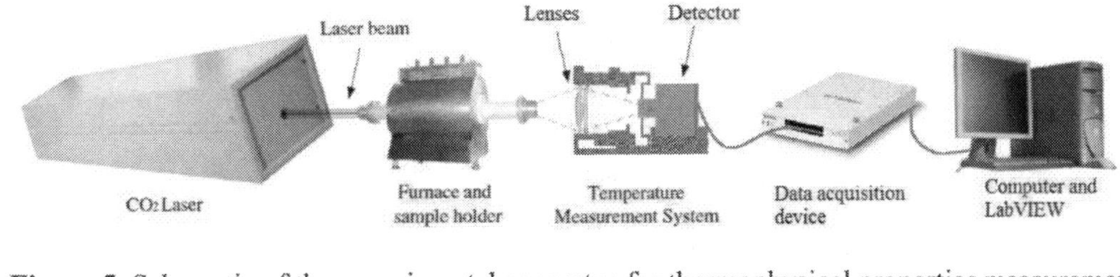

Figure 5. Schematic of the experimental apparatus for thermophysical properties measurements.

3. Results

All thermal diffusivity values presented hereafter result from the average of successive measurements carried out on repeatability conditions. The results obtained for the thermal diffusivity of Inconel 600 and Pyroceram 9606 standard samples are shown in table 2. The results show good agreement with the reference values of both samples. The maximum deviation of obtained values for thermal diffusivity using the new temperature measurement system from the certificated value is in less than 2 %, which is within the uncertainty of the value estimated in 7.5 %. The uncertainty assessment of the experimental apparatus [7] is performed in accordance with the ISO Guide to the Uncertainty in Measurement (GUM) [8]. The uncertainty of the certified thermal diffusivity values are 7.5 % at the 95 % confidence level [9].

Table 2. Thermal diffusivity of reference samples at room temperature.

Themal diffusivity / $mm^2 \cdot s^{-1}$		
Reference sample	Mean value	Reference value [9]
Pyroceram 9606	2.02	2.02
Inconel 600	3.42	3.44

4. Conclusions

In this paper was presented a new system for measuring temperature with low response time coupled to the system for determination of thermal diffusivity using the flash method. The mathematical model was developed and implemented in LabVIEW platform for signal processing, correction and conversion of electrical signal to temperature signal. The results show great potential for use in thermal imaging systems with relatively low cost and adequate accuracy, with low thermal inertia and wide measurement range. Experiments carried out with reference samples demonstrated good agreement between the reference and obtained values, with a relative deviation of less than 2 %. A more detailed investigation becomes necessary to for a better evaluation of the reproducibility of the system using others reference samples.

Acknowledgments

This work has been supported by the Fundação de Amparo a Pesquisa de Minas Gerais – FAPEMIG and by the Conselho Nacional de Desenvolvimento Científico e Tecnológico – CNPq.

References

[1] Parker W J, Jenkins R J, Butler C P and Abbott, G L 1961 *J. Appl. Phys.* **32** 1679
[2] ASTM-E-1461 2011 *Standard test method for thermal diffusivity of solids by the flash method*, vol 14.01 (Annual book of ASTM standards)
[3] Hay B P, Filtz J R, Hameury J and Rongione L 2005 *Int. J. Thermomphys.* **26** 6
[4] Migliorini F L, Silva E H C, Grossi P A, Ferreira R A N and Camarano D M 2009 *Proceedings of the XIX IMEKO World Congress 2009, Fundamental and Applied Metrology*, Lisbon, Portugal, p 2000
[5] Chrzanowski K 2001 *Non-Contact Thermometry - Measurement Errors* vol 7 Research & Development Treatises (Warsaw: Polish Chapter of SPIE)
[6] Vendt R, Juurma M, Jaanson P, Vabson V, Kübarsepp T and Noorma, M 2011 *Int. J. Thermophys.* **32** 247
[7] Camarano D M, Migliorini F L, Silva E H C, Grossi P A, Ferraz W B and de Paula J B 2010 *Int. J. Thermomphys.* **26** 6
[8] JCGM 100:2008 *Evaluation of measurement data - Guide to the Expression of Uncertainty in Measurement* (JCGM Joint Committee for Guides in Metrology)
[9] *INMETRO Test Report DIMCI 1588* 2013 - Pyroceram 9606 and Inconel 600 samples (INMETRO Instituto Nacional de Metrologia, Qualidade e Tecnologia)

Determining camera parameters for round glassware measurements

F O Baldner[1,2], P B Costa[1,2], J F S Gomes[2], D M E S Filho[2] and F R Leta[1]

[1] Universidade Federal Fluminense – UFF; Programa de Pós-Graduação em Engenharia Mecânica – Laboratório de Metrologia Dimensional e Computacional
Rua Passo da Pátria, 156 – Niterói – Rio de Janeiro – Brasil
[2] Instituto Nacional de Metrologia, Qualidade e Tecnologia - Inmetro
Av. Nossa Senhora das Graças, 50 – Xerém – Duque de Caxias – Rio de Janeiro – Brasil

E-mail: fbaldner@gmail.com

Abstract. Nowadays there are many types of accessible cameras, including digital single lens reflex ones. Although these cameras are not usually employed in machine vision applications, they can be an interesting choice. However, these cameras have many available parameters to be chosen by the user and it may be difficult to select the best of these in order to acquire images with the needed metrological quality. This paper proposes a methodology to select a set of parameters that will supply a machine vision system with the needed quality image, considering the measurement required of a laboratory glassware.

1. Introduction

Machine vision is a field in constant development. From the digital camera sensors' conception in the late 1960s to this day, their resolution has greatly increased, and it is not uncommon for cellphones to have a small sensor with a 5 MPixel resolution, or even greater.

However, traditionally, such cameras are hardly sought to be used on scientifically applied machine vision inspection or measurement systems. Are these cameras really improper for this kind of application? With digital photography equipment presenting increasing quality, one must wonder why these cameras may not be used.

Enabling users to have control over certain operational parameters (such as aperture, exposure, etc.) is a characteristic of the digital single-lens reflex (DSLR) cameras. With such control, there's a wide variety of parameters to determine in order to find the best conditions needed.

Machine vision systems in general can provide different types of measurements, direct or indirect, with the end result being a matter of choosing the correct combination of hardware (camera, lenses, lighting, etc.). Metrology is the science of measurement and is in constant evolution, and such systems may provide an interesting source of measurement results.

Ranging from dimensional measurements [1] to colorimetry [2], there are many possibilities for machine vision in metrology [3]. Fluids metrology in particular deals with a great amount of glassware in its day to day work. Devices such as viscometers require the measurement of certain dimensions to calculate the viscosity of the fluid flowing inside of it [4].

This paper has the objective of providing a methodology for choosing values for the parameters of aperture in a camera for the acquisition of glassware images that will, subsequently, be used to extract a concrete value of a dimension from.

2. Camera Definitions

This section defines important terms in the camera field that will be used throughout this paper.

2.1. Focal plane

In optics, when a light ray passes through a lens, parallel to its optical axis, it will converge to a point called the focus [5]. The focal plane is the plane which is perpendicular to the optical axes that contains this point. The objects in this plane will provide a sharp image behind the lens, as shown in figure 1 with the object to the right in the focal plane and the other two outside.

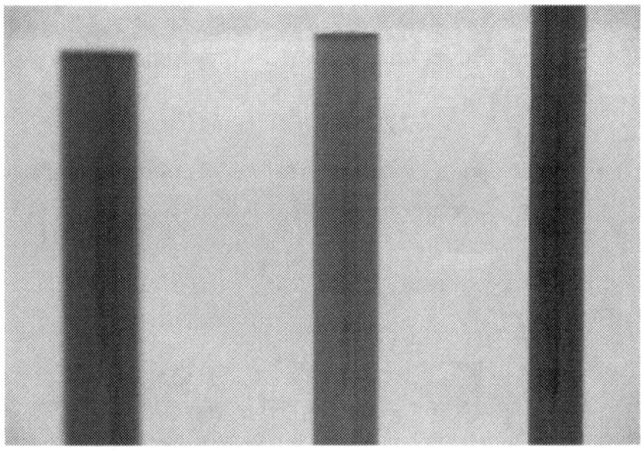

Figure 1. Focal plane example.

2.2. Exposure time

Exposure time is the amount of time in which the lens' shutter will be opened in order to expose the image sensor to the light from the object [6]. The image of an object photographed under an exposure of 1/2 s is shown in figure 2, contrasting with the object in figure 3, whose exposition was 1/8000 s.

Figure 2. Object under high exposure.

Figure 3. Object under low exposure.

2.3. Aperture

Aperture is the area of the shutter in which light will pass through. The aperture size is measured in a ratio of the focal distance of the lens and the shutter's opened diameter [6], shown in figure 4. For example, an aperture of f/2 means that the shutter's opened diameter is half the focal distance of the lens.

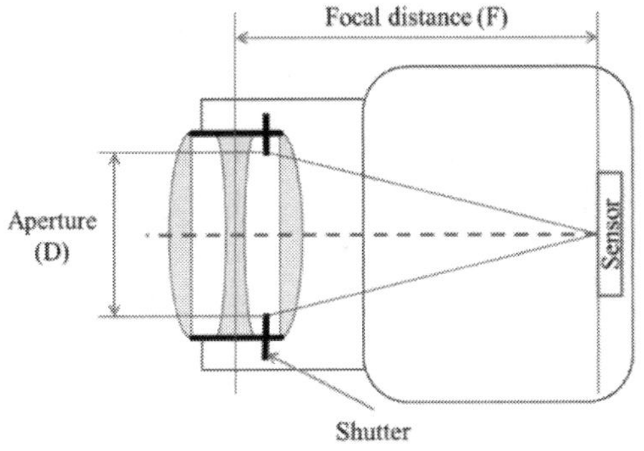

Aperture size:

$$f = \frac{F}{D}$$

Figure 4. Schematic showing the quantities involved in the aperture size determination.

2.4. ISO value

It is a system for measuring film (or sensor, in digital cameras) sensitivity. Increasing the ISO value allows for a faster shutter speed, given the same aperture size, to photograph a certain object in a darker environment [6]. This sensibility enhancement, however, creates grainy areas especially in areas of the image in which the sensor did not receive much light [7].

2.5. Exposure value (EV)

The exposure time combined with the lens aperture, in a certain object with a certain lighting condition provides the exposure value (EV) which defines the luminance (quantity of light) of the image [6]. By changing both exposure and aperture (increasing the aperture size and lowering the shutter speed, or vice-versa) the EV is kept, meaning that the same amount of light will reach the sensor, as shown by the histograms in figure 5.

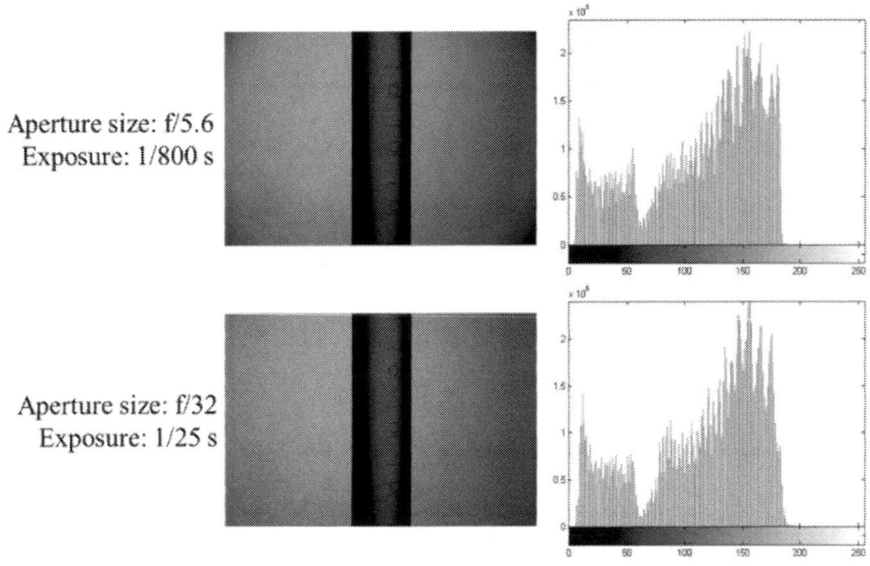

Figure 5. Two images with the same EV but different parameters.

2.6. Depth of field

In a camera lens, there is a range of values in which the light will produced a focused image of an object. This represents the depth of field, where any objects inside this range will appear in focus, and everything else out of focus. This is controlled by the focal distance, the focus point and the aperture. By setting the first two values the aperture will determine the depth of field. The greater the aperture (smallest f number), the shallowest the depth of field is [6], as seen in the example of figure 6, where, in the picture with an aperture of f/5.6 the depth of field did not encompass the two objects to the left, and with the aperture of f/9 the middle object is inside the depth of field with the rightmost one.

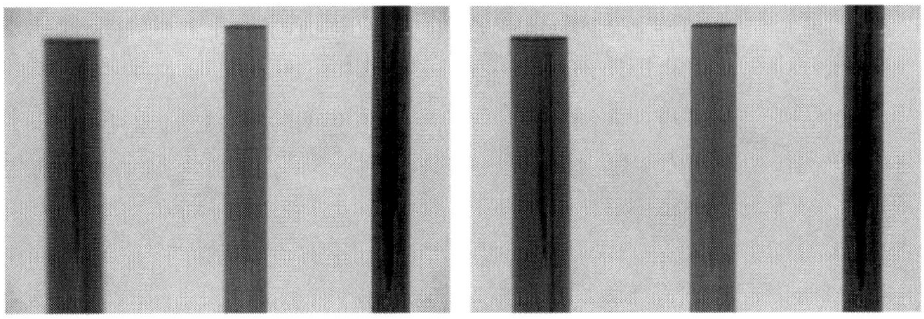

Aperture size: f/5.6 Aperture size: f/9

Figure 6. Depth of field in different apertures.

3. Materials and Methods

This section details the equipment and materials used for this experiment, as well as the methodology employed.

3.1. Materials

For this experiment, the image acquisition device used was a DSLR camera (Canon 60D) with an 18-135 mm lens connected to the computer and controlled by Canon's software (EOS Utility).

The object used was a beaker filled with red dyed water, to provide a contrast with the background. A diffuser was built with tracing paper to reduce the reflection from the light source [8]. The object and the light source (an incandescent lamp with a color correlated temperature of approximately 2700 K) were placed inside a dark box to avoid interaction with other light sources. The entire setup is shown in figure 7.

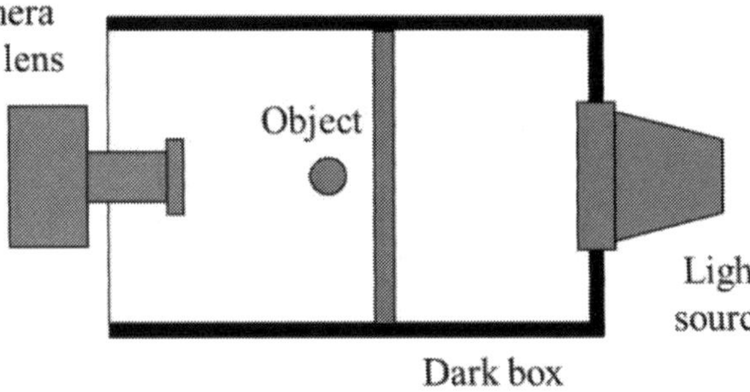

Figure 7. Image acquisition setup.

3.2. Methods

The first step in the experiment is to determine a value for the aperture size. After setting up the equipment, the lens' focal distance must be chosen so that the object will fill the image as most as possible while still being sharp around its diameter (since it's a round object). For this, the focus point must be set to the lowest possible value and then fine adjusted using the software to avoid moving the equipment.

After this, the aperture is set to the smallest value possible and the exposure is chosen so that the image is neither over-exposed (too much light) nor under-exposed (too little light). This can be achieved by looking at the image histogram, where it shouldn't be concentrated on either extremity.

This process is then repeated by decreasing the aperture size and increasing the exposition time by one step each. With this, all the images have the same EV. The images are then segmented using Otsu's algorithm [9] and everything but the center object was removed from the image. For each segmented image an average diameter is calculated and then, for the collection of images, a sample average diameter is calculated, along with the standard deviation. This is all described by the flowchart in figure 8. By analyzing these results along with each image, a proper aperture size can be chosen.

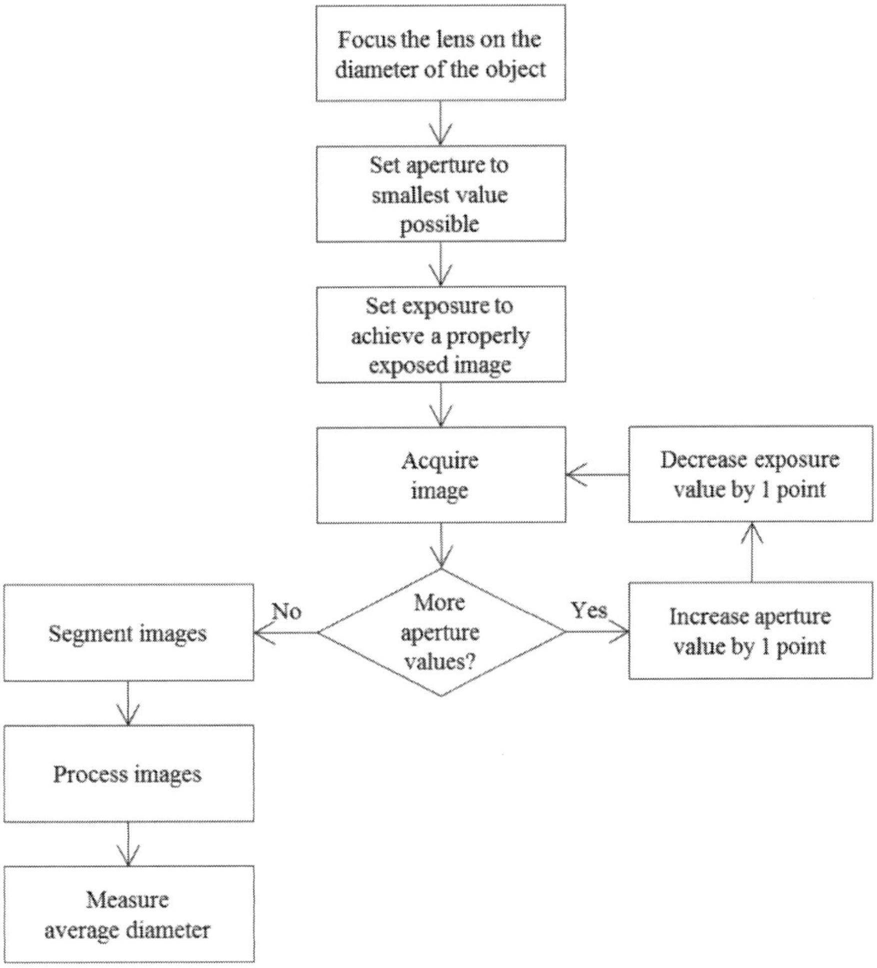

Figure 8. Flowchart for the methodology of picture processing and analyzing.

4. Experiment With Aperture

Using the methodology proposed in 3.2, the camera focal distance was set to 135 mm and the focus was adjusted via software. For the highest possible aperture for this lens' focal distance (F/5.6) the exposure that would produce a proper image is 1/800 s, generating the image shown in figure 9, alongside its histogram, in figure 10. Segmentation using Otsu's algorithm [9] produces the image shown in figure 11.

Figure 9. Acquired image.

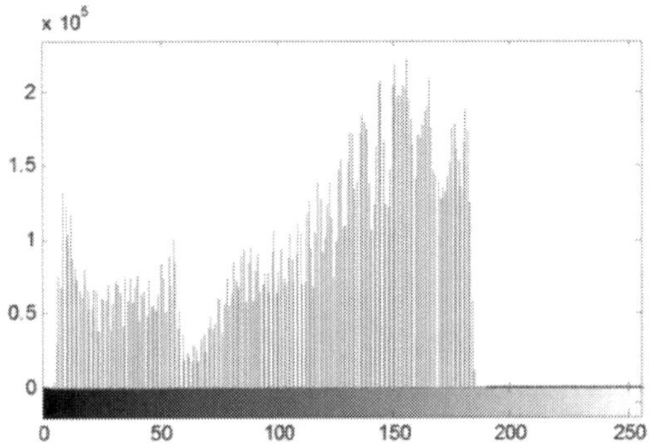

Figure 10. Image histogram.

Figure 11. Segmented image.

This process was then repeated, decreasing the aperture by one step, decreasing the exposure time by one step and acquiring an image. The resulting images would produce the exact same histogram as previously seen in figure 10. As seen in table 1, the smallest aperture possible was F/32, which had an exposure of 1/25 s. This process generated 16 images with the same EV but different aperture sizes.

Table 1. Aperture and exposure values for each image.

Image	Aperture	Exposure [s]	Image	Aperture	Exposure [s]
1	F/5.6	1/800	**9**	F/14	1/125
2	F/6.3	1/640	**10**	F/16	1/100
3	F/7.1	1/500	**11**	F/18	1/80
4	F/8.0	1/400	**12**	F/20	1/60
5	F/9.0	1/320	**13**	F/22	1/50
6	F/10	1/250	**14**	F/25	1/40
7	F/11	1/200	**15**	F/29	1/30
8	F/13	1/160	**16**	F/32	1/25

5. Results

After acquiring and processing each image, an average diameter in pixel is calculated for each image. In order to determine this diameter in millimeters, the length of the pixel must be calculated. By measuring the diameter of the beaker with a caliper, the length of the pixel is calculated using equation (1). With the length of the pixel calculated, the millimeter values for each image were then calculated with equation (2).

$$l_P = \frac{d_{\text{caliper}}}{\frac{1}{n}\sum_{i=1}^{n} d_{\text{pixel}(i)}} \tag{1}$$

$$d_{\text{mm}(i)} = d_{\text{pixel}(i)} \cdot l_P \tag{2}$$

These results are shown in the graph of figure 12, with lines marking the largest and smallest values. It can be observed that the amplitude of these measurements is of 0,04 mm.

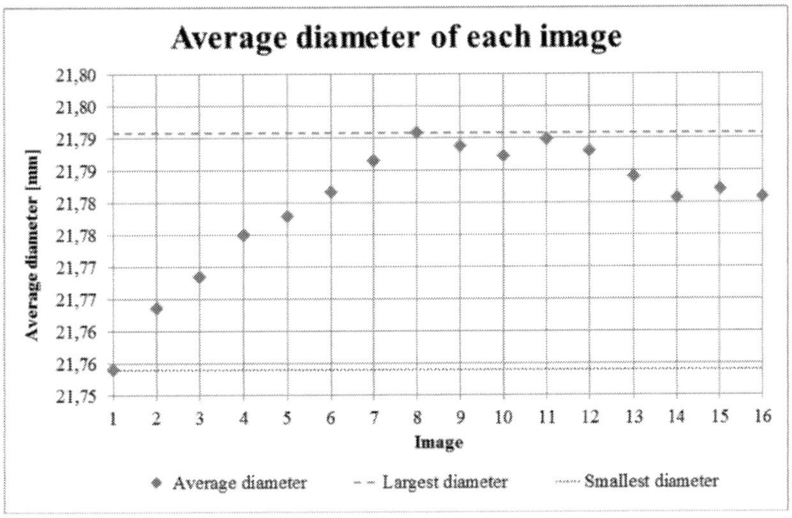

Figure 12. Average diameters for each image, alongside the maximum and minimum diameters.

6. Conclusion

Taking the average as the real value, we have a percentage error of 0,05% and -0,14%. For certain applications (for example, in Ubbelohde viscometers, the diameter needs to be within 2% of a nominal value in millimeters [10]) this can be an acceptable result.

However, how well is the average diameter a good representation of the real value? This discussion often reflects back on the application of the user.

Further analysis of the gathered data needs to be done in order to understand the influences of things such as lighting, lens' sphericity, or even the unevenness of the glassware. Future works on these subjects must concentrate on studying these error sources. Another important analysis that must be done is studying the influences for the calculation of the measurement uncertainty.

Acknowledgements

The authors would like to acknowledge FAPERJ (under grants E-26/103.591/2012, E-26/103.618/2012 and E-26/171.362/2001) for its financial support.

The authors would also like to acknowledge their colleagues from UFF and Inmetro for the support while conducting the experiments.

References

[1] Feliciano F F, de Souza I L and Leta F R 2005 Visão computacional aplicada à metrologia dimensional automatizada: considerações sobre sua exatidão *ENGEVISTA* **7** 38-50

[2] Gomes J F S 2013 Padronização de metodologia para caracterização de cor por imagem aplicada à seleção de frutas *Doctoral Thesis from Universidade Federal Fluminense*

[3] Gomes J F S and Leta F R 2012 Applications of computer vision techniques in the agriculture and food industry: a review *Eur. Food Res. Technol.* **235** 989-1000

[4] Baldner F O, Leta F R, Costa P B, Tortelote A S, Santo Filho D M E, Rodrigues C R C and Barbosa A P F 2011 Measurement of dimensional parameters in ubbelohde type viscometers using digital images *2° Cong. Int. de Metrologia Mec*

[5] Tipler P A and Mosca G 2007 *Física para Cientistas e Engenheiros* vol 2 (Rio de Janeiro: LTC)

[6] Hedgecoe J 2005 *O Novo Manual de Fotografia* (São Paulo: Editora Senac São Paulo)

[7] Trigo T 2010 *Equipamento Fotográfico* (São Paulo: Editora Senac São Paulo)

[8] Hunter F, Biver S and Fuqua P 2007 *Light Science & Magic* (Burlington, MA: Focal Press)

[9] Otsu N, A threshold selection method from gray-level histograms 1979 *IEEE Trans. Sys. Man and Cyber.* **SMC-9** 62-6

[10] ISO 3105: Glass capillary kinematic viscometers - Specifications and operating instructions 1994

Journal of Physics: Conference Series **575** (2015) 012022 doi:10.1088/1742-6596/575/1/012022

Comparative study of methods for measuring window film UV and luminous transmittance

José P Caly[1] and Antonio F G Ferreira Jr
IPT, Institute for Technological Research, Av. Prof. Almeida Prado 532 São Paulo SP 05508-901, Brazil

E-mail: pucci@ipt.br

Abstract. This work presents a comparison of window film ultraviolet transmission calculated according ISO 13837 methods A and B and window film luminous transmission calculated for illuminates A and D65. The results show a difference from 350% to 800% in ultraviolet transmission between ISO 13837 methods A and B for the evaluated window films. On the other hand, the difference in luminous transmission results for the evaluated window films is less than 1%.

1. Introduction

Based on a market research conducted by ABRAWF (Brazilian Association of Users of Window Film), 70 % of respondents wish to have solar control films installed in their vehicles [1], among the reasons that support the decision of consumers the most important are the protection against UV radiation and its damage. In solar control film data sheet [2-3], for automotive or building construction use, there is a wide range of information whose magnitudes are not in a didactic way, to reach the consumer. Example of this can be the "UV rejection rate". As shown in data sheets without a normative reference or calculation method, it is difficult to make the comparison between products and preclude a choice based on technical criteria by the consumer. Although there is no Brazilian standard headed the issue, the legislation dealing with the use of automotive films [4], Resolution 254 CONTRAN (Brazilian National Transit Department), deals only with the values of light transmittance for the illuminant A. Internationally there is a set of standards that addresses the issue with different methods of calculation. The DIN EN 410 [5] deals with the evaluation of films for building which determines the visible transmittance using illuminant D_{65} [6] and UV transmittance in the range 280-380 nm for the spectrum of the sun with an air mass equal to one. The ISO 13837 [7] defines methodology for evaluation of automotive films considering two methods for UV transmittance: the spectral range 300-400 nm with a solar spectrum air mass of 1.5 and the other in the spectral range 300-380 nm with a solar spectrum air mass of 1. For visible transmittance ISO 13837 does not mention a calculation method. The SAE J1796 standard [8] deals with automotive film application, in turn, it determines the visible transmittance using illuminant A [6] and the UV transmittance in the range of 300 to 400 nm with a solar spectrum air mass equal to 1.5, the same as the first method of ISO 13837.

[1] To whom any correspondence should be addressed.

The difference between air masses can impact the result of UV transmittance, taking in account the differences between their spectra are proportional to solar intensities, as the lower air mass, make greater the intensity of the sun. The spectral range used to calculate the UV transmittance indicates that the window film may consider the effects of blue light radiation in the 400 nm [9-10], which is not measured when this portion of the spectrum is not accounted in the spectral calculation.

The difference of illuminants in the calculation of visible light transmittance allows the evaluation of luminous transmission for different types of lamps spectral power distribution. The illuminant D_{65} resembles the sun and the illuminant A the conventional vehicles headlights. Considering these two conditions of use, the most critical situation can be compared to the minimum value of 75% for light transmittance requested by CONTRAN [4].

The objective of this study is to compare the results of calculations for 2 types of commercial window films applying the different methods of determining the luminous transmittance for illuminants A and D_{65} and the two methodologies of UV transmittance for the different spectral bands and air masses. The results will support the consumer and technical community regarding the specification criteria of choose solar control films.

2. Methodology

To develop this work, two samples of commercial sunscreen window film, designated as P1 and P2, were used. The properties of light transmittance and UV transmittance of the samples were evaluated as described in items 2.1 and 2.2. The measurements of the spectral transmittance of solar control films were performed using a double beam spectrophotometer with an integrating sphere, for a measurement spectral step and band width of 5 nm.

2.1. Evaluation of light transmittance

The light transmittance of the window films was calculated according to the procedure of SAE J1796, considering the illuminant A and a modification of this procedure considering the illuminant D_{65} as defined in DIN EN 410.

2.2. Evaluation of UV transmittance

The UV transmittance of window films was calculated according to ISO 13837 using the two methods described. The methodology A that corresponds to the spectral range from 300 to 400 nm solar spectrum for air mass 1.5 and the methodology B in the spectral range from 300 to 380 nm solar spectrum for air mass 1.

3. Results

Two window films, P1 and P2 were measured, based on the methodology described. The measurement results of the spectral transmittance of UV to visible are shown in Figure 1. Based on the results of spectral transmittance calculations for determining the values of UV transmittance and light transmittance were performed.

3.1. Measurements of the spectral transmittance

The measurement results of the spectral transmittance are shown in Figure 1. An attention shall be draw to the UV range of solar film transmittance which impacts the UV transmittance calculation.

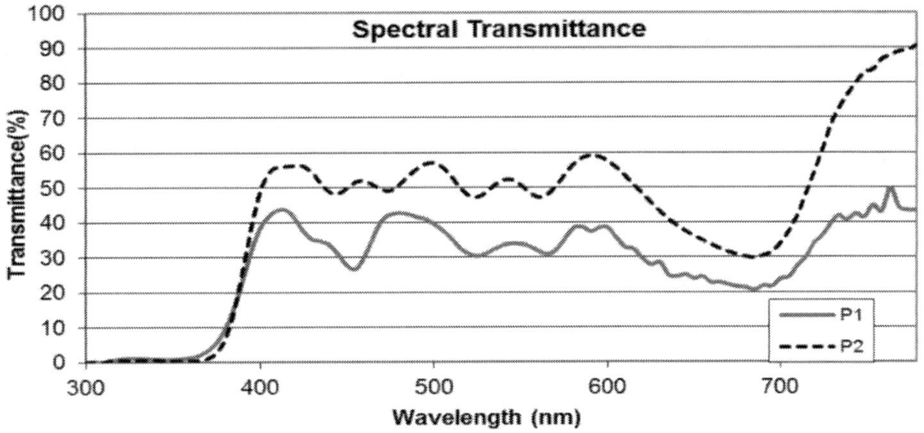

Figure 1. Spectral Transmittance of films from 300 to780 nm.

3.2. Measurement of UV transmittance

The UV transmittance (T_{UV}) was obtained from the spectral transmittance according to the methodologies of calculation (air mass equals to 1) and B (air mass equals to 1.5) of ISO 13837, and their values are presented in Table 1.

Table 1. UV transmittance with different air masses

Sample	Method A T_{UV} (%) M_{ar}=1,5	Method B T_{UV} (%) M_{ar}=1,0
P1	9,02	2,03
P2	9,18	0,96

3.3. Measurement of light transmittance

The values for light transmittance (LT) as SAE 1796 were evaluated for illuminant A and also using the same calculation methodology for the illuminant D_{65}. The results are shown in Table 2.

Table 2. Luminous transmittance for different illuminant.

Sample	LT_A (%)	LT_{D65} (%)
P1	33,38	33,82
P2	51,05	51,31

4. Discussion

The results showed different numerical values for the same sample when measured by different methods. The main reasons for it are discussed for each measured quantity.

4.1. Analysis UV transmittance

In the analysis of the values obtained from measurements of UV transmittance, it is clear that the method choice can determine significant differences in these. The data in Table 1 show that depending on the method, the tested samples had different values for T_{UV}, which presented impressive differences ranging from 350% for the sample P1 and 800% for the sample P2. The slightly greater spectral range from 300 to 400 nm in method A, compared with 300 at 380 nm in method B is a reason for the T_{UV} difference which was emphasised by window films' transmittance, despite the greater air mass method to reduce the intensity of the solar spectral power distribution [9], as seen in Figure 2.

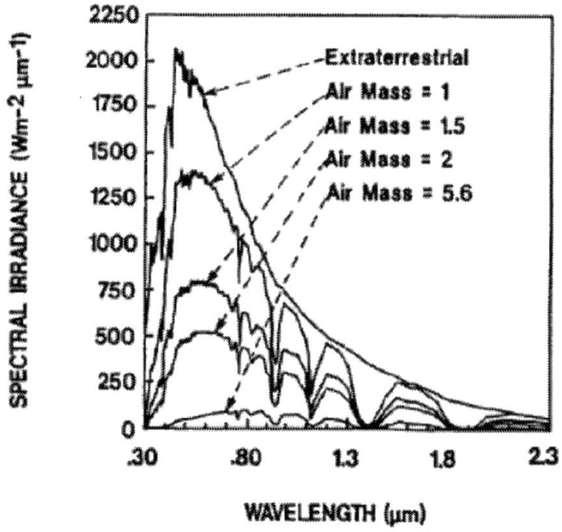

Figure 2. : Change in spectral irradiance with the air mass from [9].

4.2. Analysis of luminous transmittance

The comparison of the values of luminous transmittance showed no significant differences, when the evaluated films were subjected to various illuminant sources. Which in other words means, the use of films submitted to illuminant D_{65} (daylight) or to illuminant A (automotive headlights) does not alter significantly the vision.

The methodological difference in calculation depends on the application made. For automotive use the Resolution 254 CONTRAN establishes the illuminant A. The application of film in building construction use illuminant D_{65} as a reference to the daylight that enters the building. Although the daylight can be applied to automotive field, the use of illuminant A as a reference consider the critical condition for transmittance which lead to lower transmittance values as shown in Table 2.

5. Conclusion

Concerning the results of the UV transmittance of the solar films evaluated by two standard methods, differences from 350% to 800% were obtained for measured samples. Thus, the commercial presentation of the product shall inform the results of the UV transmittance and the standardized method used in its calculation. A better solution would be a law regulation by the responsible public agency of a specific standard method for presented the UV transmittance of the solar films in the Brazilian market.

Regarding the measurements in the visible range, the use of different illuminants (A and D_{65}), showed no significant differences, but in the case of application in the building construction environment using illuminant D_{65} is more appropriate.

6. References

[1] ABRAWF - Projeto Window Film- Relatório Final Falcão Bauer. Acessed 23th April 2013 <http://www.abrawf.com.br/new/pdf/REL-WINDOW_FILM.pdf>.
[2] Catalogo Solargard. Acessed 23th April 2013. <http://www.solargard.com.br/Comercial/Especificações_de_desempenho_2>.
[3] Catalogo 3M. Acessed 23th April 2013. <http://www.3m.com/intl/br/windowfilm/linksc0e8.html?link=boletim>.
[4] CONTRAN, Conselho Nacional de Transito, Resolução n°. 254 de 26 de outubro de 2007.
[5] DIN EN 410: 2011-04. Determination of luminous and solar characteristics of glazing. English translation.
[6] CIE 15.2. Colorimetry. 2nd Edition, 1986.
[7] ISO 13837-2008 Road Vehicles – Safety glazing materials – Method for the determination of solar transmittance.
[8] SAE J1796: 2005. Spectral transmission test.
[9] CIE 85: Solar Spectral Irradiance, 1st Edition 1989.
[10] ICNIRP. Guidelines on limits of exposure to ultraviolet radiation of wavelengths between 180 nm and 400nm (incoherent optical radiation). Health Phys. 87:171-186; 2004.

Spectral Irradiance Measurements Based on Detector

M S Lima[1], T Menegotto[1], I Duarte[1], T Ferreira da Silva[1], L C Alves[1], A D Alvarenga[1], G B Almeida[1], R N Teixeira[2] and I B Couceiro[1]

[1]Optical Metrology Division, National Institute of Metrology, Quality and Technology – INMETRO, Av. N. S. das Graças, 50, Xerém, Duque de Caxias, 25250-020, RJ, Brazil

[2]Thermal Metrology Division, National Institute of Metrology, Quality and Technology – INMETRO, Av. N. S. das Graças, 50, Xerém, Duque de Caxias, 25250-020, RJ, Brazil

E-mail: mslima-cnpq@inmetro.gov.br

Abstract. This paper presents the preliminary results of the realization of absolute spectral irradiance scale at INMETRO in the ultraviolet, visible and infrared regions using filter radiometers as secondary standards. In the construction of these instruments are used, at least, apertures, interference filters and a trap detector. In the assembly of the trap detectors it was necessary to characterize several photocells in spatial uniformity and shunt resistance. All components were calibrated and these results were analyzed to mount the filter radiometer.

1. Introduction

The realization of the spectral irradiance scale can be done using two methods, one based on the source and the other on the detector [1-3]. This second methodology was employed in this work because to be simpler and less costly and have compatible uncertainties applications for both a National Metrology Institute (NMI) and for industrial applications.

The realization of the irradiance scale based on the detector appears as a measuring chain alternative in which halogen incandescent lamps with tungsten filament is used as a source of absolute radiation. Radiometers based on reflection trap detectors in transferring the scale of absolute spectral responsivity from the cryogenic radiometer are used [4-6]. From the establishment of the spectral irradiance scale and considering all aspects of metrology, the spectral radiance scale can be developed and a new chain of traceability achieved.

2. Experimental Method

The experimental design for measuring the spectral irradiance is shown in Figure 1. A 1000 W FEL lamp was supplied with a current source of 8.2 A and it was placed at a distance of approximately 46 cm of radiometer filter (RF). A black box with an outlet aperture of about 5 cm and a shutter are used to minimize scattered light. The FEL lamp illuminates a filter radiometer which generates a photocurrent that was measured using a digital multimeter and a voltage-current converter for each filter. The idea is to construct the spectral irradiance scale using the black body model to FEL lamp and considering the calculated current to a filter radiometer. To realize the scale it was used an interactive method in MATLAB platform as described in reference [1-3].

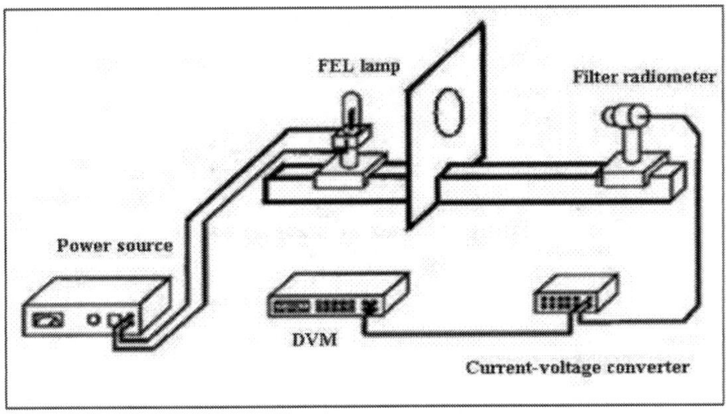

Figure 1 - Schematic of the experimental setup of spectral irradiance measurement.

The measured spectral irradiance is given by the following equation:

$$E_{med}\ \lambda\ = \frac{I_{med}}{I_{calc}} E_{calc}\ (\lambda_{eff})$$

Where: $E_{med}\ \lambda$ is the measured spectral irradiance; I_{calc} is the calculated current; I_{med} is measured current and $E_{calc}\ (\lambda_{eff})$ is the calculated spectral irradiance.

3. Experimental Results

3.1 – Evaluation of photodiodes at spatial uniformity

The spatial uniformity measurements were taken on the laser line He-Ne of 632.8 nm. The photodiode was attached to the support that is fixed on a base and it is moved in two directions by using micrometric motorized screws, allowing the measurement of the electrical signal from the photodiode surface. The measured electrical signals along the surface are normalized with respect to the electric signal measured at the center of the photodiode. Shunt resistances of these devices were also measured. Table 1 shows each set of three photocells grouped according to closer values spatial uniformity and shunt resistance for the assembly of trap detector.

Table 1. Spatial uniformity and shunt resistance of photocell S1337.

Photodiode	Uniformity	R_{shunt} (MΩ)
#1	0.998 – 1.002	258.2
#15	0.996 – 1.002	267.5
#23	0.996 – 1.000	262.2
#7	0.996 – 1.000	284.8
#8	0.996 – 1.002	287.1
#16	0.996 – 1.002	289.2
#11	0.996 – 1.000	285.1
#14	0.996 – 1.000	281.1
#24	0.996 – 1.000	280.3

Figure 2 shows the spatial uniformity of the photodiode #16 that is part of a set of three photodiodes mutually compatible. It is observed that the uniformity of the photodiode is similar to the other, as confirmed in Table 1. The uniformity variation among this set of photodiode (#7; #8 and #16) is less than 0.5 %. These evaluations indicate that this is a good set for the assembly of a trap detector.

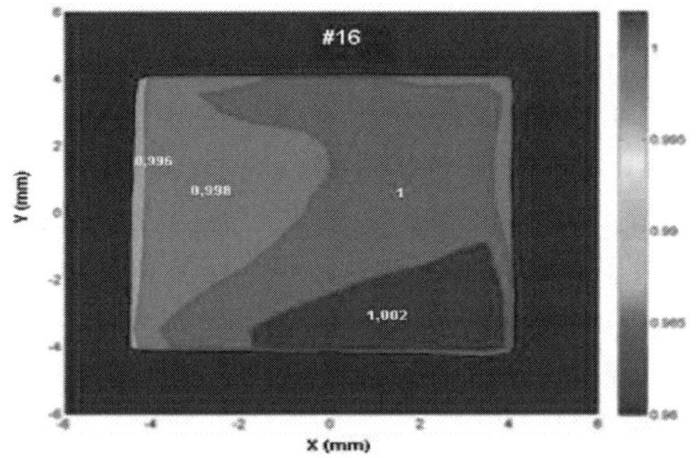

Figure 2. Spatial Uniformity photodiodes used in the assembly of the trap detectors.

3.2 – Trap Detector Spectral Responsivity Measures

Measures of absolute spectral responsivity (A/W), on the laser lines of HeNe (632.8 nm) and Argon ion (457.9 nm, 488.0 nm, 514.5 nm) were made to trap detectors by comparison with the cryogenic radiometer, reference standard of optical power (W). In Table 2 are shown these measurements.

Table 2. Spectral responsivity of a trap detector measurements.

λ (nm)	s (A/W)	u (A/W)
457.9	0.3656	2.50E-04
472.7	0.3779	1.60E-04
488.0	0.3906	1.90E-04
501.7	0.4019	1.60E-04
514.5	0.4126	2.50E-04
632.8	0.5086	3.70E-04

Figure 3 shows the result of the comparison of the spectral responsivity to three trap detectors mounted and characterized at INMETRO. It is observed that these devices exhibit very similar behavior in terms of spectral responsivity, among the other parameters.

3.3 – Evaluation of optical filters on measures of spectral transmittance

Transmittance curves were obtained as a function of wavelength for a filter set in the range of 200 nm to 1200 nm. Figure 4 shows the results of these measurements. It was selected a set of filters on the maximum wavelengths of 380, 500, 600, 700 and 800 nm. This set was chosen because it did not show a large shift in the maximum wavelength and in the transmittance when compared with the nominal values.

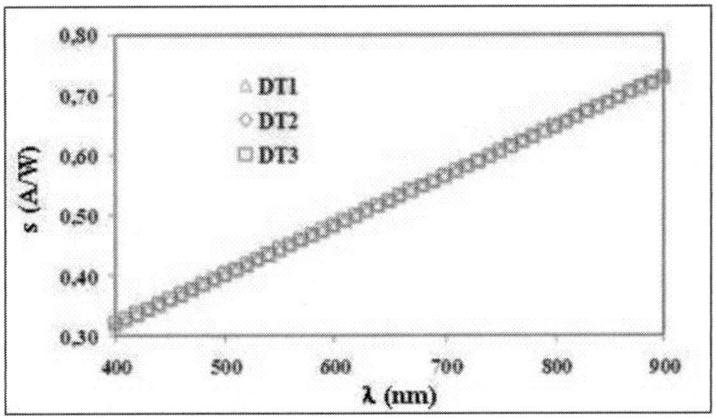

Figure 3. Spectral responsivity of trap detectors.

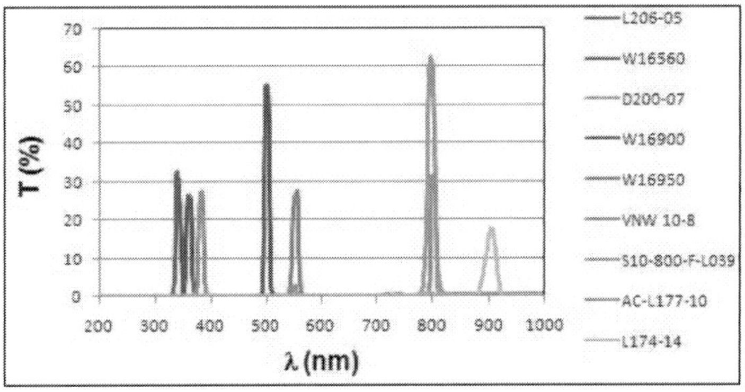

Figure 4. Evaluation of optical filters on measures of spectral transmittance.

3.4 – Measurement of circular apertures

Table 3 shows the measurements of the areas of the circular apertures and the respective expanded uncertainties. The uncertainty was initially available at approximately 4% and currently this parameter is of the order of 0.1%. These values are consistent with some National Metrology Institutes (NMI). NIST has values of the order of 0.004%.

Table 3. Measure the areas of apertures and their expanded uncertainties.

Identification code	Aperture area (mm²)	U (mm²)	u (%)
A_2	7.069	0.009	0.1
A_3	7.074	0.009	0.1
A_4	7.070	0.009	0.1

4. Final Results and Discussion

Some photodiodes were characterized for uniformity and shunt resistance. The transmittance of interferential filters was measured. This allowed the construction of filter radiometers, Figure 5.

Figure 5. Filter radiometer from INMETRO.

Some of these devices were assembled and preliminary tests were performed by determining the irradiance from the photocurrent measurement. A comparison of irradiance measurements between INMETRO (E_{teste}) and NIST (E_{NIST}) shows that these preliminary results are promising, however, some adjustments in the experimental system should be carried out and the main sources of uncertainty should be evaluated. Table 4 shows the range of wavelength of the filters used in the filter assembly of the radiometer.

Table 4. Wavelength filter, versus irradiance

λ (nm)	$E_{INMETRO}$ (mW/m^2.nm)	E_{NIST} (mW/m^2.nm)	1-$E_{INMETRO}/E_{NIST}$ (%)
380	18.40	17.50	- 5
500	83.61	81.60	- 2
600	155.52	151.00	- 3
700	219.90	207.00	- 6
800	266.31	237.70	- 12

5. Conclusions

The first step required to achieve the scale of irradiance has been practically completed. This includes calibration of the optical and mechanical components, such as filters and apertures, and characterizing several photocells on measures of spatial uniformity and shunt resistance. Preliminary tests of spectral irradiance measurements were performed and the results are consistent with what is available in the literature. The next step to be performed is the complete characterization of filter radiometers, to consolidate the realization of the scale of spectral irradiance, comparing it with the primary standard in the field of optics, the cryogenic radiometer itself.

Acknowledgments

The authors acknowledge the support of the National Council for Research and Development (CNPq / PROMETRO) and the Mechanical Metrology Division of INMETRO.

References

[1] Kärhä P, Toivanen P, Manoochehri F and Ikonen E 1997 Development of a detector-based absolute spectral irradiance scale in the 380-900 nm spectral range *Applied Optics* **36** 8909 – 8918.

Journal of Physics: Conference Series **575** (2015) 012023 doi:10.1088/1742-6596/575/1/012023

[2] Kübarsepp T, Kärhä P, Manoochehri F, Nevas S, Ylianttila L. and Ikonen E 2000 Spectral irradiance measurements of tungsten lamps with filter radiometers in the spectral range 290 nm to 900nm *Metrologia* **37** 305-312.

[3] Kärhä P, Toivanen P, Manoochehri F and Ikonen E 2010 Spectral irradiance model for tungsten halogen lamps in 340-850 nm wavelength range *Applied Optics* **49** 880-886.

[4] Issam A, Bastie J and Lepoutre F 2001 Vers la réalisation d'une nouvelle échelle d'éclairement énergétique spectrique dans le proche ultraviolet (domaine spectral 200 nm – 400 nm) *Colloque Interdisciplinaire en Instrumentation* C2I.

[5] Hoyt C C Miller P J, Foukal P V and Zalewski E F 1989 Comparison between a side-viewing cryogenic radiometer and self calibrated silicon photodiodes, *Proceedings SPIE* **1109** Optical Radiation Measurements II, 236.

[6] Fox N P Trap detectors and their properties 1991 *Metrologia* **28** 197-202.

Journal of Physics: Conference Series **575** (2015) 012024 doi:10.1088/1742-6596/575/1/012024

The interference Linnik type microscope (IM) developed at Inmetro

I Malinovski[1], I B Couceiro[1], M S Lima[1], R S França[1] and J P von der Weid[2]

[1]Optical Metrology Division, National Institute of Metrology, Quality and Technology – INMETRO, Av. N. S. das Graças, 50, Xerém, Duque de Caxias, 25250-020, RJ, Brazil

[2]Telecommunication Center (CETUC), Catolic University – PUC-Rio, Rua Marquês de São Vicente, 225, Gávea, Rio de Janeiro, 22453-900, RJ, Brazil

E-mail:mslima@inmetro.gov.br

Abstract. The interference Linnik type microscope was developed and it design was modernized to provide high resolution nanometrology facility. This microscope can perform length measurement directly relative to wavelength standards such as the frequency stabilized lasers. Thus, one can use interference microscope for primary calibration of the step heights. These step heights can be used to calibrate vertical axis of the other instruments like optical coherence tomography, atomic force microscope, confocal microscope, etc. In all cases the traceability to the National Institute of Metrology, Quality and Technology length primary standards was attempted to be established.

1. Introduction

Nano-metrology is the science and technology of measurements of the artifacts with nano-meter or sub-nanometer accuracy. In order to make such high level scientific measurements it is important to establish the chain traceability to the International System SI meter unit in this area. In this respect, INMETRO have been actively involved in this area of metrological research since 1997 [1-4] when we created new automatic type of Gauge Block Interferometer (GBI). This study has resulted in characterization of the high resolution primary GBI down to sub-nanometer level. We applied a similar methodology to measure other smaller objects, such as nanometrology secondary standards, or so called step-heights.

2. Experimental Set-up

A new primary interferometric microscope (IM) developed is based on the Linnik type interferometer [3], modernized and fully automated and with full digital processing. The light source used is the frequency stabilized He-Ne laser, calibrated against primary length. As the secondary step height standards we have used the step height triple gauge block from Mitutoyo of 2 μm and 10 μm and the step height of 100 nm, produced at Materials Metrology Division. The schematic of the Linnik type of IM is shown in Figure 1.

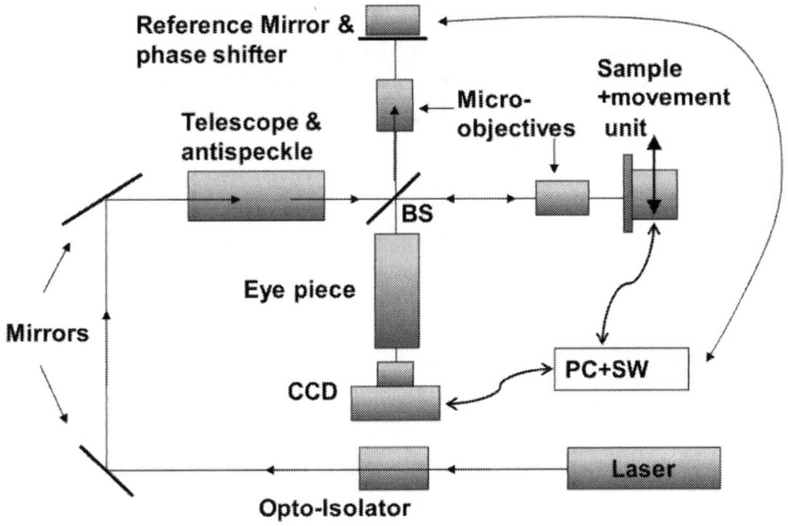

Figure 1. Set up of Linnik type interferometer

The frequency stabilized 633 nm He-Ne laser beam light is directed through a speckle removal unit consisting of two lenses of 100 mm focal length and a rotating diffuser plate. The emerging light is divided by a 50/50 beam splitter and directed into the sample and reference arms. Two identical microscope objectives (10x) are located with equal optical path length to the step height block sample and to the flat reference mirror ($\lambda/20$). The reflected beams from sample and reference are recombined at the beam splitter, passed x20 variable zoom eye-piece from HIROX and fringe is directed as image plane to the CCD camera. The CCD image is analysed using in-house developed software for evaluating the fringe fraction of the nominal step height. The interferometric frames are taken from CDD by PC and processed with specially developed and tested dedicated fringe pattern processor software (SW). The sample itself can be moved relative to the optical axis of the interferometer by motorized micrometer in X or Y directions in perpendicular to the axis direction.

3. Experimental Results

We have developed a set of hard-ware and software tools suitable for fringe image processing, phase shifting interferometry and post processing data visualization. The main fringe-pattern processing software (SW) module is based on multi-parameter iterative fit of the digitized pattern along the vertical line (or several neighbouring lines). Prior to the fit it is also possible to perform direct / reverse FFT with gaussian filter. This filter is known for not to perturbing the phase and it is used to remove pixel noise from the interferometric pattern. Additionally pixel noise was removed by averaging several frames to produce the final master interferogram used for processing. The quality of the measured fringe and the theoretical fit are in very good agreement. The minimization criterion is the least square difference between measured data and the model function. It is possible to analyze the fringe pattern along multiple lines, so that we can figure out the topography of the measured artifact. It has been previously shown that using our fringe processing algorithm the resolution of the interferometer can be as high as 0.1 nm or better. The accuracy of the measurement is determined mainly by the quality of the artifact GB and it might be as good as 1 nm the resolution of the method is well below 1 nm. One of the most serious systematic errors of IM is known to be the misalignment of the reference and measuring shoulders. While our interferometer software permits to control misalignments we have developed simple but very efficient self-calibration procedure. In this procedure we measure standard and immediately after that, we calibrate the fringe pattern on reference flat surface of the gauge block. We can optionally use flat surfaces of both GB in the master standard to perform such self-calibration. If the surface of the GB is assumed to be flat then we can remove misalignment error, by appropriate correction. Our IM instrument of the design used in set-up permits both pattern processing and phase shifting to be performed. We have developed and applied the software (SW) procedures suitable for detailed analysis of phase shifting technique. Our phase shifting unit is based on quite sophisticated 3 degree of freedom tilt-move stage. At each point the full interferogram is taken and saved into memory of the PC. After the procedure is done, it is possible to process its image stack at each position vs. distance of movement such as to reconstruct the phase for the whole measured surface. Further, to remove vibration related noise, we normally take several interferograms, averaging at each position of phase-shifter. To improve exactness of the phase stepper we can optionally make several back-forward positioning at each Z position averaging interferograms. Each individual scan produces interferometric fringe line of about 5-6 fringes at each pixel of the field of view. Each line can be Sin function fitted with an algorithm similar to the one described before. Normally we use no single pixel, but some neighbouring pixel area to average out the pixel-to-pixel noise. It is possible to process individual interferograms the way described above, something we do simultaneously with phase shifting. Thus, we have an opportunity to compare both algorithms: fringe pattern and pure phase shifting method. The experience shows that after all above precautions the signal to nose ratio permits some 0.1 nm resolution (or better) of the system. This has been confirmed by multiple repeatability testes. See example in Figure 2.

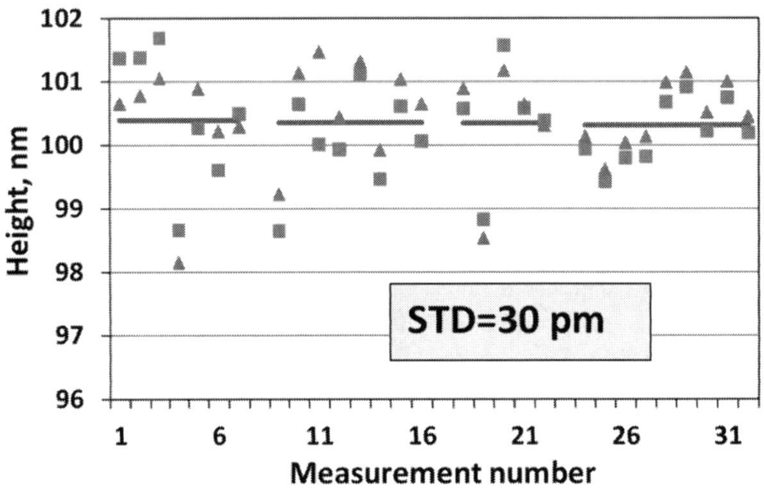

Figure 2. Reproducibility of 100 nm step height measurements taken by phase shifting technique method and corrected on field error.

Our estimation of the uncertainty of primary Linnik IM was made from the results of multiple measurements. The uncertainty budget of Linnik IM is shown in table 1. While single measurement presents relatively big phase noise, using automated system it is possible to average multiple readings at the same conditions (and positions). Next step was to apply adequate correction to the field. To do this we added vertically movable automated stage that was useful for on-flight calibration flat area of the sample. Exactness of correction is estimated to be about 2 nm. The obliquity or aperture error was theoretically estimated from the geometry of the optical Set-Up of our IM. With NIKON PLAN x10 long focus microscopic objectives the maximum incidence angle on sample is about 20 angular degrees that is significantly decreased by camera lens x20 eye-piece. The eye-piece restricts the angles, and at the end effective incident angle is about 3-4 angular degrees by estimation. This means that the possible obliquity error is about 80-90 pm on 100 nm step. To check it, we have restricted the entrance aperture of the IM down to about 2.5 The result was that the obliquity error is non-detectable within 100-200 pm uncertainty. So, we consider that at k=2 nm the uncertainty, associated to this particular error type, is no more than 0.5 nm. This is well within our target final IM uncertainty.

Table 1. Uncertainty Budget for IM.

Source of uncertainty	Uncertainty
Alignment, corrected	0.5 nm
Fringe fraction	0.3 nm
Phase noise after 20 measurements average	0.3 nm
Repeatability after 10 measurements average	0.1 nm
Reproducibility after 10 measurements average	0.8 nm
Flatness of the artefact (exactness of field correction)	2 nm
Laser Wavelength	< 0.1 nm
Combined standard uncertainty	2 nm

While single measurement presents relatively big phase noise, using automated system it is possible to average multiple readings at the same conditions (and positions). Next step was to apply adequate correction to the field. To do this we added vertically movable automated stage that was useful for on-flight calibration flat area of the sample. Exactness of correction is estimated to be about 2 nm. The obliquity or aperture error was theoretically estimated from the geometry of the optical Set-Up of our IM. With NIKON PLAN x10 long focus microscopic objectives the maximum incidence angle on sample is about 20 angular degrees that is significantly decreased by camera lens x20 eye-piece. The eye-piece restricts the angles, and at the end effective incident angle is about 3-4 angular degrees by estimation. This means that the possible obliquity error is about 80-90 pm on 100 nm step. To check it, we have restricted the entrance aperture of the IM down to about 2.5 The result was that the obliquity error is non-detectable within 100-200 pm uncertainty. So, we consider that at k=2 nm the uncertainty, associated to this particular error type, is no more than 0.5 nm. This is well within our target final IM uncertainty.

4. Final Results and Discussion

An important problem of traceability to SI unit in nanotechnology and nano-production has been discussed in several institutes and new methods were developed and applied, for example, in dimensional nanometrology. In our work we use IM Linnik using step heights to give traceability to AFM, CM, etc. Nowadays INMETRO is developing and characterizing their own step heights in sub-micrometer scale. Satisfactory results were obtained in the first artifacts but we need to improve the method of preparing and test different material to produce these secondary standards.

We have created a new interferometric microscope for primary measurements. The instrument has been shown as a powerful tool to provide the traceability for the SI meter unit length scale for several optical and opto-mechanical instruments such as AFM, CM, TEM, etc. The limits of the reproducibility of Linnik Interferometer are shown to be as high as a few tens of pico-meters. The primary measurement system for nanometrology was created and certified, so we are ready to participate in a key comparison in this area.

Acknowledgments
This work has been supported by the National Research Council of Brazil (CNPq/ PROMETRO) and also by FAPERJ.

References

[1] Titov A Malinovsky I and Massone C 2000 Gauge block measurements with a nanometer uncertainty *Metrologia* **37** 121-130
[2] Titov A Malinovsky I and Massone C 1999 Fringe-image processing gauge block comparator of high-precision *Proc. SPIE* **3477** 92-100
[3] Malinovsky I Franca R S Couceiro I B Lima M S Azeredo C L S Almeida C M S von der Weid J P 2012 Primary Imaging Interference Microscope for Nanometrology *XX IMEKO World Congress Metrology* (September 9-14, Busan, Republic of Korea)
[4] Decker J E Altschuler J Beladie H Malinovsky Prieto I E Stoup J Titov A 2007 Viliesid M and Pekelsky J R, Final Report on SIM.L-K1 (SIM.4.2) Regional Comparison. Stage One: Calibration of gauge blocks by optical interferometry, *Metrologia* **44** 04001

Traceability transfer in high accuracy contact temperature measurements for length interferometry

P A Costa[1], M S Bessa[1], J S Schutz[1], C L S Azeredo[2], R S França[1], K N Quelhas[3] I Malinovski[1]

[1]Optical Metrology Division, National Institute of Metrology, Quality and Technology – Inmetro, Av. Nª Sª das Graças, 50, Xerém, Duque de Caxias, 25250-020, RJ, Brazil
[2]Fluid Dynamics Metrology Division, National Institute of Metrology, Quality and Technology – Inmetro, Av. Nª Sª das Graças, 50, Xerém, Duque de Caxias, 25250-020, RJ, Brazil
[3]Thermal Metrology Division, National Institute of Metrology, Quality and Technology – Inmetro, Av. Nª Sª das Graças, 50, Xerém, Duque de Caxias, 25250-020, RJ, Brazil

E-mail : rsfranca@inmetro.gov.br

Abstract: Some relevant components of uncertainty in interferometric measurements are associated with temperature measurements of material standards. This paper proposes a traceability transfer chain through a high accuracy contact temperature measurement system towards other temperature measurements systems. Those are traced to the best national metrological references in appropriate configurations for gauge-block length calibrations, allowing in that way investigations of effects and uncertainty components linked to temperature measurement of those blocks as further reductions in these uncertainties. As an application example a study of temperature gradients on gauge blocks is shown

1. Introduction

Dimensional precise definitions for the length of any materialized artifact undergo exact measures its temperature at the instant of its dimensional measurement, and the uncertainty components of this magnitude is directly proportional to the sizes measured. Contact measurement systems have been specially developed in the Interferometry Laboratory of Optical Metrology Division of the Brazilian National Metrology Institute - INMETRO – as seen in references [1] and [2], aiming to significant reductions in uncertainty propagation over the entire traceability chain. This should cover all the chain since the primary standards and national references until the temperature sensor systems and readers used in synchronous temperature measurements obtained in high accuracy interferometric measurements. This method is commonly used in commercial length calibration services for gauge blocks of best classes, which are the transfer standards "par excellence" of industry for length unit. The reduction of uncertainty components in this type of measurement is linked to a detailed definition of various auxiliary quantities, and of those components with higher relative weight associated with temperature measurements.

2. Traceability Chain in Temperature & Length

To perform the realization of temperature scale standards internationally known as fixed points are used, in which the reference temperatures of phase transitions are kept stable for long times.

Whereas the dimensional measurements are performed around the reference temperature of 20 °C the standards most suitable for checking the measurement system are the triple point of water cells (reference temperature = 0.01 °C) and the melting Gallium cells (reference temperature = 29.7646 °C). Thermal Metrology Division (Diter) of INMETRO maintains cells of these types as references in international comparisons and for periodic checking of the measurement systems used in regular calibration services. The temperature reading systems used in interferometric measurements, whose measurement values affect the final result in length, should be periodically sent for calibration. However, the offered temperature calibration service is carried out through comparisons with systems directly traced to the fixed points. In those systems the sensors involved are not held in physical contact with a solid medium as in the usual operation of gauge block measurements. Instead they are measured after thermal equilibrium in a water, oil or air bath, including a possible extra insulation to prevent undesirable leakages inside the sensor body immersed in the heat transfer media. These methodological differences produce higher final uncertainties, requiring application of long time thermal exchange, thus reaching minimum conditions of thermal equilibrium for a given uncertainty after those times. Some temperature uncertainty contributions in block measurements were discriminated by Decker and Pekelsky [3].

3. High Accuracy System of Temperature Measurements

Aiming to minimize undesirable effects and to achieve best adequacy in reasonable time a contact measurement system was designed, which consists of a temperature bridge read by a locally developed automated software and the a standard resistor thermostatically stabilized with nominal value 25 Ω, and a long metal sheath 25 Ω SPRT of high quality and stability. The SPRT is kept in direct physical contact with the sensor to be compared, both being carefully positioned within a good thermal insulated Dewar. This contact is accomplished through a simple related material structure comprising a copper block with a hole for long sensors and a second block with similar dimensions to the artifact to be measured by the intended sensors. The whole system should be checked periodically against the two fixed points cited, and from these checks are produced high accuracy estimates both of their direct corrections as the coefficients of variation of the reference sensor resistance, therefore closing the traceability chain in an appropriate form for this system. The assembly is simple and fully reproducible by any metrology lab with the access to high accuracy temperature references such as the fixed points. Some images of the system can be seen in Figure 1. This system also allows the verification of other high accuracy sensors. Such sensors can, from this type of scan, be used in studies and tests of tiny effects than their direct temperature measurements. These effects are linked to both temporal / spatial evolution of artifact temperatures as also to their material propagation (for interesting case studies on these small variation scales, realized by specific type sensors, see refs. [1], [2] and [4]).

Figure 1. Temperature Measurement Systems

In Figure 2 we have the schematic view of the traceability chain, which spreads from checks against national temperature standards, having reduced uncertainty propagation, from the comparison system developed and its contact block measurement systems, until the final application steps. These steps are related to measurement of effects associated with waiting time for thermal equilibrium and gradients generated dynamically inside gauge blocks to be calibrated by interferometric measurement systems.

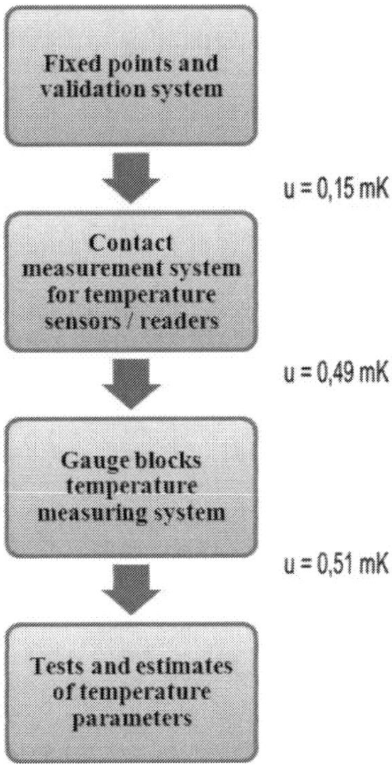

Figure 3. Traceability Chain

3.1. Uncertainty Propagation

From Table 1 we have presented a budget showing contributions from various elements of temperature measurement comparison system. From these values the greatest contributions for uncertainty propagation in all steps refers themselves to inner gradients linked to the verification process (without considering external factors such as aging of electronic components, sensors jitter / drift, large temperature differences towards the reference temperature, degradation of fixed points, etc.).

Table 1. Temperature Comparison by Contact System Uncertainty Budget

Components	u_i(mK)	c_i	u_i c_i (mK)
Uncertainties of Standards			
Triple Point of Water	8.50E-02	1	8.50E-02
Melting Point of Gallium	1.25E-01	1	1.25E-01
Measurement System Uncertainties			
Bridge Temperature	3.10E-01	1	3.10E-01
Gradient	2.50E-01	1	2.50E-01
Drift	2.50E-01	1	2.50E-01
Compared Thermometer Uncertainties			
Resolution	2.89E-02	1	2.89E-02
Stability (4 hours)	1.30E-01	1	1.30E-01
Combined Uncertainty		**0.51 mK**	

4. System Operation Proposed Methodology

Studies realized at Interferometry Lab of INMETRO show marked differences in intrinsic response time of both compared sensors (due, for example, to their physical size) that may produce anomalous results in obtained corrections if instantaneous measuring speeds are not considered (see [1] and [2]). In Figure 3 there are presented some cycles of temperature variation whose amplitude differences were due to stepped variations and of different response times in both systems. To minimize these inaccuracies for every temperature change executed inside the Dewar (this step is essential to estimate variation coefficients of the respective sensors) it should be waited a required minimum time until the difference between both systems achieves reduced values and it is stable for long enough (shown in graph lower baselines). This condition is monitored by automating readings of both sensors. Such is the case of applications of a four-channel sensor digitizer used in gradient tests or a compound temperature bridge + scanner connected to six reduced size PT100 used in a commercial interferometer Mitutoyo GBI temperature reading system, both devices are operating in Interferometry Lab.

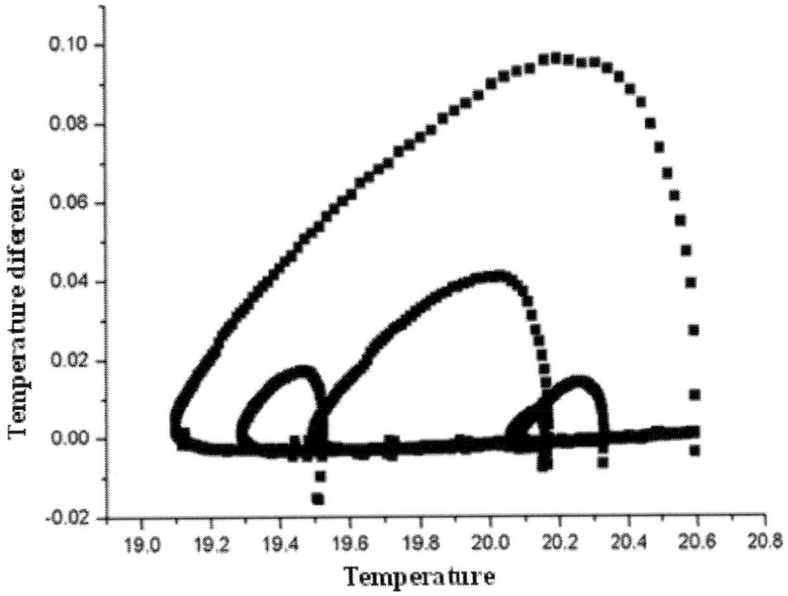

Figure 3. Evolution of the temperature difference between two sensors with different sizes and response times

5. Case Study: Gradient Measures in Gauge-Blocks

In Figure 4 two operating instances of a temperature measurement system with two PT100 connected to a gauge block of 100 mm are shown. In smaller ellipse we found reduced cyclical variations in temperature associated with night fluctuations in control room. Differences between values read from two sensors are due to their distinct deviations. In daytime, a heating caused by inner thermal sources after interferometer was switched on not only caused a progressive increase in the absolute temperature of the block as also of reading differences between both sensors, indicating the presence of dynamic gradient effects to be considered.

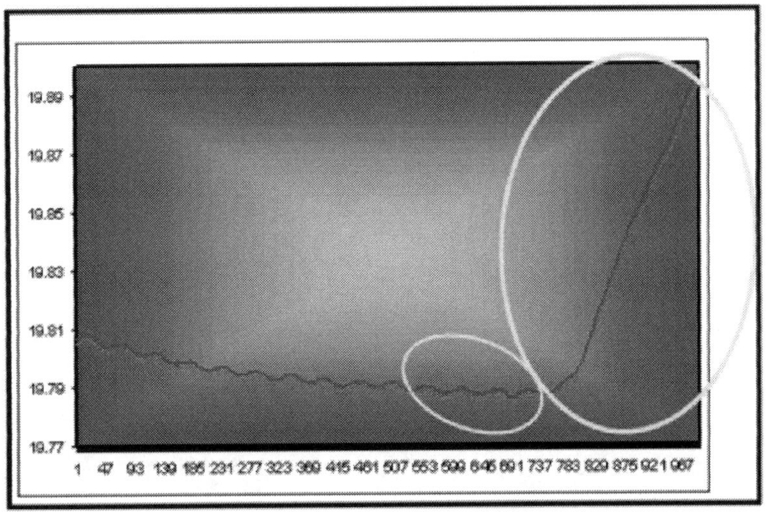

Figure 4. Temperature evolution at two instances for a 100 mm gauge block

6. Conclusions

The proposed system was internally characterized and adequately compared to similar systems and fixed points in Thermometry Metrology Division (Diter) of INMETRO, for its validation. Given the requirements for high accuracy temperature measurements in interferometric systems, Diter staff can still easily reproduce an identical system, intending future mutual periodic comparisons and, therefore, a good warranty of long-term quality for these best defined results. Some isolated uncertainty components of the system must still be evaluated, as well as certain effects and additional contributions, such as electrical noise, characterization of immersion depth in fixed-point cells thermometric wells, and setting limits on waiting times for thermal stabilization. Calculation routines for computing effects associated with possible dynamic variations in compared systems were developed, settling it in a measurement form very similar to that used in gauge blocks measurements.

Acknowledgments

This work has been supported by the National Research Council of Brazil (CNPq/ PROMETRO).

References

[1] Titov A and Malinovsky I 2005 Precise certification of temperature measuring system of the original Kösters interferometer and ways of its improvement *Proceedings of SPIE* **5879** 42-52

[2] Titov A and Malinovsky I 2005 Nanometrology and high-precision temperature measurements under varying in time temperature conditions *Proceedings of SPIE* **5879** 11-22

[3] Decker J E and Pekelsky J R 1997 Uncertainty evaluation for the measurement of gauge block by optical interferometry *Metrologia* **34** 479-93

A Methodology for Obtaining the Mean Glandular Dose in Mammography Exams

J C Feital[1], J U Delgado[2] and R T Lopes[3]
[1] Senior Technologist, Radioprotection and Dosimetry Institute (IRD / CNEN), Brazil
[2] Senior Technologist, Radioprotection and Dosimetry Institute (IRD / CNEN), Brazil
[3] Professor / COPPE / UFRJ, Brazil
[1] e-mail: jfeital@ird.gov.br

Abstract. This study proposes to validate measurements of mean glandular dose quantity (D_G) for a given thickness and glandularity of compressed breast based in a PMMA phantom. 10 exposures were made in an ionization chamber in conventional mammographic equipment for a research. It obtained an average value for the incident air kerma, Ka,I = 9.59 mGy leading to a D_G = 1.82 mGy. Experimental results too were obtained for the acquisition of this quantity in other mammography clinics and these results were also reported and discussed. After the comparison of the results the methodology was validated.

1. Introduction

In Brazil, estimates for 2014, reported the appearance of about 57,000 new cases of breast cancer [1], being estimated that approximately 30 thousand of these findings still are found late, which in the year of 2010 has led to almost 13 thousand deaths among the female population. According to publication of the International Commission on Radiation Units (ICRU) [2], the incidence of breast cancer in the age between 45 to 64 and more than 65 years is 51 % and 39 %, respectively. Early diagnosis, through equipment's adjusted in relation to their basic operation [3,4] parameters, together with the radiographic image quality, allows besides the detection of diseases at an early stage with lower radiation doses to the patient. Even if this dose is from recall due to an inappropriate image.
Absorbed dose (D) at the entrance of the skin is a very important in the evaluation of quality control in mammography [4] and is obtained from the measurement of incident air kerma -**Ka,i**, multiplicand by the backscattering factor. This quantity is also known as Entrance surface air kerma - **Ka,e** [5]. Already the mean glandular dose - **D_G**, is dosimetric quantity which best characterizes the carcinogenic risk induced by X rays in mammography exams; it is recommended by the International Commission on Radiation Protection ICRP, and adopted in the European Protocol of Dosimetry in Mammography [6.7]. This study aims to validate a representative average value for measurements of D_G to the phantom measurements in mammography equipment's.

[1] To whom any correspondence should be addressed.

Journal of Physics: Conference Series **575** (2015) 012026 doi:10.1088/1742-6596/575/1/012026

2. Methods and Measurements

It were used for the measurements in the laboratory: a mammography equipment Siemens mammomat 1000 model, a dosimeter manufacturer Radcal (model 9015, 910506 series), coupled to ionization chamber (modelo10X56M), calibrated by the National Laboratory of Metrology of Ionizing Radiation (LNMRI - IRD /CNEN) and aluminum (Al) filters (purity of 99%) with thickness of 0.1 mm for measurement of half-value layer (HVL), getting to this basic parameter the value of 0.38 mm in Al. The methodology adopted here was to associate quality control procedures of the radiographic image with experimental determinations and other mathematical models for the dosing quantities **Ka,i**, D (**Ka,e**) and D_G [8.9]. For the determination of **Ka,i**, 10 exposures were carried out yet in the ionization chamber.
This experimental arrangement occurred in accordance with the sequence shown in Figs. 1 and 2, respectively. It was used the radiation quality with x-rays tube voltage of 28 kV, for a < 2 % reproducibility. In turn, the product current x milliamperage (mAs) ranged from 56 to 63 to manual and automatic modes. For the set Molybdenum-Molybdenum (Mo-Mo) target - filter this remained constant throughout the experiment, so as not to significantly change the beam spectrum.
The "Phantom mama", as adopted by the Brazilian College of Radiology, is composed of acrylic with dimensions for width, height and depth of 12, 5 and 16 cm, respectively, where are found the objects related to the quality of radiographic image[10], was positioned next to and in the same plane of ionization chamber positioned perpendicular to the central axis of the primary x-ray beam in the following dimensions: Focus to the center of the ionization chamber - Camera Center (FCC= 59 cm) and radiation field size of 24 x 18 cm on the surface of the Bucky, as can be seen in these images (Figs. 1, 2 and 3). These parameters will not have too changed.

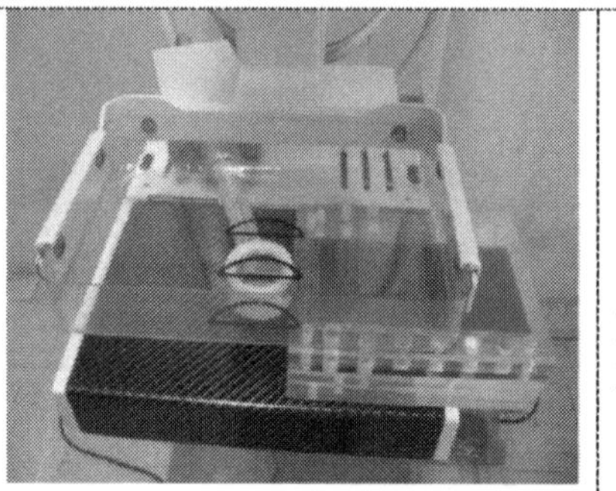

Figure 1. Ionization chamber positioned adjacent to and in the same plane of the phantom of 5 cm height and centralized along with "Bucky" of mammography equipment.

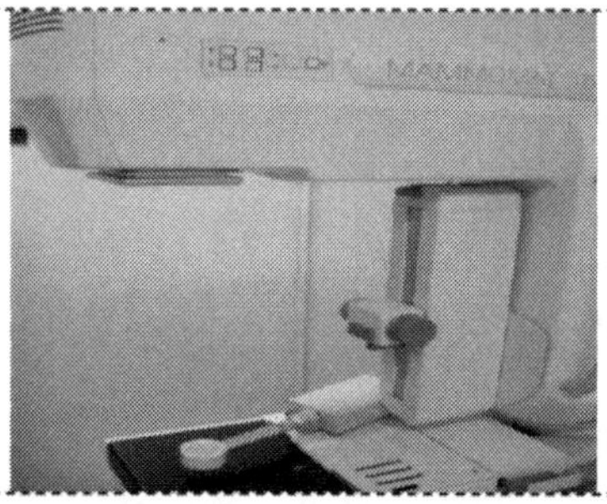

Figure 2. Ionization chamber positioned in the center of the field of radiation (FCC= 59 cm) for the Ka,i measurements.

Radiation field.

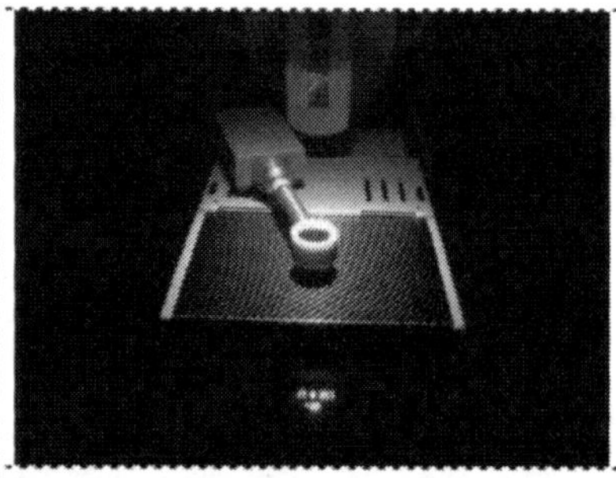

Figure 3. Ionization chamber positioned in the centre of the field of radiation (FCC= 59 cm) for the Ka,i measurements.

Light field.

For the determination of HVL and D_G, with proper procedures, methodologies were already described in the literature [9, 11] fot the quantity incident air kerma, the estimated uncertainty has been reported for k = 1. For the calculation of the D_G were used: one computerized mammography equipment (CR) and four screen-film systems .

3. Results and Discussion

The value **Ka, i** = 9.59 ± 0.1% mGy was obtained from ten to exposures in the ionization chamber. This result was multiplied by the scattering factor (1.09), in accordance with the TEDOC-1517, IAEA, 2006, being obtained 10.45 mGy ± 0.1% for the value of the D (**Ka,e**).

In this study we obtained the calculation of D_G, 1.82 ± 0.2, based in a PMMA phantom where the conversion factor (g.c = 0.190) [9] and the fibro-glandular percentage equivalent to 20% and 5,0 cm of the compressed thickness, adopting: S =1; to set target anode (Mo-Mo); **Ka,i** = 9.59 mGy ± 0.1 %, which was interpolated to a HVL 0.38 mm of Al (obtained experimentally).

Respect to Table 1, the D_G values are in accordance with the reference level below of 3.0 mGy to an acrylic phantom of 4.5 to 5.0 cm [12]. This shows the validate the calculation model adopted to the D_G values obtained in this study.

For the level of uncertainty associated with the D_G, this proved to be up to 5 times greater than the estimated uncertainties for measurements in each clinic, based on quality control tests. However, this result was expected because here were considered other components of significant uncertainty in its composition: so it was consistent with the practices and negligible in relation to the ultimate given quantity.

Table 1. D_G for four screen-film systems and one CR equipment based in a PMMA phantom with 5 cm of compressed thickness equivalent.

DATE ORIGIN	D_G (mGy ± %)
This study	1.82 [a] ± 0.2 (k=2)
Clinic A[h]	2.33 [b] ± 0.04 [e]
Clinic B	1.63 [c] ± 0.08 [f]
Clinic C	1.53 [d] ± 0.08 [g]
Clinic D	1.83 [j] ± 0.07 [i]

[a] Value calculated from the obtaining Ki and HVL obtained experimentally in mammography equipment. In this calculation were also considered: S=1 and the conversion factor (product c.g = 0.190) [9].

[b]. Value obtained in accordance with report methodology [8].

[c, d, j] Values obtained from these same tests [8]. In the clinics (B) and (C) are conducted about 15.600 and 8.500 annual exams, respectively.

[e, f, g, i] Expanded uncertainty (%) obtained by combined standard uncertainty value multiplied by the supply factor (k=2), to a level of confidence of 95.45%. Were here considered the uncertainty components due to the Ka,i values measured in each clinic (0.02, 0.01, 0.02, 0.14) % and the values for the uncertainties of the certificate (0.02, 0.04, 0.04, 002) %.

[h] Computadorized Radiography (CR).

Already, for the absorbed dose at the entrance of the skin (D), so named by technical regulation 453 of the Ministry Health 453, it should be recalled that, as long as they do not take effective measures in relation to quality control practices, as well as the mistakes made in the positioning of the breasts during the exam, news exposures due to repetitions of the following may occur, leading to unnecessary radiation dose to the patient.

4. Conclusions and Recommendations

According to the results shown in Table 1, the D_G values obtained in this study are consistent with those reported in the available literature, in relation to reference levels for such quantity, which validates the methodology presented here.

This study is relevant due the necessity to contemplate others phantom with different features (glandularity and thickness of compression), also increasing the number of measurements in the mammographic equipment with other technologies (digital radiography and tomosynthesis). Furthermore, it is observed an increase in the D_G due to the changes on the mammography equipment, i.e. of conventional technology to the CR, mainly in state of Rio de Janeiro - Brazil. It is known that such technology has been causing an increase of at least 20 % between the values found for such quantity.

5. References

[1] Regional Tables 2012 http: //www.inca.gov.br .
[2] International Commission on Radiation Units & Measurements 2009 *Mammographyin clinical practice.* Journal of ICRU; **9**, pag.09-14.
[3] Instituto Nacional do Câncer 2007 *Mammography: the practice to control. Recommendations for health care professionals - INCA*, Rio de Janeiro.
[4] Ministério da Saúde 1998. *Guidelines for radiological protection in medical and dental diagnostic radiology*, 453 Technical Regulation Report Official Governmental. Brasilia, DF .
[5] Ramos M M O, Tauhata L 2011 *Quantities and Units for Ionizing Radiation.* Institute of Radiation Protection and Dosimetry / Laboratory of metrology of ionizing radiation. Rio de Janeiro.
[6] *European Protocol on Dosimetry in Mammography 1996.* Report EUR EN Luxembourg: European Communities.
[7] Almeida C D, Coutinho C M, Dantas B M, Peixoto J E, Koch H A 2012 *A new mammographic dosimetric phanton.* Radiation Protection Dosimetry, pp. 1-3.
[8] Atomic International Energy Agency 2006. *Quality Control in Mammography*, TECDOC – 1517, Viena.
[9] Atomic International Energy Agency 2011. *Quality Assurance programme for digital Mammography.* Human Series 17. Viena.
[10] Menezes, C J M, Hwang, S F, Castro W A , Peixoto J E 2002. *Development of a dosimetric card for mammography.* Journal of research and development. Vol. 4, N.3.
[11] Atomic International Energy Agency 2007. *An international code of practice.* Technical Reports Series. Publication 457. Dosimetry in Diagnostic Radiology. Vienna .
[12] European Comission, 2006. *European Guidelines for Quality Assurance in Breast Cancer Screening and Diagnosis.* 4th edition. In PERRY N., BROEDERS M., DE WOLF C. *et al.* (eds): Office for Official Publications of the European Communities, Luxembourg.
[13] Dance D R 1990. Monte Carlo calculation of conversion factors for the estimation of mean glandular breast dose, Phys. Med. Biol., v35, pp. 1211–1219.
[14] Dance D R, Skinner C L, Young K C, Beckett J R, Kotre C J 2000. "Additional factors for the estimation of mean glandular breast dose using the UK mammography dosimetry protocol", Phys. Med. Biol., v. 45, pp. 3225–3240.

.

Acknowledgments

Authors have also to express our thanks to Bárbara Fernandes Gonçalves Cristiano for the suggestions and comments.

Determination of uncertainty components for a system in Radiation Protection Dosimetry

F Lopez[1], **TS Cabral**[2], **JG Peixoto**[2]
[1]Medical Physicist, Researcher: Instituto de Radioproteção e Dosimetria
[2]IRD/CNEN, Salvador Allende av, Recreio dos Bandeirantes, Rio de Janeiro, RJ, Br
E-mail flopez@unan.edu.ni
tschirn@ird.gov.br
guilherm@ird.gov.br

Abstract. This work is about the theoretical calculation of uncertainties associated to the dosimetry of photons of a ^{137}Cs source that will be used in a Dosimetry Laboratory. In this case recognition of the influence quantities that provide most uncertainty and the right choice of resolution of auxiliary equipment to obtain the smallest uncertainties according to the laboratory.

1. Introduction

In Latin America there are several SSDL which were established with cooperation by the IAEA and WHO [1] and currently started a more in this area in Nicaragua, Central America. Based on all the experience of other laboratories LNMRI such as Brazil, which has a long track record in implementing ISO 17025 , what this work for the purpose of determining prior to the installation more feasible resolutions to reduce of uncertainty. It is through a system of quality management and the implementation of ISO 17025 that is successful written documentation that describes many procedures [2, 3], necessary condition requirement for traceability of other laboratories recently as the SSDL Protection Center and HRPC Radiation Hygiene of Cuba , who in 2011 initiated the calibrations in the field of radiology [4] although in somewhat different areas the basic idea is the same as the definition of the measurand and establishment of correct method and procedure to ensure repeatability and reproducibility. These two LSCD examples demonstrate that each will have its own characteristics in terms however traceability should be traceable to a primary laboratory. To achieve this it is not enough just to calibrate the equipment needed but to know the entire system of the laboratory, this includes air conditioning, positioning system, radiation protection, tests and trials main and auxiliary equipment with appropriate resolutions for uncertainties in the extent be less than 5% not to mention that there must be a system of quality management and the information to the highest standard of quality for customer satisfaction. This motivates a careful strategy intended to give the end product a calibration certificate with a value of the measurand and its associated uncertainty [5, 6, 7].

2. Photon Beam Dosimetry

2.1. Measuring definition

Before calibration on an Dosimetry Calibration Laboratory should be performed first on beam dosimetry using [5]. In this case the beam corresponds to a ^{137}Cs source, with an activity of 20 Ci for calibration of area and personal dosimeters. The measurand is the air kerma rate at a point of interest (test point) on a volume of an ionization chamber at reference conditions and considering the input and output time of the source [1], the effect of positioning layer it and needed the camera to reproduce the magnitude air kerma rate.

The equation for our measurement model is:

$$\dot{K}(1m) = \left(\frac{\bar{Q}}{T} - \frac{Q_f}{T_f}\right)\varphi(\vartheta, P) N_K f_s K_Q C_f \tag{1}$$

Where Q is the reading made with the electrometer, φ is the correction factor for pressure and temperature, N_k is the calibration factor for the ionization chamber, f_e is the factor of the electrometer, K_Q is the radiation quality and a conversion of C_f that is equivalent to 3600 X 1000 to produce the units mGy/h.

2.2. Influence quantities

2.2.1. Pressure and temperature correction

The symbol φ represents the correction factor of the environmental conditions in the chamber which depends on the relative humidity, temperature and barometric pressure place. The expression for this correction is

$$\varphi(\vartheta, P) = \left(\frac{a + \vartheta}{b}\right)\left(\frac{P_0}{P}\right) \tag{2}$$

Where $a + \vartheta$ is the thermodynamic temperature expressed ϑ in ^{0}C, P is the barometric pressure of the room, P_o is the atmospheric pressure, $a = 293.15\ K$ and $b = 273.15K$ respectively.

2.2.2. Correction for the transit time from the source

The transit time from the source given by the expression

$$\Delta \dot{T} = \frac{(K_n - K_i)}{(nK_i - K_n)} \tag{3}$$

Where Δ T represents the time it takes the source to return to their safe position, K_n is determined for n kerma irradiations, k_l is the kerma for one irradiation, and T the duration of irradiation.

2.2.3. Correction for electrometer' leakage

The leakage current will affect the reading of the electrometer as shown in (1), is represented by Q_f *divided by* T_f; the readings for leakage could be decreased. It depends of leakage of the camera, sometimes the effect is negligible in this case a correction is not important.

2.2.4. Correction for chamber positioning

This point is crucial because the standards indicate that the camera should be placed according to the manufacturer's specifications [2, 3] in terms of the geometry of it. Yet another point is the rangefinder on the rail or in this case a very interesting point is the position at 1m once this is achieved must register a benchmark with light system thus any deviation will be determined according to that point this detail is not written in standard however it is very important to know how to establish reference conditions in positioning.

Given a resolution of 1mm in terms of positioning from source to point of interest if we estimate the relative error percentage would get maximum value calibrations ± 0.2% for radiological protection equipment is very small.

3. Determination of uncertainty combined for air kerma rate at 1 m.

To determine the uncertainty associated with our measure must recognize the sources of our uncertainties make the necessary corrections and determine the uncertainty that these effects contribute to it. Very important is to recognize which is the distribution of our variables. To give an idea of these distributions are shown in Table 1 which shows the mean and its corresponding standard deviation.

Once recognized each variable and its distribution determines the combined standard uncertainty associated with our measurand [6] for this we use

$$U(y)_c = \sqrt{\sum_i^n \left(\frac{\partial y}{\partial x_i}\right)^2 u(x_i)^2} \tag{4}$$

Where, y represents the average value of the measurand and x_i the estimated values of the variable input. An alternative expression assuming a measurand in the form

$$Y = cX_1^{p_1} X_2^{p_2} \ldots X_n^{p_n} \tag{5}$$

It is given by

$$\left(\frac{U_c(y)}{y}\right)^2 = \sum_{i=1}^{N}\left[P_i \frac{U(x_i)}{x_i}\right]^2 \tag{6}$$

Table 1. Probability distributions, their means and standard deviations.

Distribution	Mean and standard deviation
Rectangular	μ y $U(xi) = \frac{a}{\sqrt{3}}$, where **a** is the half interval of distribution.
Triangular	μ y $U(xi) = \frac{a}{\sqrt{6}}$, where **a** is the triangular base
Normal	μ is the mean $\bar{X} = \left(\frac{\sum_{i=1}^{n} xi}{n}\right)$, and **σ** is the variance, where **s** is $s = \sqrt{\frac{\sum_1^n (X_i - \bar{X})^2}{n-1}}$.

As we can see, finally can be expressed in relative units i.e. a percentage, however it should be mentioned that in (1) variables really are not all look exactly this way (3) but for which if the calculation is correct and to the other is calculated according to (4) but to express uncertainty divided the measurand combined standard uncertainties remain expressed in relative units and relative sensitivity coefficients are 1. The expression that allows us to calculate the relative combined uncertainty is given by:

$$U_C(rel) = \sqrt{\left[\frac{U(L)}{L}\right]^2 + \left[\frac{U(T)}{T}\right]^2 + \left[\frac{U(P)}{P}\right]^2 + \left[\frac{U(\vartheta)}{\vartheta}\right]^2 + \left[\frac{U(f_e)}{f_e}\right]^2 + \left[\frac{U(N_k k_Q)}{N_k k_Q}\right]^2} \quad (7)$$

N_k factor multiplied by K_Q is due to their relationship to these factors and its associated uncertainty.

4. Conclusions

We observe that for such an uncertainty less than or equal to 5% the thermometer should have a resolution of 0.01 ° C, the time in the order of 0.01 s or better, and smaller time would be better for the millisecond in the case of humidity is important to maintain the conditions of camera calibration and environmental conditions within the conditions of reference [4, 5]. It is important to assess the effects of moisture and the uniformity of the radiation field.

Acknowledgements

First thank God for this opportunity, to the International Atomic Energy Agency, who sponsored my stay in Rio de Janeiro, Brazil and therefore allowed the realization of this article. Similarly greatly appreciate LNRMI director, Mr. Jose Delgado Ubiratan and Master Tânia Schrin Cabral for accepting my presence in her laboratory, also Dr. José Guilherme Pereira Peixoto by the direction of this work and to other colleagues that collaborated with supporting material, productive information to enrich this essay.

Reference

[1] Einselhor Horst, Red de laboratorios de dosimetría con patrones secundarios, OIEA Boletín 19 No. 2.

[2] ISO 4037 Part 2, X and gamma reference radiation for calibrating dosemeters and doseratemeters and for determining their response as a function of photon energy —Part 2: Dosimetry for radiation protection over the energy ranges 8 keV to 1,3 MeV and 4 MeV to 9 MeV

[3]ISO 4037 part 3, X and gamma reference radiation for calibrating dosemeters and doserate meters and for determining their response as a function of photon energy —Part Calibration of area and personal dosemeters and the measurement of their response as a function of energy and angle of incidence. P 35.

[4]Walwin-Salas Gonzalo, Martínez G. Alina, González Rodríguez Niurka, Vergara Gil Alex, Gutiérrez Lores Stefan, Establecimiento y diseminación en Cuba de patrones dosimétricos aplicados a radiología diagnostica. BCT INIMED No 1 de 2011, La Habana, Cuba.

[5]Dosimetría das fontes radioativas de radiopreoteção, Código: PE2B-002, revisão 03, P 1, Instituto de Radiproteção e Dosimetria, Comssão Nacional de Energia Nuclear, Rio de Janeiro, 2003

[6] Guia para a expresssão da incerteza de medição, 3ra Ediçao revisada, INMETRO, agosto 2003

[7] Spiegel M. Estadísticas y probabilidades, New York 1978.

Journal of Physics: Conference Series **575** (2015) 012028 doi:10.1088/1742-6596/575/1/012028

Estimated of the associated uncertaintes with the linearity test of activimeters

C H S Sousa[1], G J Teixeira[2] and J G P Peixoto[3]
[1]Medical Physicist, PhD student at Instituto de Radioproteção e Dosimetria
[2]Radiology Technologist, PhD student at Instituto de Radioproteção e Dosimetria
[3]Physicist, PhD at Instituto de Radioproteção e Dosimetria
IRD/CNEN, Salvador Allende av, Recreio dos Bandeirantes, Rio de Janeiro, RJ, Br

E-mail: chenrique@ird.org.br, gt@ird.gov.br, guilherm@ird.gov.br

Abstract. Activimeters determine the activity of radioactive samples and are validated by performance tests. This research determined the expanded uncertainties associated with the linearity test. We used two activimeters and a source of $^{99}Tc^{m}$ for testing recommended by the IAEA protocol, which consider the decay of radioactive samples. The expanded uncertainties evaluated were not correlated with each other and our analysis considered a rectangular probability distribution. The results are also presented in graphical form by the function of normalized activity measured in terms of conventional true value.

1. Introduction

The knowledge of measurement, either by adjustments, conditionings or indications of instruments, is used for obtaining the value of a phenomenon, body or substance that can be qualitatively distinguished and quantitatively determined called quantity [1]. To determine it, it is necessary to mention the associated uncertainty to the result, which characterizes the dispersion of values around a mean and can be expressed as a standard deviation or half of an interval corresponding to a confidence level stated.

The activimeter is an instrument used in nuclear medicine services for measuring the activity of radioactive materials. It is basically constituted by a well ionization chamber and an associated electronics, which measure and record the ionization current produced by the interactions created from the radiation emitted from samples. They are adjusted at the factory, using a set of standard reference sources, certified and tracked, enabling the determination of the calibration curve of different radionuclides.

The characteristic response of a activimeter depends on the intrinsic nature and pressure of the gas, the material and thickness of the wall chamber [2,3]. The detector consists of two cylindrical coaxial electrodes, contained in a sealed space, filled with a gas under pressure that serves as the ionizing medium. The associated electronics convert the ionization current measured in signal voltage, amplifies it, processes and displays in a digital readout in units of activity. However, the instrument response depends on other factors that may influence the outcome of the measurements, such as different energies and emission rates of photons of various radionuclides to be measured.

According to the technical recommendations [4], all the instrumentation used in diagnostics or therapeutics nuclear medicine services must periodically pass through performance tests, which will attest the functional conditions of the instruments, focusing on the precision of the measurement and consequent reduction in probabilities of radiobiological interactions.

The linearity test is an acceptance test, which should define the measurement range in which the instrument has a linear response as a function of activity [4], however, their result can not be submitted separately, since it depends directly or indirectly another parameters that can influence the final result of the test. Therefore, the objective of this work was the analysis of the metrological of the linearity test for activimeters focusing on determining uncertainties.

2. Materials e methods

2.1. Materials

As suggested by the literature, radionuclide Technetium-99m ($^{99}Tc^{m}$) [4], was used because it is an element gamma emitter with a relatively short half-life. The sample was obtained through a single elution of $^{99}Tc^{m}$ generator system, resulting in an initial activity of 66.6 gigabequereis to a declared volume of 6 ml. The material was collected in a vial P6 obtained from the Instituto de Pesquisas Energéticas e Nucleares – IPEN[5].

Two activimeters were selected for these study doses, the first detection system is provided by an ionization chamber [6], of approximately 6.400 cm cubic filled with argon in a pressure of 13 psi and a holder made in polymethylmethacrylate (PMMA). The other one a detector system, equipped with two tubes Geiger Muller [7] in opposed positions to measure the radioactive sources between them. The most relevant features of these instruments are listed in Table 1.

Table 1. Most relevant features to this study, as range, resolution, accuracy, precision, linearity, response time, dimensions of the detectors, number of digits of the display and the shield thickness.

		ACTIVIMETERS	
		A[a]	**B**
Range (MBq)	Min	$7{,}0\times10^{-4}$	0,37
	Max	$7{,}0\times10^{5}$	$3{,}7\times10^{4}$
Resolution (MBq)		0,001	0,370
Accuracy (%)		± 2,0	Ni
Precision (%)		± 0,1	Ni
Linearity (%)		± 2,0	3,4
Response time (s)		≈ 10	≈ 50
Dimension (cm)		17 X 26	2 GM
Display		4	4
Shield		3,0	6,4
Detector		IC[b]	GM[c]

[a]Dosimetric reference
[b]Ionization chamber
[c]Geiger-Müller

2.2. Methods

The method defined and applied was the source decay, where measurements of the sample activities were compared to the true conventional values calculated for radionuclide decay. Three measurements series were performed every 3 hours during 105 hours and their standard deviations calculated [8].

Before starting the measurements there was an adjustment to zero for background radiation and input voltage were checked. The background radiation was subtracted from the value measured for evidence net metering. The vial, which containing radioactive sample, was positioned in the holder over the chamfered mark. The time set for each measurement was 15 seconds the same defined for system stabilization between the measurements.

The values were recorded and entered into a spreadsheet to make a log-linear graph, which already contained the conventional true values (CTV) of the activity corrected for source decay. The analysis also considered the measurement value standardization, which was defined as the ratio (*R*) result of the measured activity (*Aa*) by source reference activity corrected by the time of measurement (*Ca*), (equation 1). The average was applied to the end of the test to obtain the percentage deviation.

$$R = \frac{Aa}{Ca} \tag{1}$$

This study did not consider the environmental conditions of the laboratory, because sealed ionization chambers were used, and factors such as temperature, relative humidity and atmospheric pressure are not significant [9].

Type A uncertainties were defined as the percentage standard deviation of the measurement of linearity tests (σ_L) and background radiation (σ_b) because their repeatability. The probability distribution of these tests was described as rectangular due to marking ± 10% as determined by the technical document.

All the other parameters and contributions such as accuracy (σ_a), precision (σ_p), reproducibility (σ_r), the resolution (σ_{res}), radionuclides impurities (σ_i) and volume (σ_v), had their standard deviations and percentages recorded considered as type B uncertainties, emphasizing that testing accuracy, precision and reproducibility have been previously applied to the linearity test.

3. Results and discussions

The results of the performance tests of the instruments analyzed were described as percentages for a posterior determination of the value uncertainties uncorrelated to the tests. The amplitude of the measures of the reproducibility test performed with the standard reference source 137Cs was 1.8% for the calibrator A and 6.8% for the calibrator B, but these value are not so certain as the uncertainties associated with testing.

The calculation of the average measurements and standard deviations for the linearity test found that both instruments have a linear response within the acceptance limits recommended. After the application of Equation 1, the results were easier to be viewed graphically (figure 1).

Performance tests on both instruments showed concordant results with the limits of acceptability and are presented in Table 2. The combined standard uncertainty for the calibrator, equipped with IC was estimated at 3.802% with an expanded uncertainty (U) of 8.802% to a degree of freedom (k) equal to 2.32 (Table 2). The instrument B, with 2 GM tubes, presented a combined standard uncertainty of 9.02% with an expanded uncertainty (U) of 19.69% for a degree of freedom (k) was 2.18 (Table 2).

4. Conclusions

After analyzing the results, we conclude that volume and radionuclide impurities were considered important in estimating of associated uncertainties of the linearity test. The activimeter A, showed an uncorrelated expanded uncertainty lower than 4.0% and the activimeter B an uncorrelated expanded uncertainty below 10.0%. The possibility of correlation between the quantities can influence the final results. The results obtained in this study refer exclusively related to this research instruments and under the same conditions, it is not our goal, set or recommend the use of any type of instrument.

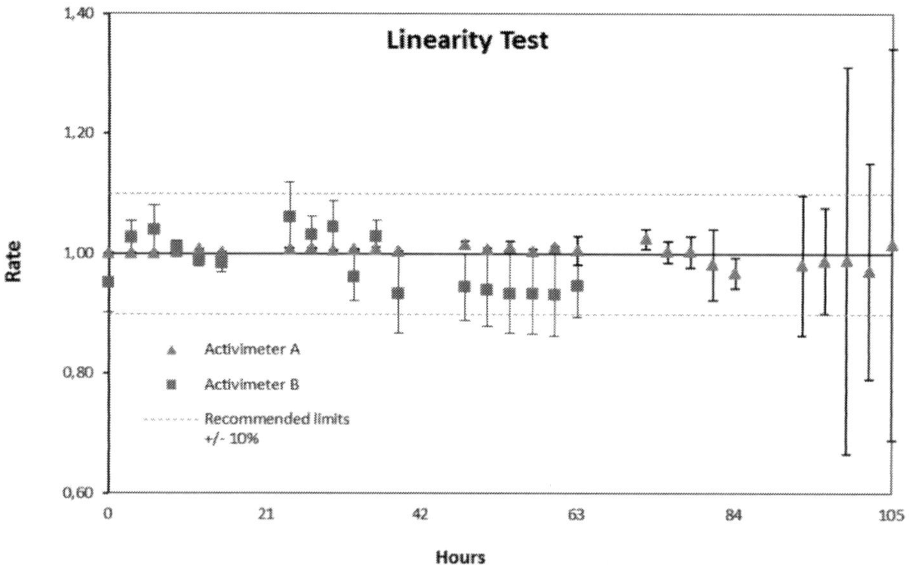

Figure 1: Graph linearity tests and standard deviations of activimeters A and B, showing percentage results in the dotted area representing the lower and upper bounds for this parameter.

Table 2: Results of the percentual standard deviation from performance tests, types of probability distributions adopted for each test and the results of the estimation of uncertainties.

Quantity	Distribution	N	ACTIVIMETER A		ACTIVIMETE B	
			Type A	Type B	Type A	Type B
Background	normal	10	0,052		3,82	
Accuracy	normal			0,097		0,46
Precision	normal			0,020		0,48
Reproducibility	normal			0,046		1,21
Volume	normal			0,043		0,04
Impurities	normal			0,003		0,01
Resolution	normal			0,004		1,62
Linearity	rectangular	10	3,8		7,9	
combined standard uncertainty			3,802		9,02	
expanded uncertainty			$U = 8{,}820$		19,69	
			$k = 2{,}32$		2,18	

results in percentage

5. References

[1] INMETRO 2012, Vocabulário Internacional de Metrologia – Conceitos fundamentais e gerais e termos associados. 94p – 1ª ed

[2] Schrader H 1997, Activity Measurements with Ionization Chambers. *Monographie BIPM - 4*

[3] Schrader H and Svec A 2004, Comparison of ionization chamber efficiencies for activity measurements. *Applied Radiation and Isotopes* ***60*** *369-78*

[4] IAEA 1991, Quality control of nuclear medicine instruments, *International Atomic Energy Agency TECDOC* ***602***

[5] IPEN 2012, Catálogo de produtos e serviços do IPEN

[6] CAPINTEC 2003, Operator´s Manual Digital Isotope Calibrator model CRC -127R, *CAPINTEC NJ*

[7] VICTOREEN 1982, Operating Instructions CAL/RAD Digital Isotope Calibrator model 34.061, *Nuclear Associates NY*

[8] SOUSA C H S 2013, Estimativa das Incertezas Associadas aos Testes de Desempenho de Calibradores de Doses, *Instituto de Radioproteção e Dosimetria/CNEN, Mestrado stricto sensu*

[9] IEEE 200, American National Standard Peformance Criteria for Hand-Held Instruments for the Detection and Identification Radionuclides, *ANSI N42.34/2006:47*

The role of time and frequency metrology inside the new definitions of SI quantities

L V G Tarelho, F A Rodrigues, A C Baratto, G A Garcia, R S L Souza

Division of Metrology for Information and Communication Technology (Dmtic)
INMETRO, Av. Nossa Senhora das Graças, 50- Xerém, Duque de Caxias, RJ – ZIP: 25250-020

lvtarelho@inmetro.gov.br

Abstract. There is an on-going discussion, initiated in 2011, which proposes several changes to the official definitions of the base units of the International System of Units (SI). This will also entail changes in the relationships between the primary quantities of SI and some universal constants. One result of this process points in the direction of giving greater importance and centrality to unit of time (s), increasing the practical importance of the realization of the quantities frequency and time. So, it is strongly recommended that the metrological maintenance of these quantities, in Brazil, be done in the Campus of INMETRO.

1. Introduction

The CGPM (Conférence Générale des Poids et Mesures) adopted a Resolution and encouraged communication and debate on the possible future revision of the SI (Systéme International de Unites) during its 24th meeting in October 2011. In order to disseminate the discussion it was given universal access at BIPM (Bureau International des Poids et Mesures) site to key documents on the redefinition of certain SI units, which are currently at draft stage and will be amended in the coming years [1].

In the "New SI", four of the seven SI base units, namely the kilogram, the ampere, the kelvin and the mole, will be redefined in terms of invariants of nature [2]. This new definitions will be based on fixed numerical values of the Planck constant (h), the elementary charge (e), the Boltzmann constant (k), and the Avogadro constant (N_A), respectively. Some different constants were evaluated [3], namely the Compton Length (λ_e), the magnetic constant (permeability of the vacuum - μ_0), the constant of Rydberg (R_∞) and the mass of electron (m_e), but they did not created a complete set to form a unit basis.

The choice for these four numerical constants was made based on the algebra associated with units´ exponents which was used both to display the relationships between different sets of units and to provide simple tests of whether a particular set of constants provides combinations of units that forms a complete set [3]. This approach gave rise to defining units in a quantum based SI, about which all NMIs (National Metrology Institutes) should be aware.

In the "New SI" the definitions of all seven base units will also be uniformly expressed using the explicit-constant formulation, and specific *mises en pratique* (practical realizations) will be drawn up to explain the realizations of the definitions of the base units in a practical way.

Sometimes making predictions and forecasts are only mental exercises about different future scenarios that should not be taken so seriously into account due to the ephemeral aspect of the future

Journal of Physics: Conference Series **575** (2015) 012029 doi:10.1088/1742-6596/575/1/012029

itself, but the redefinition of the SI is not from the same futurology studies: it is imminent and is already technically discussed in consultative committees in BIPM. Of course a major diplomatic discussion is still to be made at CIPM (Comité International des Poids et Mesures) because there are several economic and technological aspects of these redefinitions that should be addressed before the change take place and a world consensus be achieved, but the change is coming.

2. Time Definition

The redefinition of SI will not change the definition of time (and frequency) because this quantity is already explicitly connected to a constant of nature or fundamental constant, ν_{Cs} - the frequency of the transition between the two hyperfine levels of the Cesium ground state.

The proposed definition - "*The second, s, is the unit of time; its magnitude is set by fixing the numerical value of the ground state hyperfine splitting frequency of the cesium 133 atom, at rest and at a temperature of 0 K, to be equal to exactly 9 192 631 770 when it is expressed in the unit* s^{-1}, *which is equal to Hz*" - changes the words without changing the sense of the previous definition: "*The second is the duration of 9 192 631 770 periods of the radiation corresponding to the transition between the two hyperfine levels of the ground state of the cesium 133 atom*".

All other SI units will be redefined in order to be explicitly linked to a fundamental constant. The constants of nature linked to them will be:

- *the ground state hyperfine splitting frequency of the cesium 133 atom* $\boldsymbol{\nu_{Cs}}$ *is exactly 9 192 631 770 hertz, Hz,* s^{-1},
- *the speed of light in vacuum* **c** *is exactly 299 792 458 meter per second,* $m\,s^{-1}$,
- *the Planck constant* **h** *is exactly* $6.626\,06X \times 10^{-34}$ *joule second, J s,*
- *the elementary charge* **e** *is exactly* $1.602\,17X \times 10^{-19}$ *coulomb, C,*
- *the Boltzmann constant* **k** *is exactly* $1.380\,65X \times 10^{-23}$ *joule per kelvin,* $J\,K^{-1}$,
- *the Avogadro constant* $\boldsymbol{N_A}$ *is exactly* $6.022\,14X \times 10^{23}$ *reciprocal mole,* mol^{-1},
- *the luminous efficacy* $\boldsymbol{K_{cd}}$ *of monochromatic radiation of frequency* 540×10^{12} *hertz is exactly 683 lumen per watt,* $lm\,W^{-1}$.

With the seven base units it is possible to describe all the derived quantities because they are chosen to form a complete set.

Using the BIPM recommendations expressed at the 8th edition of SI-International System of Units [4][5], we could easily describe the dimension of a derived quantity Q as expressed in Equation 1,

$$\mathrm{im}\, Q = T^{\alpha}\, L^{\beta}\, M^{\gamma}\, I^{\delta}\, \theta^{\varepsilon}\, N^{\zeta}\, J^{\eta} \tag{1}$$

where the exponents $\alpha,\beta,\gamma,\delta,\varepsilon,\zeta$,and η are small integer numbers, positive, negative or null, and are called dimensional exponents.

In this work, to make easier to recall the quantities involved in the algebraic formalism, we will express the derived quantity directly by its unit symbol and the dimensional analysis will be made using matrix formalism.

One can describe the SI units defined in the 8th edition of SI [4] using the constants described there, where the mass, electric current, thermodynamic temperature and amount of substance are related to the mass of IPK (International Prototype of Kilogram), the magnetic permeability constant, the temperature of triple point of water and the mass of isotope 12 of carbon. Table 1 presents the dimensional exponents for each fundamental constant as stated in 2006 – SI Brochure, and Table 2 presents the same content as proposed in 2011 for discussion.

Table 1 – Dimensional exponents for each fundamental constant as stated in 2006.

	s	m	kg	A	K	mol	cd
ν_{Cs}	-1	-1	0	1/2	0	0	1
c	0	1	0	1/2	0	0	2
m(IPK)	0	0	1	1/2	0	1	1
μ_0	0	0	0	-1/2	0	0	0
T_{PTE}	0	0	0	0	1	0	0
$M(^{12}C)$	0	0	0	0	0	-1	0
K_{cd}	0	0	0	0	0	0	1

Table 2 - Dimensional exponents for each fundamental constant as stated in 2011

	s	m	kg	A	K	mol	cd
ν_{Cs}	**-1**	**-1**	**1**	**1**	**1**	**0**	**2**
c	0	1	-2	0	0	0	0
h	0	0	1	0	1	0	1
e	0	0	0	1	0	0	0
k_B	0	0	0	0	-1	0	0
N_A	0	0	0	0	0	1	0
K_{cd}	0	0	0	0	0	0	1

When comparing both tables one can realize that the previous definition of mass and temperature were related to material artifacts and in the redefinition these quantities are now related to frequency definitions. Only the amount of substance definition is independent of the frequency realization.

Time and frequency standardizations by using the frequency of the cesium atom has become a strategic technology and a very important metrological knowledge in order to assure the practical realization of six from the seven SI units according to the new definition.

3. Time and Frequency Standards at Inmetro

LAORT, one of the laboratories of Dmtic at Inmetro, through the measurement of the Local Time Scale at Campus Xerém, participates in the BIPM Circular T since October 2012. This monthly report presents the results of a time and frequency continuous key comparison. They can be used to demonstrate the degree of equivalence between all NMIs clocks (~400 clocks at 68 laboratories and 13 primary standards at 8 laboratories).

The name chosen for the Local Time Scale at the Campus of Inmetro in Xerém is INXE (INmetro XErém) and it is available for dissemination inside the Campus. The Universal Coordinated Time - UTC(INXE) - has the technical recognition because it is validated by the key intercomparison but it did not have any CMC (Calibration and Measurement Capability) assigned to this effort.

At the moment Inmetro´s Time Scale UTC(INXE) is composed by a commercial Cesium clock and it will be expanded in the near future to two commercial Cesium clocks and one research Cesium primary frequency standard employing a thermal beam and optical laser pumping.

The behavior of the availability of UTC(INXE), compared to the UTC, can be seen in Figure 1.

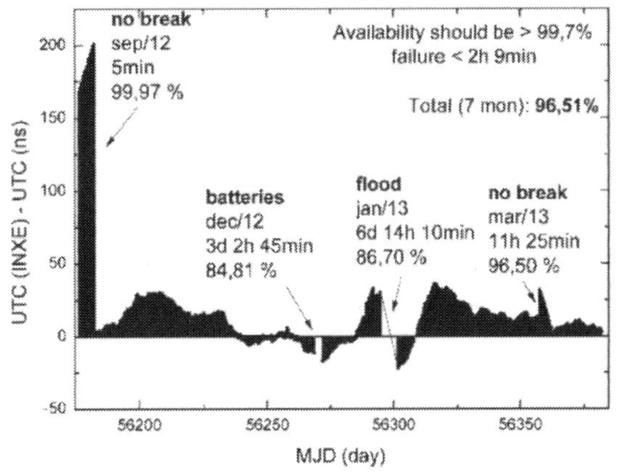

Figure 1. Difference between UTC(INXE) and UTC in nanoseconds

4. Conclusions

Some of the efforts to create an infrastructure for time and frequency metrology at Inmetro are now giving rise to more fruitful possibilities in developing and realizing other SI units' standardizations.

The availability of the Local Time Scale, as can be seen in the Figure 1, is almost in the range of 3σ, with some episodes of 2σ month availability due to power failure and flood.

Some of the discussions presented here should help to trace new routes for time and frequency metrology at the Campus of Inmetro.

5. Acknowledgements

The authors would like to thank the discussions with Dr. Luc Erard during his presentation at Inmetro and Dr. Richard Davis during his presentation at SYRTE-LNE.

6. References

[1] Documents available at: http://www.bipm.org/en/si/new_si/
[2] Draft Chapter 2 for SI brochure - following redefinitions of the base units. (draft for 9th edition of SI - The International System of Units)
[3] Peter J Mohr, "Defining units in the quantum based SI", Metrologia 45 (2008) 129–133.
[4] SI Brochure – The International System of Units, 8th ed, BIPM, Paris, 2006.
[5] Sistema Internacional de Unidades: SI. — Duque de Caxias, RJ: INMETRO/CICMA/SEPIN, 2012. Available at http://www.inmetro.gov.br/infotec/publicacoes.asp

New generation of a mobile primary frequency standard based on cold atoms

S T Müller [1], J de Martin Júnior [1], R D Pechoneri [1], P Santa Catharina[1], V S Bagnato[2] and D V Magalhães[1]

[1] Departamento de Engenharia Mecânica, Escola de Engenharia de São Carlos, Universidade de São Paulo, São Carlos, Brazil
[2] Departamento de Física e Ciência dos Materiais, Instituto de Física de São Carlos, Universidade de São Paulo, São Carlos, Brazil

E-mail: stella@ifsc.usp.br

Abstract. We have constructed a mobile primary frequency standard using intra-cavity cold cesium atoms and the results have shown the advantages of using this kind of system compared to cesium beam standards. Based on the first setup with an expanding cloud of atoms, we can plan the construction of a clock more compact since it has no strict size limitations. In order to assemble the new system even smaller, the development of a system containing lasers, microwave source and cavity in a single box has already begun. The mobile atomic standard based on cold atoms is a possible strategic product with a broad range of applications and an important contribution to a primary standard of high relevance.

1. Introduction

With exceptional accuracy, atomic clocks have applications in several fields of modern fundamental physics [1]. They allow not only to investigate the macroscopic scales theories like relativity or gravitation, but also to know better the microscopic world, in atomic scale, thanks to quantum mechanics. Two of the main applications are navigation and telecommunication systems, and atomic references are the heart of the most advanced navigation system, satellite constellations positioning like GPS, GLONASS, and Beidou [2].

Most of the high performance time standards require an extremely complex construction and operation, but the size of these systems makes them incompatible with some applications, which require a compromise between size and performance. These include space and industrial applications, such as telecommunication lines, mobile telephone networks or internet, calibration of instruments, dissemination of local time references with good quality, etc.

The Atomic Time and frequency Standard Program USP São Carlos [3] started a few years ago with the construction of a thermal beam atomic frequency standard, today our research is focused on the development of two frequency standards based on cold atoms: an atomic fountain [4] and a compact system based in a cold atomic cloud in free expansion [5]. Besides those experimental standards, our facility has commercial atomic clocks and a hydrogen maser.

The atomic frequency standards developed up to now provided us the theoretical and experimental knowledge for the development of a mobile atomic frequency standard with the same level of precision and stability of the previous setups. The new diode laser sources, microwave synthesizer, vacuum and control systems are being reduced in order to ensure the portability of the system.

2. Experimental setup and temporal sequence

The aim of this project is the construction of a cold atoms compact frequency standard that works in an innovative way, since the steps of the sequence occur in the same place, within a microwave cavity. For the physical package we use a cylindrical cavity, sculpted in a stainless steel vacuum chamber and resonant in 9.2 GHz. This operation mode allows us to design a much more compact system than a conventional cold atoms clock, where the various interactions are in different parts of the instrument. Therefore, the compactness of the clock is achieved through a sequence purely temporal. The most relevant steps can be seen in Figure 1.

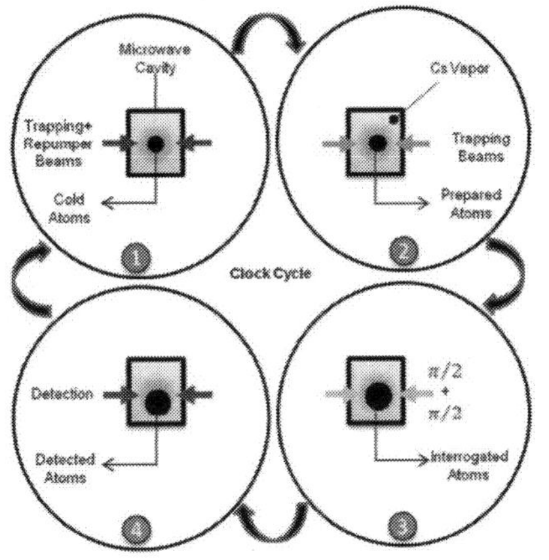

Figure 1. Temporal sequence of a clock cycle performed inside the microwave cavity.

The temporal sequence of the compact frequency standard can be divided in four well defined steps (see figure 1), as follows:

(Step 1)Trapping atoms in a magneto optical trap (MOT)

In this stage is possible to capture 10^8 atoms using magnetic field and three counter-propagating pairs of laser beams (MOT) [6]. This is important to catch up the atoms together in the same spatial region, forming a cold cloud, and to remove a huge portion of the cloud's kinetic energy. In this step the atomic cloud attains very low temperatures and becomes more stable. The $6S_{1/2}|F=4\rangle \rightarrow 6P_{3/2}|F'=5\rangle$ cycling transition is used, and repumping is done using the $6S_{1/2}|F=3\rangle \rightarrow 6P_{3/2}|F'=4\rangle$ transition.

(Step 2)Sub-Doppler Cooling and the Preparation

After the first reduction of the kinetic energy, the system passes from MOT configuration to optical molasses by switching off the current supply of the magnetic field. Simultaneously, the atomic cloud is cooled even more, changing the intensity and detuning of the laser beams with acousto-optical modulators, reaching levels below 10 μK. In order to prepare the atoms in their ground electronic state $6S_{1/2}|F=3\rangle$, the repumping light is switched off 5 ms before shutting off the cooling light. During this interval, optical pumping efficiently transfers the atoms to the required ground state.

(Step 3) The clock transition interrogation

Now, the cold atoms are in a single initial state and after the total shutdown of the lasers beams, the cloud starts a free expansion. During the expansion the atoms are interrogated in a Ramsey sequence

[7], consisting of two coherent microwave pulses $\pi/2$ with 1ms of duration and separated by 8ms. These pulses excite the clock transition between the two hyperfine levels $6^2S_{1/2}|F=3, m_F=0\rangle \rightarrow 6^2S_{1/2}|F=4, m_F=0\rangle$.

(Step 4) The Detection
To detect the atoms that have transitioned to the $6S_{1/2}|F=4\rangle$ level after the two microwave pulses, the light beams originally used as a cooling laser are turned back on for 40 ms, and the fluorescence signal is collected.

Once the resonance is observed, it can be used as a frequency discriminator. The transition probability difference of two successive measurement acts as an error signal used to control the local oscillator frequency.

3. Results
Figure 2 presents a typical scan of the microwave frequency across the clock resonance in our device [5]. Using the Ramsey fringe of Figure 2, we can scan around the central fringe to determine the linewidth with a theoretical fitting (Figure 3), obtaining a value of 47 Hz. The contrast is better than 80% (where the contrast is defined as a difference between resonance amplitude and background).

To measure the frequency stability of the system is necessary to lock the central resonance observed on figure 3 to the microwave chain (macroscopic oscillator). The microwave chain on the other hand is phase locked to the 5 MHz output of the Hydrogen Maser. The modulation of the frequency is controlled by a computer, which also register the introduced correction to keep the interrogation signal at its maximum.

The stability observed is $\sigma_y(\tau) = (5 \pm 0.5)x10^{-13}\tau^{-1/2}$, as presented in Figure 4, demonstrating that this compact cold atom frequency standard can reach performances better than commercial beam clocks, which typically have short time stabilities on the range of 10^{-11} - 10^{-12}.

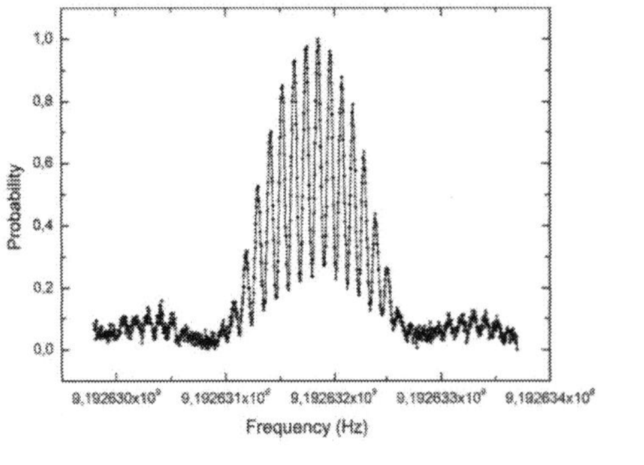

Figure 2. Ramsey Fringes obtained by an interrogation performed in the cavity by two microwave pulses of 1 ms each and separated by 8 ms. The linewidth of the central fringe is 47 Hz.

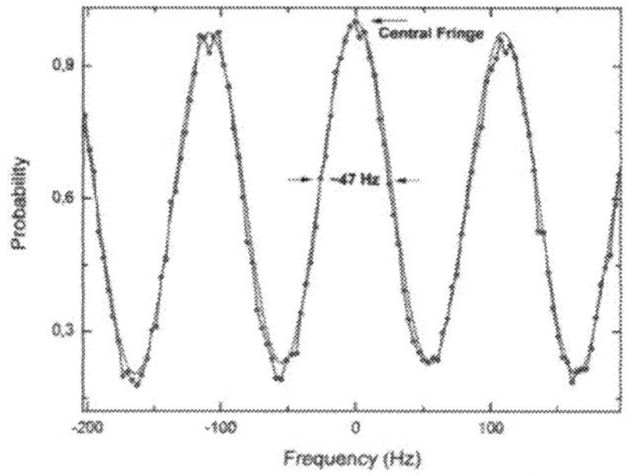

Figure 3. Scan over the central Ramsey fringe (figure 2) giving a 47 Hz linewith with 80% contrast.

The advantage of cold atoms in such compact system can be easily observed if compared with commercial thermal Cs beam clocks. The long term stability already surpasses the currently used commercial standards and the clock transition linewidth is on the order of 50 Hz for our system, meanwhile the same resolution can be achieved only with a thermal system that has some meters in the interrogation region (26 Hz for a 4 m long interrogation region [8]).

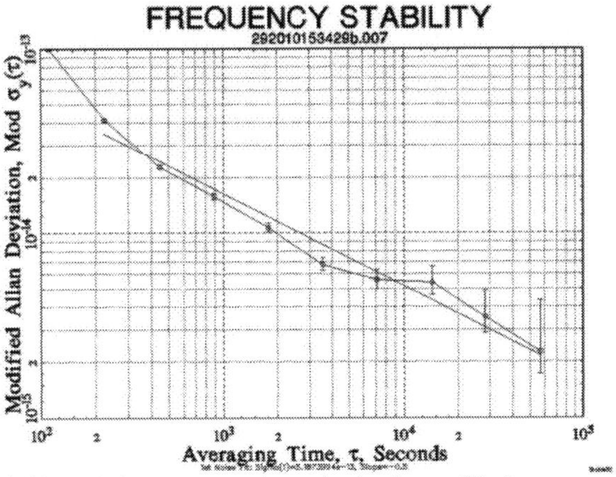

Figure 4. Frequency stability of the system phase locked to a Hydrogen maser calculated by Allan standard deviation.

4. Reducing the system volume

We actually envisage a more compact system, mounting the apparatus in a single block. This should be done keeping all the necessary processes to produce a frequency standard signal.

4.1. The macroscopic oscillator

A fundamental piece of the atomic standard, it corresponds to the signal that generates the time reference. The feedback from the atoms interrogated with its signal closes the loop for the atomic frequency reference. It should be as small as possible, but still taking into account some parameters as temperature deviations [9, 10]. To achieve reliable performance, we will use PID (Proportional-Integral-Derivative) control and peltier elements as actuators.

4.2. Lasers Diode

The use of laser diodes is fundamental in the system. They should keep a very low effective linewidth, in order to not degrade the SNR of the clock transition. In the limit of high resolution spectroscopy, the laser noise can add a significant parcel to the overall budget of noise contributions.

4.3. General Control of the System

Careful control of all the steps of the clockwork (trap, sub-doppler cooling, interrogation and detection) is mandatory. In each cycle the microwave source is corrected, and the steps consist of well synchronized events, such as laser light frequency shifts, pulsing of the microwave signal, triggered signal detection and processing of the atomic fluorescence [11].

5. Conclusions

We have developed a compact system to be used as a mobile reference for frequency signals. We managed to cool 10^8 atoms at 10 μK in a microwave cavity and after the Ramsey interrogation the detection by fluorescence is applied. The core idea is the use of cold atoms as a high performance and reliable clock source. We have been working in the reduction of volume to make an embeddable prototype, in order to maintain the high quality characteristics necessary for the operation of such device.

References

[1] Jespersen J and Fitz-Randolph J 1999 *From sundials to atomic clocks* (Mineola: Dover Publications)

[2] Ramsey N F 2002 *Application of atomic clocks* in: Laser physics at the limits, ed Figger H. *et al* (Berlin: Springer) p 3
[3] Ahmed M, Magalhães D V, Bebeachibuli A, Müller S T, Alves R F, Ortega T A, Weiner J and Bagnato V S 2008 The brazilian time and frequency atomic standards program *An Acad Bras Cienc* **80**, 2
[4] Magalhães D V 2004 Construção de uma fountain atômica para utilização como padrão primário de tempo *Doctorade thesis*
[5] Müller S T, Magalhães D V, Alves R F and Bagnato V S 2011 Compact Frequency Standard based on an Intracavity Sample of Cold Cesium Atoms *J. Opt. Soc. Am. B* **28** 11 p 2592
[6] Metcalf H and Van Der Straten P 2003 Laser cooling and trapping of atoms *J.Opt. Soc. Am. B* **20** p 887
[7] Vanier J, Audoin C 1989 *The quantum physics of atomic frequency standards* v2 (Adam Hilger)
[8] Glaze D J, Hellwig H, Allan D W and S. Jarvis 1977 NBS-4 and NBS-6: The NBS primary frequency standards *Metrologia* **13** p 17-28
[9] L. E. Schnurr 1967 The transient thermal characteristics of quartz resonators and their relation to temperature-frequency curve distortion *21st Annual Symposium on Frequency Control* (1967)
[10] Ji Wang, Yook-Kong Yong and Tsutomu Imai 1998 High-order plate theory based finite element analysis of the frequency-temperature relations of quartz crystal oscillators *Frequency Control Symposium* Proceedings of the 1998 IEEE International (1998)
[11] J. Levine 1999 Introduction to time and frequency metrology *Review of scientific instruments* **70**, 6

Journal of Physics: Conference Series **575** (2015) 012031 doi:10.1088/1742-6596/575/1/012031

Method validation for methanol quantification present in working places

E D M Muna[1], C H B Bizarri[1], J R M Maciel[1], G P da Rocha[1] e I O de Araújo[1]
[1] Centro de Tecnologia SENAI Ambiental, FIRJAN – 53 Rua Morais e Silva- Rio de Janeiro - Brazil

E-mail: edobloni@firjan.org.br

Abstract. Given the widespread use of methanol by different industry sectors and high toxicity associated with this substance, it is necessary to use an analytical method able to determine in a sensitive, precise and accurate levels of methanol in the air of working environments. Based on the methodology established by the National Institute for Occupational Safety and Health (NIOSH), it was validated a methodology for determination of methanol in silica gel tubes which had demonstrated its effectiveness based on the participation of the international collaborative program sponsored by the American Industrial Hygiene Association (AIHA).

1. Introduction

Methanol is widely used as an industrial solvent in the production of other chemicals, such as paint and varnish removers and in automotive antifreeze and defrosting solutions as well as liquid windshield wipers. Many other products such as plastics, paints and explosives are derived from the main product of methanol, formaldehyde [1] [2] [3].

Once absorbed, methanol is slowly eliminated by human body, therefore it should be regarded as a cumulative poison. Although short exposures to vapors should not cause adverse effects when they become daily may result in methanol accumulation sufficient to cause health damage [5] [7]. Chronic exposure to methanol vapor has been reported to produce symptoms such as headache, dizziness, blurred vision, eye irritation, nausea and gastrointestinal upset [2].

A review of data on environmental levels and human exposure notes that most emissions to the environment arise from the production and use of methanol as a solvent in industrial processes and, to a lesser extent, from a variety of other industrial processes and consumer applications [2].

Regarding the toxicological effects of methanol, a regulatory standard enacted by the Ministry of Labor and Employment of Brazil established the methanol exposure limit that a worker may be submitted in production environment. For working hours of 48 hours per week, the tolerance limit specified for methanol is 156 ppm [4]. There is an obvious need to develop an inexpensive, sensitive, rapid, and reliable alternative GC (gas chromatography) method to determine methanol in low level presented in working environments.

Based on NIOSH 2000 method [6], a method was optimized and validated using gas chromatography with flame ionization detection and headspace injection in order to quantify methanol presented into silica gel samplers.

2. Materials and Methods

2.1. Materials
Reverse-osmosis type quality water (purified with a Milli-RO plus Milli-Q station from Millipore, Milford, MA) and HPLC quality were used throughout.

2.2. Extraction step
Methanol is collected form the air by a silica gel tube, containing two sections of 20/40 mesh silica gel (front = 100 mg, back = 50 mg) separated by a 2-mm urethane foam plug.
After the forced passage of air through the tube using of a vacuum pump, the adsorbent is transferred to a vial headspace with 2 mL of ultrapure water. The sealed vial is placed in an ultrasound bath for 30 minutes before the injection.

2.3. Chromatographic conditions
A headspace injection system (Perkin Elmer Turbo Matrix 40 Trap) coupled to gas chromatograph with flame ionization detector (Perkin Elmer Clarus 500) were used to quantify methanol extracted from silica gel adsorbent. The analyses were performed under the following chromatographic conditions: Column, RTX-1301 (Crossbond® cyanopropylphenyl dimethyl polysiloxane), 15 m × 0.32 mm i.d., DF = 1 mm, from Restek (Bellafonte, PA). The temperature of the FID was 240°C, and the injector temperature was 200°C. The oven temperature was programmed to 40°C (for 6 min), Helium was the carrier gas with a flow of 0.5 mL/min for two minutes followed by an increase of 100mL/min to 1mL/min. The parameters of headspace injection were: incubation time of 30 minutes, temperature of 90 °C, needle temperature 200 °C and time of pressurization of 1 minute. The retention time for methanol in the chromatographic system was observed at 2.567 minutes.

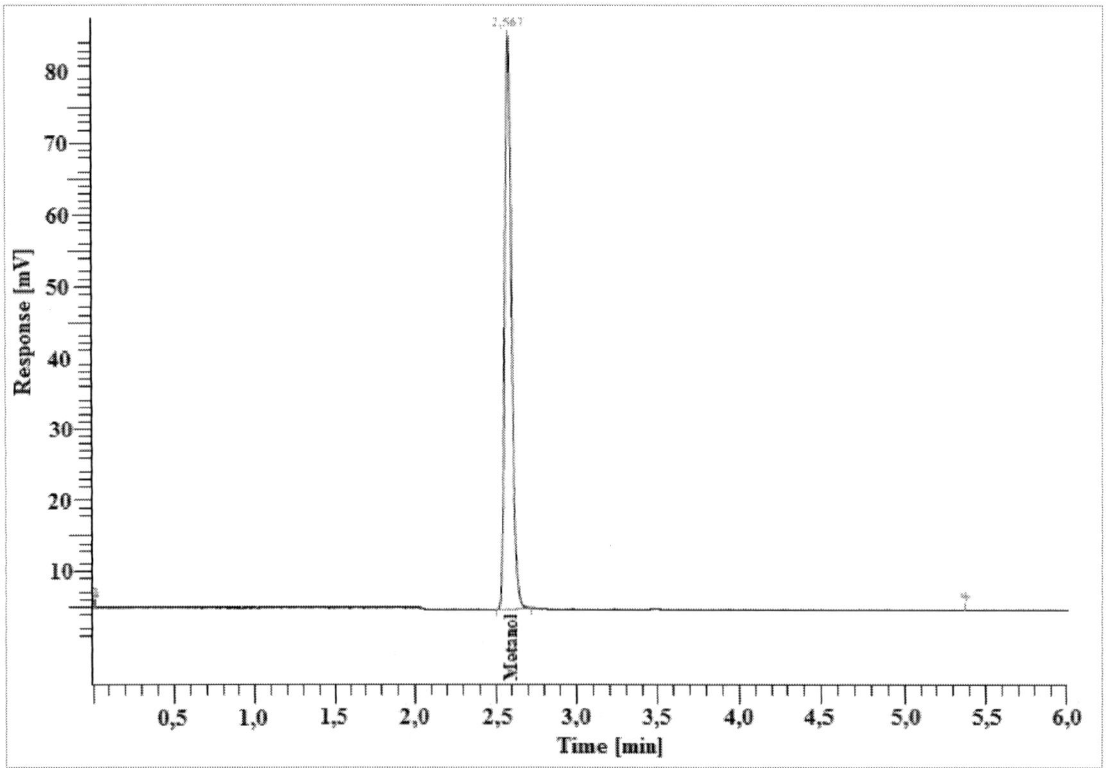

Figure 1. Chromatogram of an aqueous solution with 50 mg/L methanol (2.567 minutes).

Journal of Physics: Conference Series **575** (2015) 012031 doi:10.1088/1742-6596/575/1/012031

2.4. Method Validation

Validation of the method was made in terms of selectivity, linearity, recovery, repeatability and inter-day precision, quantification and detection limits, according to the guidelines of INMETRO (Brazil Metrological Institute) [8].

3. Results and Discussion

3.1. Method Validation

3.1.1. Selectivity

The t test was used to compare the slope of the analytical curves of the analytes in solution and with fortified matrix (silica gel tubes), which presented $t_{calculated} < t_{table}$ (Table 1), showing that there was no matrix effect for this analyte.

Table 1. Comparison of analytical curves of methanol in solution and extracted from fortified matrix

Concentration (mg/L)	Analytical curve in solution Area	Analytical curve extracted from fortified matrix Area
	30544	30760
50	29649	28858
	29029	29068
	55522	56494
100	55925	54393
	57357	56306
	11386	113114
200	119023	110210
	112304	113773
	239995	221744
400	225010	231476
	234666	229870
	336433	339304
600	346917	340823
	338986	352482
	460882	472138
800	467082	477241
	460962	475651
	552981	575848
1000	558220	566656
	567105	584156
Angular Coefficient	**568**	**580**
Linear Coefficient	**1067**	**-855**
R^2	**0.999240**	**0.999203**
t_{calc}	**0,419072**	
t_{tab}	**2,024394**	

3.1.2. Linearity

Method linearity was determined by evaluating the regression curve and is indicated by the square correlation coefficient (R^2). The line of best fit for the relationship between the ratio of peak area and internal standard area and concentration of analytes in the samples was determined by linear regression performing calibration curves in the considered concentration ranges (50-1000 mg/L), depicted in Figure 2. The slope was calculated taking into account calibration curves in triplicate prepared in methanol aqueous solution. Linearity was achieved with a minimal R^2 of 0.99.

The residue graph (Figure 3) showed randomly distributed around the center line, without a trend, confirming the absence of random errors. Cochran test showed (Table 2) that the residues are homocedastic ($C_{calculated} < C_{table}$). Regression analysis of calibration data achieved satisfactory linearity over the considered concentration range (50-1000mg/L). Grubbs test was applied in the results of analytical curve and no outlier value was observed (Table 3).

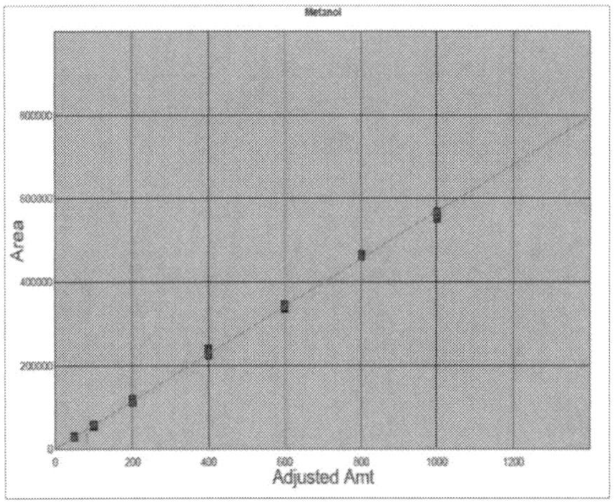

Figure 2. Analytical curve of methanol in aqueous solution.

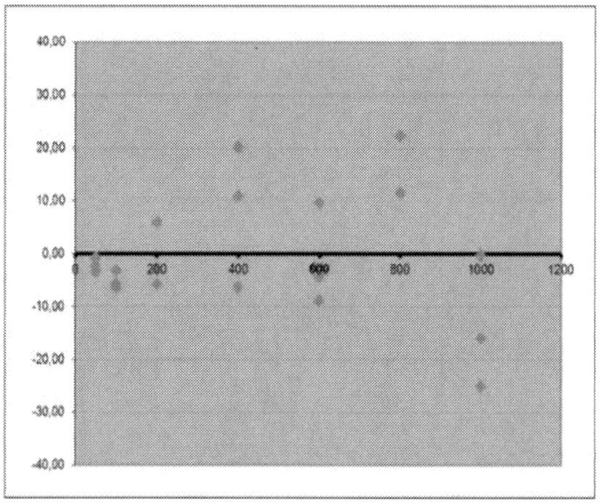

Figure 3. Residuals graphic of analytical curve.

Table 2. Cochran test for metanol analytical curve

Cochran values	
$C_{calculated}$	0.349
C_{table}	0.561

3.1.3. Limit of Detection (LOD) and quantification (LOQ)

In order to determine the limits of detection and quantification it was performed an evaluation of 10 fortified matrix at low concentration value of LOD was obtained by multiplying the standard deviation by t_{table} (n= five samples, IC = 95%). LOQ value was based on multiplying LOD value by 3.3.

The obtained LOD was 2.6 mg/L methanol, while LOQ presented a calculated result of 8.5 mg/L methanol (Table 4). Fortified matrices were tested at the calculated concentrations of LOQ and the results of signal / noise ratio were greater than five and the relative standard deviation of less than 5%.

Table 3. Grubbs test for outlier evaluation

Nominal Methanol Concentration (mg/L)	Experimental Methanol Concentration (mg/L)	Average Conc. (mg/L)	Standard Deviation	Variance	G calculated		G table
					Superior	Inferior	
50	51.9 50.3 49.2	50.5	1.340	1.796	1.055	0.934	
100	95.8 96.5 99.1	97.1	1.697	2.880	1.129	0.774	
200	198.5 207.6 195.7	200,61	6.187	38.283	1.126	0.785	
400	420.5 394.1 411.1	408,54	13.367	178.666	0.891	1.081	1.156
600	590.2 608.6 594.7	597,81	9.621	92.556	1.123	0.795	
800	809.2 820.1 809.3	812,85	6.259	39.170	1.155	0.589	
1000	971.2 980.4 996.1	982,59	12.564	157.866	1.074	0.904	

Table 4. Limit of Detection (LOD) and quantification (LOQ) of Methanol

Parameters	Experimental concentrations of methanol (mg/L)
50 mg/L (nominal concentration)	50.8 49.8 49.7 49.6 50.0 51.1 52.2 49.3 52.2 49.3
Average	50.4
Standard Deviation	**1.1**
t_{table}	**2.2622**
LOD	**2.6 mg/L**
LOQ	**8.5 mg/L**

3.1.4. Recovery

Fortified matrices were prepared in three known concentrations (low, medium and high) over the working range as shown in table 5. The recovery of this method was determined by analysis of nine replicates of samples spiked with methanol at three different concentrations and the results were quite satisfactory the average recovery of 99.9% was calculated.

Table 5. Methanol recovery study.

Nominal Concentration (mg/L)	Recovery (%)
	99.2
50	99.5
	99.5
	101.2
400	98.0
	101.3
	99.6
1000	100.6
	100.2
Average	**99.9**

3.1.5. Repeatability

Nine determinations were carried along the working range at three concentration levels (triplicates per concentration level) in matrix fortified. Then calculated the standard deviation per concentration level studied and all values were below 2% (Table 6).

Table 6. Method repeatability

Level	Concentration (mg/L)	Average	Standard deviation	Repeatability limit	RSD%
	49.7				
1	49.8	49.7	0.0856	0.2398	**0.2**
	49.7				
	404.7				
2	392.0	400.6	7.4309	20.8065	**1.8**
	405.0				
	996.1				
3	1005.8	1001.5	4.9287	13.8004	**0.5**
	1002.5				

3.1.6. Inter-day Precision

Different analysts analyzed samples under the same sample preparation and equipment conditions (Table 7). After comparison of variances (F test) and means (t test) it was not observed statistically significant difference between the two assessments by different analysts. On tables 8 are depicted the results of F test and t test.

3.1.7. Reproducibility

Given the lack of certified reference material, the participation in an interlaboratory program (IP) was crucial to evaluate the effectiveness of the methodology validated, which had as provider of the study the American Industrial Hygiene Association (AIHA) All results were satisfactory in PI, as shown on Table 6, with z-scores bellow 2.0.

Table 7. Values obtained from the two analysts.

Nominal Concentration (mg/L)	Experimental concentration (mg/L)	
	Analyst 1	Analyst 2
	51.9	50.7
50	50.3	50.9
	49.2	51.4
	198.5	202.8
200	207.6	200.7
	195.7	209.7
	590.2	599.6
600	608.6	600.9
	594.6	605.7
	1049.8	1009.6
1000	971.2	1011.0
	980.4	997.2

Tabela 8. Result of F and t test for the two analysts

Parameters	Analyst 1	Analyst 2
Average	465.8	462.3
Variance	150212.5	149297.5
Evaluations number	12	12
Freedom level	11	11
F	1.00613	
P(F<=f)	0.496048	
F critical	2.817930	
Stat t	0.022236	
P(T<=t)	0.49123	
t critical	1.717144	

Table 9. Interlaboratory Program results (AIHA)

Samples	Laboratory Results	Designated value	z-Score
1	0.617 mg	0.618 mg	0. 0
2	1.683 mg	1.523 mg	1.5
3	0.068 mg	0.081 mg	-1.9

4. Conclusions

In conclusion, the headspace injection coupled with GC–FID method using capillary column presented here is a highly sensitive, rapid, and reliable procedure to determine methanol in tubes with silica gel adsorbent.

The limit of quantitation obtained by this method (15ppm) is bellow more than ten times of the defined exposure limit (156 ppm), considering 5 L of collected air. Therefore, the validated methodology meets the specifications of regulatory standard number 15 of Ministry of Labour and Employment of Brazil.

5. References

[1] Methanol Institute http://www.methanol.org/ Methanol - Basics / Methanol - Applications.aspx, access on june 2013

[2] Environmental Health Criteria, No 196 - World Health Organization

[3] Environmental Protection Agency (EPA) - Toxicological Review of Methanol (noncancer) - In Support of Summary Information on the Integrated Risk Information System (IRIS) May 2013

[4] Ministério do Trabalho e Emprego (MTB) Portaria nº 3.214/78 - NR15 – anexo 11 Avaible at: <http://www.mte.gov.br/legislacao/normas_regulamentadoras/nr_15.pdf>, access on june 2013

[5] Pereira P A P e Andrade J B 1998 Quím. Nova 21 6 744-754

[6] NIOSH Manual of Analytical Methods (NMAM), Fourth Edition - METHANOL: METHOD 2000, Issue 3, dated 15 January 1998

[7] Barceloux DG, Bond GR, Krenzelok EP, Cooper H, Vale JAJ *2002 Toxicol Clin Toxicol.* 40 (4) 415-46

[8] INMETRO - DOQ-CGCRE-008 04 Rev - Jul/2011

Acknowledgments

We thank Serviço Social da Indústria – Federação das Indústrias do Esatado do Rio de Janeiro (FIRJAN) for the research funding to carry out the present study.

Validation and Uncertainty of the method for multiresidue analysis of 35 pesticides in melon using Gas Chromatography Coupled to Quadrupole Mass Spectrometry (GC-QP/MS)

M A L Milhome[1,2], J S Sousa[1], G A Andrade[2], R C de Castro[2], L K Lima[2], C G Lima[2], R F Nascimento[1]

[1]Departamento de Química Analítica e Físico-Química, Universidade Federal do Ceará, R. Humberto Monte, S/N, Campus do Pici, Bl. 935, CEP: 60451-970, Fortaleza-CE-Brazil

[2]Fundação Núcleo de Tecnologia Industrial do Ceará -NUTEC, R. Rômulo Proença S/N, Campus do Pici, CEP:60451-970, Fortaleza-CE, Brazil

E-mail: maria.milhome@nutec.ce.gov.br

Abstract. QuEChERS method and detection in GC/SQ-MS were validated for the analysis of 35 pesticides in melon. Validation parameters (selectivity, linearity, LOD, LOQ, accuracy and precision) were determined according ABNT NBR 14029:2005. The recoveries rates for all the pesticides studied were from 63% to 117% with relative standard deviation (RSD) lower than 15% in the concentration range of 0.05 - 0.20 mg/kg. The limit of quantification (LOQ) for most compounds were below the MRLs established in Brazil. The combined relative uncertainty (*Uc*) and expanded uncertainty (*Ue*) was determined using repetitivity, recovery and calibration curves data for each pesticide.

1. Introduction

Multiresidue analysis of complex matrices such as fruit requires constant modifications and optimizations methodologies and therefore need to be validated. Agencies like ANVISA and INMETRO, require validation of analytical methods as a key requirement in the accreditation for quality assurance and demonstration of technical competence.

According to ANVISA, validation should ensure through experimental studies that the method meets the requirements of analytical applications, ensuring the reliability of the results [1]. The analytical parameters normally determined for validation of methods of separation are: selectivity, linearity, precision, accuracy, limit of detection, limit of quantification and robustness. One of the requirements of ABNT NBR 14029:2005, is the determination of the uncertainty associated with the results [2].

According EURACHEM the term uncertainty is "a parameter associated with the result of a measurement, which characterizes the dispersion of the values that could reasonably be attributed to the measurand [3]. In estimating the overall uncertainty, it may be necessary to take each source of uncertainty and treat it separately to obtain the contribution from that source. Each of the separate contributions to uncertainty is referred to as an uncertainty component. When expressed as a standard deviation, an uncertainty component is known as a standard uncertainty.

The gas chromatography coupled to quadruple mass spectrometers (GC-QP/MS) has been used for multiresidue analysis due to its good resolution, selectivity and sensibility expected given purpose, in addition to its versatility, featuring condition in different matrices [4]

Recently several articles related to multiresidue method validation in fruits has been published [5-6]. However, few articles report the determination of measurement uncertainty as parameter validation [7].

The work aims at validating and determining the uncertainty of multiresidue method in the matrix of melon and quantification by GC-QP/MS. The implementation of the method of analysis will expand the scope of laboratory analysis, allowing quality control of fruit, by monitoring the level of residues.

2. Materials and Methods

2.1. Chemicals, reagents and samples extraction

Certified standards for 35 pesticides studied were purchased from Sigma-Aldrich (Brazil) and Dr. Ehrenstorfer (Brazil). All the standards and reagents had purities exceeding 97.0 %. Melon samples were analyzed based on QuEChERS method described by Anastassiades *et al* [8].

A blank matrix of the melon were prepared by adding spiking solution to blank extract, to produce a final concentration of 0.02, 0.05, 0.1, 0.2, 0.5 and 0.75 $mg.Kg^{-1}$

2.2. Chemicals, reagents and samples extraction

It was used a gas chromatograph coupled to single quadrupole mass spectrometers (GC-SQ/MS, DSQII model, Thermo, USA), of the *Fundação Núcleo de Tecnologia Industrial do Ceará - NUTEC.* A Vertibond 50 m x 0.25 mm i.d. capillary column with a 0.10 μm film was used in the pesticide separation and helium as carrier gas at a constant flow of 1 $mL.min^{-1}$. The temperature program was the following: initial temperature 100 °C held for 1 min, 15 $°C.min^{-1}$ rate to 180 °C, then 4 $°C.min^{-1}$ rate to 280 °C and held for 14 min. The injection temperature was 250°C, and 1 μL volume was injected in splitless mode (45.3 minutes). The mass spectrometer was operated in electron impact (EI) mode, ion source temperature 200°C, MS Transfer Line 270 °C, electron multiplier voltage 1295 V, scanning from m/z 50 to 500 at 2.0 s per scan; solvent delay 6.0 min.

Quantitative analysis was performed in the selected ion monitoring mode (SIM) based on the use of one ion target (T) and two or three ions qualifiers (Q1, Q2 and Q3).

2.3. Validation and Uncertainty

A blank matrix of the melon were prepared by adding spiking solution to blank extract, to produce a final concentration of 0.02, 0.05, 0.1, 0.2, 0.5 and 0.75 $mg.Kg^{-1}$

Limit of detection (LOD) and Limit of quantification (LOQ) were estimated in the SIM mode analysis as the lowest concentration injected that yielded S/N ratio of 3 and 10, respectively.

Accuracy and precision of the method were tested with recovery experiments, performed with five replicates of blank samples spiked with pesticides at 0.05, 0.10 and 0.20 $mg.kg^{-1}$ and determination of relative standard deviation (RSD, %).

Uncertainty was determined for all of the pesticides, according to the procedures recommended by EURACHEM/CITAC Guide CG 4 [3]. The uncertainty of measurement was obtained at the level of 0.05 $mg.Kg^{-1}$ using validation data. Three sources of uncertainty were taken into account: uncertainty associated with precision (u_1), uncertainty associated with bias (u_2) and uncertainty associated with the calibration curve (u_3). The combined relative uncertainty (*Uc*) was calculated using the expression:

$$Uc = (u_1^2 + u_2^2 + u_3^2)^{1/2} \quad (1)$$

The expanded uncertainty (*Ue*) was obtained by multiplying *Uc* by a coverage factor $k=2$ (confidence of 95%).

3. Results and Discussion

3.1. Method Validation

The standard matrix was used for determining the linearity range of pesticides studied. The correlation coefficients obtained were all above 0.99 and the linearity range from 0.02 to 0.75 mg Kg^{-1} was considered satisfactory.

The LOD and LOQ values obtained ranged from 0.005 - 0.050 $mg.Kg^{-1}$. The limit of quantification (LOQ) for most compounds were below the MRLs established in Brazil.

The recoveries for all the pesticides studied were from 63% to 117% with relative standard deviation lower than 15%. According to ABNT NBR 14029:2005, the range of acceptable recovery level of concentration of 0.01 $mg.kg^{-1}$ is 60 - 115%, and 0.1 $mg.kg^{-1}$ is 80-110% [2]. The ANVISA recommended recovery percentages of 70 - 120% (> 0.01 - 0.1 $mg.kg^{-1}$) and RSD $\leq$ 20% [1].

The QuEChERS method was used by Barakat *et al,* in the extraction of 36 pesticides in honey samples, using the GC-ECD and GC-NPD systems. Afforded recovery percentage between 70 and 120% for RSD $\leq$ 22% [9]. Nguyen *et al,* achieved satisfactory recoveries (80% to 115%) for 107 pesticides in cabbage and radish [10]. Hernández-Borges *et al,* using the method QuEChERS and GC-NPD for the determination of 11 pesticides in banana obtained recovery percentages between 67 and 118% with RSD $\leq$ 16% [11].

3.2. Measurement of uncertainty analyses

Uncertainty of measurement represents a quantitative indicator of the reliability of the analytical results, expressed as a range which is estimated to be the real value, usually associated with a confidence level [3]. In an analytical procedure, the uncertainty about the result may arise from many possible sources, including sampling, matrix effects, environmental conditions, glassware, method of measurement and random variation. In this work, the combined relative uncertainty (*Uc*) and expanded uncertainty (*Ue*) was determined using repetitivity, recovery and calibration curves data for each pesticide. The *Uc* and *Ue* values ranged from 4.0 - 18.1% and 7.9 - 36.1%, respectively. On the basis of the expanded uncertainties values, the pesticides could be classified into three groups: group I (*Ue*<10%), group II (*Ue*:10–20%) and group III (*Ue* >20%). The results are showed in Figure 1.

About 74% of the pesticides analyzed showed *Ue* ranging from 10-20% (Group II). The pesticides chlorpyrifos, pyraclostrobin, deltamethrin and azoxystrobin had low uncertainty values (*Ue* <10%) and were classified in Group I. *Ue* values considered high (>20%) were observed in 14% of pesticides. Chloroneb and metalaxyl showed *Ue* >30%.

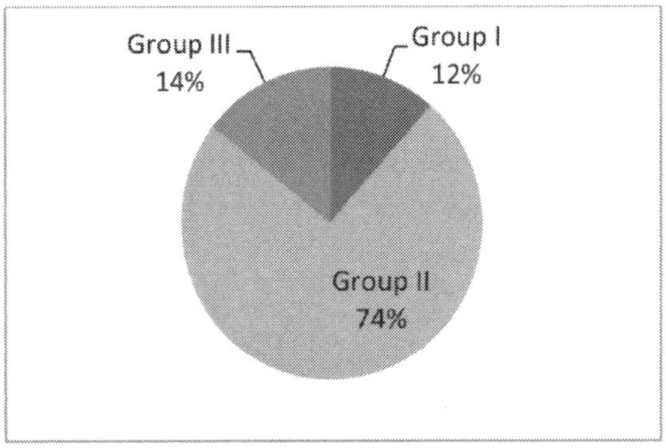

Figure 1- Classification of pesticides into groups based on the values of expanded uncertainty (*Ue*) group I (*Ue* <10%), group II (*Ue*:10-20%) and group III (*Ue* >20%).

The repeatability was the largest contribution to the measurement uncertainty. Other sources, such as purity of the reference standard feature small contribution in uncertainty values, not exerting considerable influence on the final value.

Frenich *et al* used LC-MS-MS to validate a method for the analysis of 31 multiclass pesticide residues. The uncertainty associated to the analytical method was lower than 23% for all compounds tested [11]. Kanrar *et al* obtained measurement uncertainty of less than 20% of method validation of 42 pesticide residues in tea [12].

Conclusion

The proposed multiresidue method was successfully validated and can be applied to monitoring of 35 pesticides in melon. The method offers low measurement uncertainty (≤ 20%), indicating the adequacy of the requirements by accrediting agencies. The repeatability was the largest contribution to the measurement uncertainty.

Acknowledgements

The authors gratefully the FUNCAP e CNPq for financial support of this search and the NUTEC for providing the infrastructure of laboratories for chromatographic analysis

References

[1] ANVISA- Agência Nacional de Vigilância Sanitária; Guia para o controle da Qualidade para análise de resíduos de agrotóxicos em alimentos para os laboratórios integrantes do PARA (2007) http://www.anvisa.gov.br/reblas/para/CONTROLE_QUALIDADE.pdf.>Accessed: 10.05.2012.

[2] ABNT- Associação Brasileira de Normas Técnicas, Agrotóxicos e Afins – Validação de métodos Analíticos, NBR 14029, 2005.

[3] EURACHEM/CITAC, Guide CG 4, Quantifying Uncertainty in Analytical Measurement, second ed. 2000, URL http://www.measurementuncertainty.org/mu/QUAM2000-1.pdf. Accessed: 10.06.2012.

[4] Yang X, Zhang H, Liu Y, Wang J, Zhang Y C, Dong A J, Zhao H T, Sun C H and Cui J 2011 *Food Chemistry* **127** 855-65

[5] Menezes Filho A, Santos F N and Pereira F A P 2010 *Talanta* **81** 346-354

[6] Koesukwiwat U, Lehotay S J, Miao S and Leepipatpiboon N 2010 *Journal of Chromatography A* **1217** 6692-6703

[7] Kanrar B, Mandal S and Bhattacharyya A 2010 *Journal of Chromatography A* **1217** 1926-33

[8] Anastassiades M, Lehotay S J S, Tajnbaher D and Schenck F J 2003 *Journal of AOAC International* **86** 412-431

[9] Barakat A A, Badawy H M A, Salama E, Attallah E and Maatook G J 2007 *Journal of Food, Agriculture & Environment* **5** 97-100

[10] Nguyen T D, Yu J E, Lee D M and Lee G 2008 *Food Chemistry* **110** 207–213

[11] Hernández-Borges J, Cabrera J C, Rodríguez-Delgado M A, Hernández-Suarez E M and Saúco V G 2009 *Food Chemistry* **113** 313-319

[12] Frenich A G, Vidal J L, Lopez T, Aguado S and Salvador I 2004 *Journal of Chromatography A*, **1048** 199–206

Comparison of methods for accurate end-point detection of potentiometric titrations

R L A Villela[1,4], P P Borges[2] and L Vyskočil[3]

[1] Acoustics and Vibration Metrology Division, National Institute of Metrology, Quality and Technology - Inmetro. Av. Nossa Senhora das Graças, 50, Duque de Caxias, RJ, CEP 25250-020, Brazil.
[2] Chemistry Metrology Division, National Institute of Metrology, Quality and Technology - Inmetro.
[3] Slovak Institute of Metrology-SMU. Karloveská 63, SK-84255 Bratislava 4 Slovak Republic.

E-mail: rlvillela@inmetro.gov.br

Abstract. Detection of the end point in potentiometric titrations has wide application on experiments that demand very low measurement uncertainties mainly for certifying reference materials. Simulations of experimental coulometric titration data and consequential error analysis of the end-point values were conducted using a programming code. These simulations revealed that the Levenberg-Marquardt method is in general more accurate than the traditional second derivative technique used currently as end-point detection for potentiometric titrations. Performance of the methods will be compared and presented in this paper.

1. Introduction

Titration is one of the most utilized techniques for measuring the amount of an analyte by considering the stoichiometry relationship in a chemical reaction.

Potentiometric titration curve are represented by a plot of the indicator-electrode potential (y) versus the volume (or mass) of the titrant. For obtaining the end point of a titration, the first and second derivative plots of the potential as a function of the volume (or mass) of the titrant is normally used [1]. This method has no good accuracy due to its high susceptibility to measurement uncertainty of y and insensitivity to the number of data points around the end point.

In this paper is proposed a non-linear regression of measurement data from potentiometric titrations by the Levenberg-Marquardt algorithm [2] for end point determination. This non-linear regression can be used to determine the end point with better accuracy. For example, in potentiometry a 0.25 mV error represents a 1% relative error in the concentration of the analyte [1]. This can be a huge problem for characterization of reference materials, since it has influence in the accuracy of the property value measured due to experimental errors.

A code was developed in LabVIEW® (a proprietary development environment for programming) to determine the end point from acid-base coulometric titration with potentiometric detection. This paper aims to present the performance and comparison of the two methods, namely the traditional second derivative (SD) technique and the Levenberg-Marquardt (LM) approach.

[4] To whom any correspondence should be addressed.

2. Implementation of the methods

2.1. Potentiometric titration data

The coulometric titration data were obtained by using the primary system of coulometry from Inmetro [3]. The experimental data are related both to the initial and to the final acid-base coulometric titration. The end points were calculated through the software developed using both the SD method and the LM algorithm.

2.2. Second derivative method

This method consists of the estimation of the end point by the second derivative that is obtained numerically. It was implemented in the software that reads the n experimental data points, time (t_i), in seconds, and pH (y_i), where i = 1, 2, ..., n. The n–2 points of the second derivative (y_j'') are used to estimate the value of the end point (t_e), that is computed considering that $y''(t_e) = 0$ and using a linear interpolation of t_j and t_{j+1}, where $t_j \leq t_e \leq t_{j+1}$ and $y''(t_{j+1}) \leq y''(t_e) \leq y''(t_j)$. The respective value of pH, $y(t_e)$, was calculated from the linear interpolation of $y(t_i) \leq y(t_e) \leq y(t_{i+1})$ and $t_i \leq t_e \leq t_{i+1}$ (figure 1).

Unfortunately, in many cases this technique cannot provide a measurement result with low uncertainty because numerical derivatives are very prone to measurement noise (random error). Furthermore, this method requires two successive differentiations and uses only two points to estimate t_e from the second derivative data. Therefore a better approach to reduce the uncertainty of end-point estimations is reasonably expected.

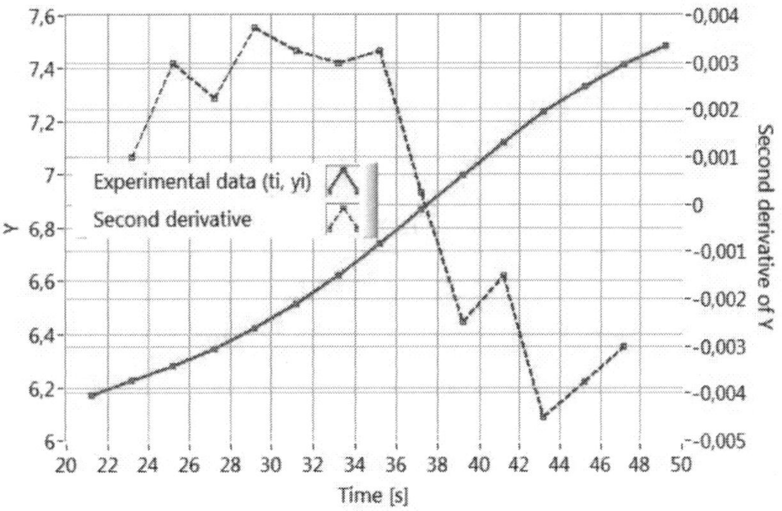

Figure 1. Experimental data and second derivative both linearly interpolated.

2.3. The Levenberg-Marquardt Method

An empirical equation (1) presented at the first time by Vyskočyl and Mathiasová [4] was used in the LabVIEW® code to model the electrochemical reaction occurring in coulometric titrations when the detection of the end point is obtained by potentiometry.

$$y = \frac{a}{1 + e^{k(t-q)}} + b(t - q) + c \tag{1}$$

Where: y is the electrode potential; t, time for the titration reaction (may be volume or mass); c, position of the titration curve; b, slope of the titration curve; q, position of the end point; k, width of the titration curve and a amplitude of the titration curve.

An initial guess of the q value can be obtained from the maximum value of the first numerical derivative. It can be easily demonstrated, according to equation (2), that $t_e = q$ is the end-point value.

$$\frac{\partial^2 y(t_e)}{\partial t^2} = \frac{ak^2 e^{k(t_e+q)}\left(e^{kt_e} - e^{kq}\right)}{(e^{kt_e} + e^{kq})^3} = 0 \qquad (2)$$

The program implemented reads the n experimental data points, time (t_i) and pH (y_i), where $i = 1, 2, \ldots, n$, computes numerically the first derivative to specify the initial guess for the parameter q (maximum value) and uses the LM algorithm to determine the set of parameters that best fit the set of experimental data points. The best fit parameters minimize the weighted mean squared error between the observations y_i and the best nonlinear fit. From the parameters a, c and q, the end point (t_e) and the respective value of pH (y_e) are obtained from equations (3) and (4).

$$t_e = q \qquad (3)$$

$$y_e = a/2 + c \qquad (4)$$

The stopping conditions for the fitting process were set to 5×10^5 for the maximum number of iterations of the fitting routine and the value of 10^{-10} for the tolerance that specifies the relative change in the weighted distance between y_i and the fit curve. The best fit curve and the experimental data are shown in figure 2.

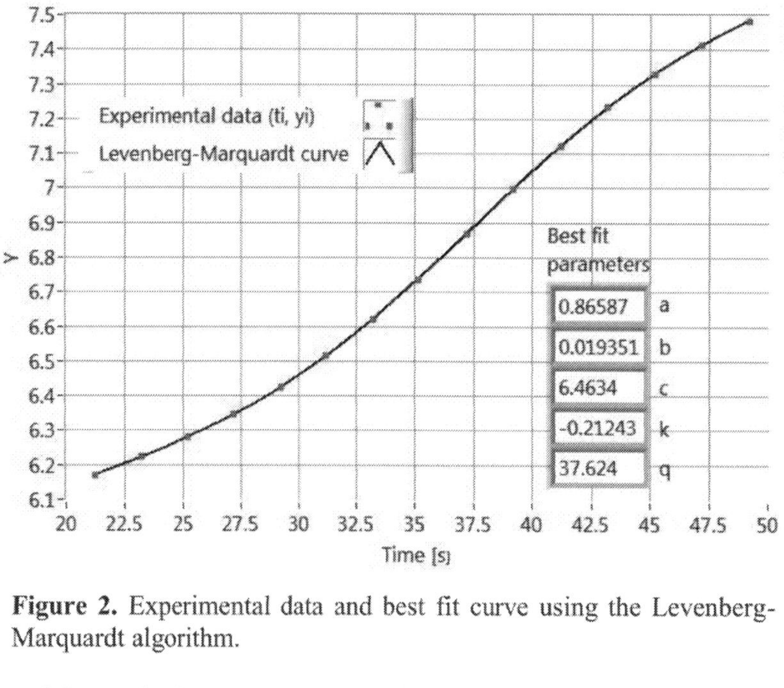

Figure 2. Experimental data and best fit curve using the Levenberg-Marquardt algorithm.

3. Comparison of the methods

3.1. General procedure

The comparison between the SD technique with the LM method was carried out taking into account their performance. This comparison was undertaken through the evaluation of result deviations obtained from simulated data points t_i and y_i.

It was considered that the equation (1) describes the electrochemical reaction and the measurement data y_i can be mathematically modeled by equation (5).

$$y_i = \frac{a}{1 + e^{k(t_i - q)}} + b(t_i - q) + c + e_i \tag{5}$$

Where e_i is the measurement error of y_i and a normally distributed random variable with zero mean and standard deviation σ_e, where $i = 1, 2, ..., n$.

When the σ_e and all parameters of the equation (5) are specified, the value of each y_i can be computed (simulated data) and the true values of t_e and y_e can be calculated from equations (3) and (4) for a given set of data points t_i and y_i generated.

Accordingly, the generated values of t_i and y_i can be used by the SD and LM methods for estimation of t_e and y_e and finally to compare with the true values of t_e and y_e. The assessment of method accuracy for a specific data set t_i , y_i where $i = 1, 2, ..., n$, can be carried out computing the method errors $e(t_{eSD})$, $e(y_{eSD})$, $e(t_{eLM})$ and $e(y_{eLM})$ presented at the flow diagram in figure 3.

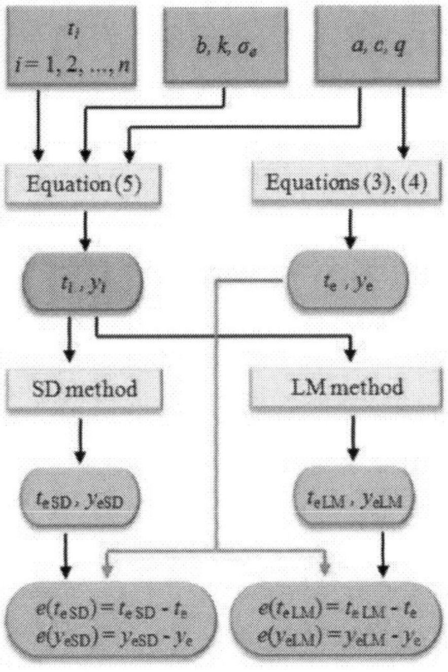

Figure 3. Flow diagram for calculation of the resultant errors of t_e and y_e due to the errors e_i.

In summary, a simulation of experimental data consisting of n points t_i , y_i , where $i = 1, 2, ..., n$, provides the errors of t_e and y_e for both methods: $e(t_{eSD})$, $e(y_{eSD})$, $e(t_{eLM})$ and $e(y_{eLM})$. Consequently, if a large number of simulations are performed the probability distribution, mean and standard deviation of the errors can be computed for an error analysis.

3.2. Parameter specifications

If a simulation for an error analysis is conducted using only a specific set of values of a, b, c, k, q, the conclusion of the analysis (the best method for end-point detection) will be very particular to the case tested. In order to achieve some generality, 1×10^5 simulations were performed, where each simulation s has its own parameters a_s, b_s, c_s, k_s, q_s, t_{is} (n is constant) and resultant errors $e_s(t_{eSD})$, $e_s(y_{eSD})$, $e_s(t_{eLM})$,

$e_s(y_{eLM})$. Moreover, to obtain representative values of a_s, b_s, c_s, k_s, q_s, t_{is} and σ_e for all simulations (s = 1, 2, ..., 1×10^5) many data sets of titration experiments were used to determined their minimum and maximum values. This was computed through the LM method and was crucial to avoid unrealistic simulations. In each simulation, the parameters a_s, b_s, c_s, k_s, q_s were generated as random variables according to a uniform distribution with their respective minimum and maximum limits (table 1).

Table 1. Parameter limits used for error analyses.

Parameter	Min value	Max value
a	0.76	0.99
b	0.015	0.028
c	6.4	6.5
k	-0.3	-0.21
q	27	49

3.3. Error analyses

Each error analysis consisted in 1×10^5 simulations of n experimental data points to compute the errors $e_s(t_{eSD})$, $e_s(y_{eSD})$, $e_s(t_{eLM})$ and $e_s(y_{eLM})$ 1×10^5 times and the corresponding means and standard deviations. It was performed four error analyses to reveal the behavior of the methods in different conditions of σ_e and n (table 2).

After some examination of the analysis results the following conclusions were reached for the cases tested:

- For both methods, an increase in the value of σ_e causes an increase in the standard deviations of the errors of t_e and y_e.
- The means of the errors of t_e and y_e are negligible compared to their respective standard deviations. Thus, an unbiased estimation of the end-point value can be obtained by application of any of the two methods.
- The performance of the LM method was markedly superior than the SD method for most of the simulations performed. For all error analyses of t_e using fourteen data points, the absolute value of LM error was smaller than the respective absolute value of SD error in approximately 95% of the simulations.
- The LM method exhibited improved performance with an increase of the number of data points (n) while the SD method did not show this behavior.

Table 2. Error analyses of t_e for different conditions.

Error analysis		1	2	3	4
Standard deviation of e_i (σ_e)		0.0008	0.0016	0.0024	0.0008
Number of data points (n)		14	14	14	10
LM method	Mean of $e(t_e)$	6.48×10^{-6}	8.60×10^{-5}	9.30×10^{-5}	1.09×10^{-4}
	Standard deviation of $e(t_e)$	0.0145	0.02896	0.0429	0.02523
SD method	Mean of $e(t_e)$	2.30×10^{-4}	1.87×10^{-4}	-2.14×10^{-4}	-6.66×10^{-5}
	Standard deviation of $e(t_e)$	0.2163	0.3909	0.5088	0.2161
Percent of simulations $\lvert e_s(t_{eLM})\rvert \leq \lvert e_s(t_{eSD})\rvert$		95.0	94.8	94.5	91.9

4. Conclusions

The error analyses revealed that in most cases the Levenberg-Marquardt method compared to the second derivative method reduces drastically the error of the end-point detection from potentiometric titrations. In consequence, the application of the Levenberg-Marquardt method can increase the accuracy of the end-point value from potentiometric titration data and contribute to a better characterization of reference materials by using coulometric titration with potentiometric end-point detection.

References

[1] Sawyer D T, Sobkowiak A and Roberts Jr J L 1995 *Electrochemistry for Chemists* 2nd edition (New York:John Wiley & Sons).
[2] Marquardt D W 1963 *J. Soc. Indl. Appl. Mathematics* **11** 431.
[3] Borges P P, Sobral S P, Silva L, Araujo T O and Silva R S 2011 *J. Braz. Chem. Soc.* **10** 1931.
[4] Vyskočyl L and Mathiasová A 1986 *8. Medzinárodné Sympózium Metrológie*. INSYMET'86.

Evaluation of sampling inhalable PM10 particulate matter (≤ 10 μm) using co-located high volume samplers

R R S Rajoy[1,2], J W C Dias[2], E C P Rego[2], A D Pereira Netto[3]

[1] National Institute of Metrology, Quality and Technology - INMETRO, Duque de Caxias, Rio de Janeiro, Brazil
[2] Energética Indústria e Comércio Ltda, Rio de Janeiro, Brazil
[3] Federal Fluminense University - UFF, Niterói, Rio de Janeiro, Brazil

Email: qualidade@energetica.ind.br

Abstract. This paper presents the results of the determination of the concentrations of atmospheric particulate matter ≤ 10 μm (PM10), collected simultaneously by six PM10 high volume samplers from two different manufacturers installed in the same location. Fifteen samples of 24 h were obtained with each equipment at a selected urban area of Rio de Janeiro city. The concentration of PM10 ranged between 10.73 and 54.04 μg m^{-3}. The samplers were considered comparable to each other, as the adopted methodology presented good repeatability.

1. Introduction

Aerosol may be defined as a stable suspension of solid and liquid particles in a gas. The atmospheric aerosol is formed by particles with a wide range of aerodynamic diameters (10^{-3} to 10^{2} μm) [1]. These particles result from innumerable sources and processes; therefore, the atmospheric particulate matter (APM) is a complex mixture containing a large number of organic and inorganic species, belonging to a broad variety of families of chemical compounds [2].

A well accepted classification of atmospheric particles employs the aerodynamic diameters as a parameter. This classification takes into consideration the penetration of the particles with 50% efficiency through inlets with cut points of 2.5 and 10 μm, so that two fractions results: PM2.5 and PM10, respectively. This classification procedure considers that the smaller the particle the greater will be their ability to penetrate deeper in the human respiratory tract, where they remaining for long periods, thereby causing high impacts on the human health [1].

The standards for ambient air quality (in terms of μg m^{-3}) were defined for some atmospheric pollutants in Brazil, by the Resolution number 3 of the Brazilian CONAMA (Conselho Nacional do Meio Ambiente) [3]. These standards include total suspended particles (TSP) and inhalable particles (PM10). This Resolution also states that the sampling method for the PM10 sampling should be "the high volume method coupled to a system for inertial separation of particles". The high volume sampler described in the method is non-automatic and contains a specially designed inlet with a cut point of 10 μm.[4] There are equivalent methods, and only those approved in compliance with criteria of the US EPA (Environmental Protection Agency of the US) are considered.

This study aimed obtaining metrological data of PM10. With this purpose, six PM10 samplers were located in the same area and operated simultaneously, for the same period. Therefore, this study allowed obtaining a broader knowledge of the process as well as of the variables, which mostly influence the measurements of PM10 concentrations. Although the study was not intended to assess the air quality in the sampling region, the obtained results can be used with this purpose, although they represent a short period campaign.

2. Materials and methods

2.1. PM10 samplers

Six PM10 high volume samplers were used in this study. Three samplers consisted of Thermo Scientific® PM10 inlets mounted on Energética® bases (motors) and three samplers consisted of Energética® PM10 inlet and bases. The inlet is the component through which air enters the sampler and where the particles carried by the air flow are fractioned by inertia at the design cut point of the inlet (i.e. 10 μm). The inlet contains several jet nozzles with appropriate inside diameter and an impaction plate for the retention of particles larger than 10 μm (Figure 1). The base contains – following the direction of the air flow – a filter holder, a Venturi type volumetric flow rate controller, and a motor blower for suctioning the air through the sampler.

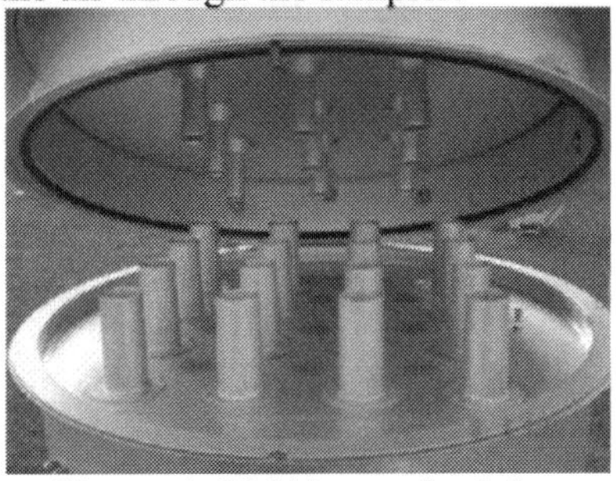

Figure 1. PM10 sampler inlet.

2.2. Sampling site description

The six samplers were installed in an open urban area in the neighborhood of Jacaré, in Rio de Janeiro City, Rio de Janeiro State, Brazil (22°53'30.S 43°15'04.W). The samplers were installed on wooden platforms (50 cm high), and distributed in a circular form, with a distance of approximately 2 meters from each other (Figure 2), without physical obstacles such as buildings within a radius of 2 m. In such way, the effects of sampling site differences and wind direction were minimized, following the recommendations of US EPA [5].

Figure 2. Sampling site layout.

2.3. Sample collection

The samples were simultaneously collected between October 16 and November 29 of 2009, on glass fiber filters (Energética E558X10IN), previously equilibrated during 24 h in conditioning chamber

with controlled temperature and humidity. They were weighed using an analytical balance with a sensitivity of 0.0001 g (Shimadzu, model AX200). The samplers were calibrated and operated according to the ABNT standard method (NBR 9547:1997) [6]. The samples were collected during 24 h periods. The meteorological data were obtained at every hour using a meteorological station installed near the sampling site (Met One Instruments, model 466A).
The mass of PM10 was determined as the difference between the average weights of each filter before and after the sample collection. The average weights were obtained from three repeated weight measurements and the volumes of sampled air were determined according to NBR 9547. The concentrations of PM10 were calculated by dividing the net weight of particulate by the volume of sampled air.

3. Results and discussions

3.1. PM10 concentration

The concentrations of PM10 during the studied period varied from 10.73 to 54.04 μg m^{-3}. The individual measurements obtained with the 6 samplers were always below the daily average primary standard (240 μg m^{-3}) and the daily average secondary standard (150 μg m^{-3}), adopted by the Brazilian CONAMA. These results indicate that, as far as PM10 is concerned, the sampling site is submitted to a little impact by particulate pollutants. The concentrations of PM10 measured with the samplers and in the different sampling dates are shown in figure 3. The result of November 2nd for sampler identified as 3-TS was considered an outlier, and, this data was excluded from the subsequent statistical analysis.

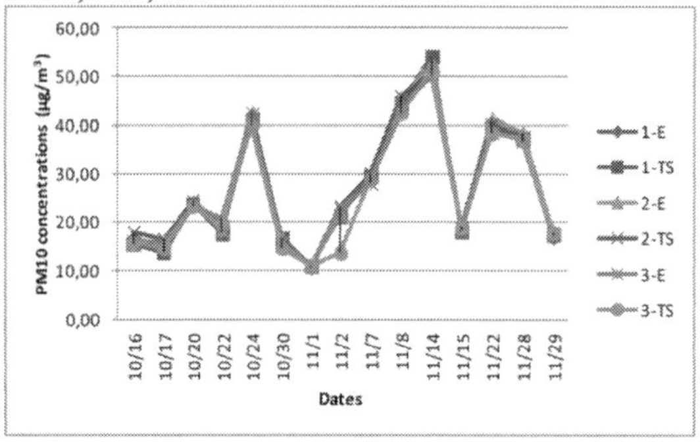

Figure 3. PM10 concentration measurements.

The PM10 concentrations measured with the six samplers were quite similar in the different days and showed the same pattern of variation throughout the period of testing (Figure 3). This demonstrated that the sampling process is repetitive, even in the different conditions (different days) that take place at the testing site, especially taking into account that the PM10 concentration can only be assessed with the different samplers when the environmental conditions remain unchanged.

The results obtained are fundamentally important as they indicate that the sampling tests with only one sampler at the site, as it is commonly done, may lead to reliable results.

3.2. Comparison among the PM10 samplers

The performances of the two groups of equipment were verified using parameters established by the US EPA [5]. In such comparison, the instruments manufactured by Thermo Scientific® (designated as TS) were considered as reference, since they are certified by the US EPA, and those manufactured by Energética® (designated as E), as candidate.

The results obtained for both groups of equipments are presented and compared in figure 4. The values of the slope (a), the intercept (b) and the determination coefficient (R^2) comply with the prerequisites ($a = 1 \pm 0.1$; $b = 0 \pm 5$; $R^2 > 0.97$), thus indicating that there is no difference in the results obtained with the two sets of samplers.

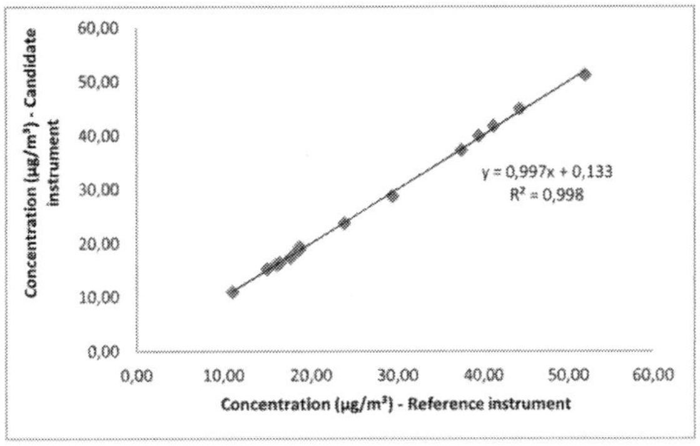

Figure 4. Evaluation of data correlation between the two groups of samplers.

The study was conducted in only one site due the difficulty of finding appropriate and safe locations, thereby limiting the range of concentration measurements. However, since the lower concentrations of particulate matter are more affected by the variability of the sampling systems and of the collection procedure, it was inferred that the resulting range of concentrations have actually strengthened the adequacy of the procedures. Therefore, it was concluded that the measurements with reference group and those with the candidate group turned out on equivalent results.

3.3. Comparison among the dates of sampling tests

Considering that the samplers provide equivalent results, these results were then compared by a single factor analysis of variance (ANOVA), in which the tested factor was the difference between the dates. The value obtained for the calculated F (1344.2) turned out to be larger than the tabulated F (1.86), thus indicating that the results between the dates are different. This outcome was expected as there is variation in the environmental conditions along the days of sampling tests. The relative standard deviation of each sampling day varied from 1.25 to 6.18%, thereby remaining below the maximum value established by the US EPA (7%). This indicated that the results are repetitive on each sampling day, when the non-homogeneous factors of the sampled environment and the variations inherent to the operator are not significant, and that the differences detected by ANOVA among the sampling days ensue from the differences taking place among the concentration values.

4. Conclusions

The results obtained in this study indicate variations among the data of PM10 during the collection days in a short period of time (15 days), since the environmental conditions undergo variation. However, an evaluation of this parameter at a particular point of collection, with six samplers, indicated good repeatability of the results, and, thus, the adequacy of the methodology used for the sampling of this pollutant, as well as the little influence of the operator. This study serves as a support for the estimation of uncertainties in sampling tests using high volume samplers with inertial separation, thus helping researchers and other professionals in the area. Furthermore, it shows that the results are highly comparable on every day, indicating that the results obtained with a single sampler may be considered reliable. Finally, the study demonstrates that the domestic sampling equipment lead to data comparable to those obtained using a certified and imported equipment.

References

[1] Colbeck I and Lazaridis M 2010 *Naturwissenschaften* **97** 117
[2] Alves C 2005 *Quim. Nova* **28** 859
[3] CONAMA – Resolução 003 1990
[4] ABNT – NBR 13412:1995
[5] USEPA 40 CFR Part 53 Subpart C
[6] ABNT – NBR 9547:1997

Journal of Physics: Conference Series **575** (2015) 012035 doi:10.1088/1742-6596/575/1/012035

Quality control of the analysis of histamine in fish by proficiency test

Warlley P. Evangelista[1], Patrícia A. S. Tette[1] and Maria Beatriz A. Gloria[1]

[1]Laboratório de Bioquímica de Alimentos – LBqA, Faculdade de Farmácia, Universidade Federal de Minas Gerais, Av. Antônio Carlos, 6627, Belo Horizonte, MG, 31270-901, Brasil

E-mail: warlley.pinheiro@yahoo.com.br

Abstract: The analysis of histamine is required by the European Union for the importation of tuna and other Scombroid fish. The aim of this study was to investigate the quality of the analysis of histamine in fish, by means of reference material (RM) and of proficiency test (PT). Sample analysis carried out using RM provided 89.4% recovery. During the proficiency test, the histamine content of the sample was 311.9 mg/kg and the z-score was zero. These results assure the good performance of the laboratory in the analysis of histamine in fish, asuring reliability of results to clients.

1. Introduction

Accredited laboratories and non-accredited ones which care about the quality of the testing results, often participate in national and/or international quality control programs. The analytical quality control can be undertaken by means of Certified Reference Materials (CRM) or by means of proficiency testing programs. These activities are very important to obtain formal recognition of the technical competence of the laboratory and they are also essential to confirm reliability and accuracy of the results. These tests are required by a Brazilian Government Institute named Coordenação Geral de Acreditação do Instituto Nacional de Metrologia, Qualidade e Tecnologia (Cgcre/Inmetro) as a requirement for laboratory ISO 17025 accreditation by NIT DICLA – 026 [1, 2]. The proficiency tests are intended to provide an independent assessment of the competence of participating laboratories. Along with the use of validated methods, proficiency testing is an essential element to the laboratory as a control quality activity [3, 4].

In the academic environment, there is a great discussion on the applicability of the principles of Good Laboratory Practices (GLP), because to obtain the accreditation a series of detailed and costly, evaluations are required, making it difficult. However, the use of the concepts of GLP by universities can support in the quality of the generated data. The principles don't need be implemented in its totality, but can guide the research. In theory, research and development, if properly documented and validated, can be accredited. Consequently, the research achieve higher reliability and data integrity through proper documentation [5].

Histamine is a neuro and vasoactive biogenic amine which, at high concentrations, can cause adverse effects to human health. The analysis of histamine in tuna is a requirement in the EU for the importation of fish. Recent studies indicated that tunas from the Brazilian coast which are used for

Journal of Physics: Conference Series **575** (2015) 012035 doi:10.1088/1742-6596/575/1/012035

exportation are of excellent quality. However, histamine poisoning outbreaks have been reported in the country [6]. CRM for analysis of histamine in tuna are not available in the national and international market. In this case, Reference Materials producted by the Food Analysis Performance Assessment Scheme (FAPAS®) can be used in proficiency tests [4].

The aim of this study was to investigate the quality of the analysis of histamine in tuna by means of reference material and proficiency test.

2. Material and methods

2.1. Samples

To ensure the effectiveness of the method through the recovery percentage and check the levels of histamine, there were analyzed a RM (canned tuna, FAPAS®) with known amounts of histamine. A sample of canned tuna with unknown histamine content was sent by FAPAS® for the proficiency test.

2.2. Reagents and solvents

Histamine (HIM) dihydrochloride and *o*-phthalaldehyde were purchased from Sigma Chemical Co. (St. Louis, MO, USA). The reagents were of analytical grade, except for the HPLC solvents (acetonitrile and methanol), which were chromatographic grade. The solvents were filtered through 47 mm diameter and 0.45 mm pore size membrane (Millipore Corp., Milford, MA, USA) for aqueous and organic solvents (HAWP and HVLP, respectively). For the preparation of the solutions, ultrapure water from Milli-Q Plus System (Millipore Corp., Milford, MA, USA) was used.

2.3. Determination of histamine

Histamine was extracted from the sample (5 g) with 7 mL of trichloroacetic acid (TCA) 5 %. The sample was homogenized on a vortex mixer (Biomatic, Brazil) for 70 sec and centrifuged (MR23i, Jouan SA, Saint Herblain, France) at 11,250 x *g* for 3 min at 0 °C. After centrifugation the supernatant was filtered on paper quality and the filtrate was collected in a 25 mL flask. The extraction was repeated two more times and the filtrates were combined. An aliquot of the extract was filtered through a HAWP membrane (13 mm, 0.45 mm, Millipore Corp., Milford, MA, USA) prior to HPLC analysis. Histamine was quantified by ion–pair HPLC using a reverse phase column (Nova-Pak® C18, 300 x 3.9 mm, 4 µm) and a guard column (Nova-Pak® C18, 20 x 3.9 mm, 4 mm) and fluorimetric detection at 340 nm excitation and 450 nm emission after post-column derivatization with OPA [7, 8].

The identification of histamine was possible by comparison of the retention time of the peaks found in the samples with that of standard solution of the amine in hydrochloric acid 0.1 mol/L. The quantification of histamine was performed by interpolation in an external calibration curve and multiplication by a correction factor [7, 8]. The analyzes were carried out in triplicate and the mean values and standard deviations were calculated.

3. Results

The mean levels of histamine found in RM and their respective percentages of recovery can be seen in table 1. The percent recovery ranged from 89.1% to 100.9%, since the mean concentration was certified 226.4 mg/kg. These results are within the acceptable range of recovery factor that is 80% to 110% for concentrations ≥ 10 μg/kg. Recovery aims to correct the result of the analysis of the systematic errors arising from the effects of extraction or digestion and losses from all stages of the analysis, performed by reading the instrumental response [9].

Table 1. Histamine contents and respective recoveries of reference materials

Reference Material	Date	HIM content (mg/kg)	Mean concentration certified (mg/kg)	Recovery (%)
1	January 10, 2013	213.4	226.4	94.2
2	January 23, 2013	201.6	226.4	89.1
3	April 03, 2013	202.4	226.4	89.4
4	May 08, 2013	313.9	311.1	100.9

Table 2 presents the mean level of histamine obtained in the analysis of the sample for proficiency test. This test was performed on April 3, 2012. The mean content was found to be 278.81 mg/kg. We also performed analysis of the RM. This had recovery of 89.4%. Considering this recovery, the histamine content of the sample corrected for proficiency test was 311.9 mg/kg.

The histamine content of the proficiency test sample was subsequently released by FAPAS® through report (FAPAS® Report 27101). According to the report, the expected result of the proficiency test was 311.1 mg/kg.

Table 2. Histamine content in the sample for the proficiency test (FAPAS®)

Sample FAPAS® (Proficiency Test)	HIM Content (mg/kg)
152/12 – Sample 1	282.06
152/12 – Sample 2	276.48
152/12 – Sample 3	277.88
Mean value	278.81

Among the participating laboratories, the Laboratório de Bioquímica de Alimentos (LBqA) had the best performance ($-2 \leq z \leq 2$), since the z-score obtained was 0.0. The z-score technique is adopted when you want to evaluate the performance of technical laboratories, especially in proficiency tests [10].

Of the 98 participating laboratories 56% achieved satisfactory performance ($-2 \leq z \leq 2$) and 44% had poor performance being: $|z| > 2$.

4. Conclusion

The Laboratorio de Bioquimica de Alimentos, LBqA - UFMG achieved satisfactory performance through RM test and had the best result in FAPAS® proficiency test for the analysis of histamine in fish. Therefore, the analysis of histamine in fish performed by LBqA is of excellent quality, ensuring confiability of results and customer confidence.

References

[1] Jardim, W F, and Sodré, F F 2009 Desempenho analítico de laboratórios prestadores de serviço na determinação de metais em água *Quim. Nova* **32** 1083-1088

[2] Inmetro. Instituto Nacional de Metrologia, Qualidade e 2013 *http://www.inmetro.gov.br* Accessed on 08/26/2013

[3] Nogueira, R and Soares, M A 2013 Accreditation and recognition programs in Brazil: current situations and perspectives *Accred Qual Assur* **18** 217-223

[4] FAPAS®. Food Analysis Performance Assessment Scheme - Proficiency Test Report 27101 2012 Histamine in Canned Fish

[5] Rodrigues, N R, Souza, A P F and Watanabe, M 2012 Implantação e implementação das normas das boas práticas laboratoriais (BPL) no laboratório de análise de resíduos da Universidade Estadual de Campinas *Quim. Nova* **35** 1276-1280

[6] Evangelista, W P Prevalence of histamine in scombroid fish poisoning and histamine in Brazil from 2007 to 2009 2010 Belo Horizonte – Faculdade de Farmácia - UFMG (Dissertation, MSc in Food Science) 70 p

[7] Vale, S R and Gloria M B A 1997 Determination of biogenic amines in cheese *J. AOAC Int* **80** 1006-1012

[8] Silva, T M Optimization and validation of a method for determination of histamine in fish 2008 Belo Horizonte – Faculdade de Farmácia - UFMG (Dissertation, MSc in Food Science) 103

[9] Ministério da Agricultura Pecuária e Abastecimento (MAPA) 2011 Manual de garantia da qualidade analítica – Secretaria de Defesa Agropecuária, Brasília – MAPA/SDA

[10] Chui, Q S H, Bishop, J M A and Iamashita, C O 2004 O papel dos programas interlaboratoriais para a qualidade dos resultados analíticos *Quim. Nova* **27** 993-1003

Matrix effect on the analysis of amphenicols in fish by liquid chromatography-tandem mass spectrometry (LC-MS/MS)

L R Guidi[1], P A S Tette[1], W P Evangelista[1], C Fernandes[2] and M B A Glória[1]
[1]LBqA - Laboratório de Bioquímica de Alimentos - Faculdade de Farmácia, UFMG, Av. Antônio Carlos, 6627, 31270-901, Belo Horizonte, MG, BRAZIL
[2]Laboratório de Controle de Qualidade de Medicamentos - Faculdade de Farmácia, UFMG, Av. Antônio Carlos, 6627, 31270-901, Belo Horizonte, MG, BRAZIL
E-mail: leticiaguidi@yahoo.com.br

Abstract. Matrix effect is an important parameter to be investigated during the development and validation of a method for the quantitative determination of contaminants in food. The objective of this study was to evaluate the matrix effect, through statistical tests, in the quantification of amphenicols in fish by HPLC-MS/MS. The study was performed by comparing the standard curves prepared in solvent solutions and in a fish sample previously known to be free of amphenicols. Since matrix effect was observed for the three analytes, calibration curves for quantification of chloramphenicol, thiamphenicol and florfenicol should be constructed using the matrix.

1. Introdution

Chloramphenicol and its analogues thiamphenicol and florfenicol are antibiotics with broad spectrum activity against bacteria. However, the clinical use of chloramphenicol was banned in many countries because of their severe side effects in humans such as aplastic anemia. Thiamphenicol and florfenicol appear to be viable substitutes for chloramphenicol in veterinary medicine. In order to ensure consumers' safety, it is necessary to develop sensitive and reliable methods for routine monitoring of these compounds.

Several chemical analyzes have been performed routinely in laboratories for the determination of a wide variety of substances. Frequently, the determination of the exact amount of the substance to be analyzed is necessary, and, in such cases, a quantitative analysis is performed. In these analysis calibration curves, which relate analyte concentration to the response in the detection system, are needed. This ratio corresponds to a performance parameter in the validation of an analytical method called linearity [1, 2].

It is possible to confirm linearity and to determine the range by the construction of analytical curves relating analyte concentration and analyte peak area obtained in the detection system. The range is defined based on the Maximum Residue Limit (MRL) or the Minimum Required Performance Limits (MRPL), choosing equidistant concentration levels, above and below these limits.

The experiments described for evaluation of linearity often involve preparation of calibration curves in solvents or in the matrix, mostly with five to six concentration levels, with a minimum of two to seven replicates per level [3].

A method proposed by [4] for evaluating the linearity consists on the following steps: (a) define the range of interest, where the midpoint of the range should correspond to the expected concentration in the sample; (b) prepare calibration solutions in solvent or in matrix (in case the matrix effect is significant), at least, six equally spaced concentration levels in three independent replicates (c) measure the response of the calibration solutions in a random order.

The sample matrix may contain components that interfere with the performance of the measurement by the detector selected, without causing a visible signal in the selectivity test. The interferences can increase or decrease the signal, and the magnitude of the effect can also depend on the concentration. Different samples, extracts and matrix concentrations may exhibit different levels of matrix effect. Then, a matrix which represents all these different levels of matrix effects should be selected during calibration curve construction [5]. When there is no significant matrix effect, the standard curve can be directly prepared in solvent [6]. If an effect is detected, the analytical curves should be constructed on the matrix. Several tests and their corresponding statistics can be used to study this effect.

The aim of this study was to evaluate the matrix effect, by statistical tests, on the quantification of amphenicols in fish by LC-MS/MS.

2. Material and methods

2.1. Material

Samples of *Tilapia* were collected in the consumer market of Itabirito, MG, Brazil.

Reagents were from analytical grade, except LC-MS/MS solvents, which were chromatographic grade. The ultrapure water used was obtained from Milli-Q Plus system (Millipore Corp., Milford, MA, USA). The internal standard used was deuterated chloramphenicol (CAP-d5). Standard solutions of thiamphenicol (1000 ppb), florfenicol (1000 ppb), chloramphenicol (10 ppb) and deuterated chloramphenicol (100 ppb) were prepared in methanol and stored at -20 °C.

2.2. Methods

2.2.1. Determination of the linearity

To determine the linearity tests were carried out with standard solutions in solvents and in the matrix using peak area [4]. The standard curves of amphenicols in solvent (water:methanol - 80:20 v/v) were built in the following concentration levels: chloramphenicol - 0.30, 0.90, 1.50, 2.10, 2.70 and 3.30 ng/mL; thiamphenicol - 30.0, 60.0, 90.0, 120.0, 150.0 and 180.0 ng/mL; florfenicol - 50.0, 100.0, 150.0, 200.0; 250.0 and 300.0 ng/mL. A graph was constructed relating analyte peak area/internal standard peak area versus the analyte concentration. Regression analysis was performed, providing the linear equation and the correlation coefficient for each analyte.

For each concentration level, three independent replicates were prepared and analyzed randomly to evaluate the behavior of the variances along the analytical curve. After construction of the standard curve, the regression parameters (intersection and slope) were calculated by the Method of Ordinary Minimum Squares, following the method proposed by [4].

2.2.2. Determination of the matrix effect

The study of matrix effect was achieved by comparing the analytical standard curves prepared in solvent and in the fish matrix. The average response of the analytes from the two calibration curves (in the same concentration range), obtained in the same experiment, were compared using the statistical tests F (Snedecor) of variance homogeneity and *t test* (Student) of averages comparison. To accept that the matrix effect is not significant, there should be no matrix effect at any concentration of the calibration curve [3].

3. Results

3.1. Determination of the linearity

The calibration curves of chloramphenicol, thiamphenicol and florfenicol in the solvent and in the matrix were constructed and evaluated for linearity. The curves were obtained from the relationship between the analyte concentration and the analyte peak areas/internal standard peak area. Outliers were treated and all assumptions regarding the Method of Ordinary Minimum Squares (normality, homoscedasticity and independence) were confirmed for the curve constructed in the matrix. The significance of the regression and the non-significant deviations from linearity confirmed the linear model and indicated the possibility of comparing the inclinations and the intersections of both curves by *t test* to evaluate the matrix effect (Table 1).

Table 1. Evaluation of model assumptions and linearity for the standard curves of chloramphenicol, thiamphenicol and florfenicol in fish.

	Values / Curve in					
	CAP		TAP		FF	
Statistics	Solvent	Matrix	Solvent	Matrix	Solvent	Matrix
Number of observations (n)	17	16	15	16	18	16
Normality						
R	0.992	0.997	0.974	0.991	0.984	0.967
P	p > 0.10	p > 0.10	p > 0.10	p > 0.10	p > 0.10	p > 0.10
Homoscedasticity						
tL	-1.176	-1.756	-1.827	-0.679	0.924	0.784
P	0.258	0.101	0.091	0.508	0.369	0.446
Independence						
D	2.278	1.525	2.018	2.337	1.766	1.635
P	p > 0.10	p > 0.10	p > 0.10	p > 0.10	p > 0.10	p > 0.10
Regression						
F	9591.08	6419.97	562.90	979.37	1178.03	878.57
P	1.81×10^{-22}	4.84×10^{-20}	4.36×10^{-12}	2.33×10^{-14}	2.05×10^{-16}	4.93×10^{-14}
Linearity deviation						
F	1.591	1.501	1.077	2.106	4.660	25.282
P	0.245	0.274	0.423	0.155	0.168	0.327

n = number of observations, R = Ryan-Joiner correlation coefficient, p = significance, tL = t-statistics of Levene, d = statistics of Durbin-Watson, F = variance ratio.

3.2. Determination of the matrix effect

The calibration curves for the three amphenicols in solvent and in the fish matrix are shown in Figure 1.

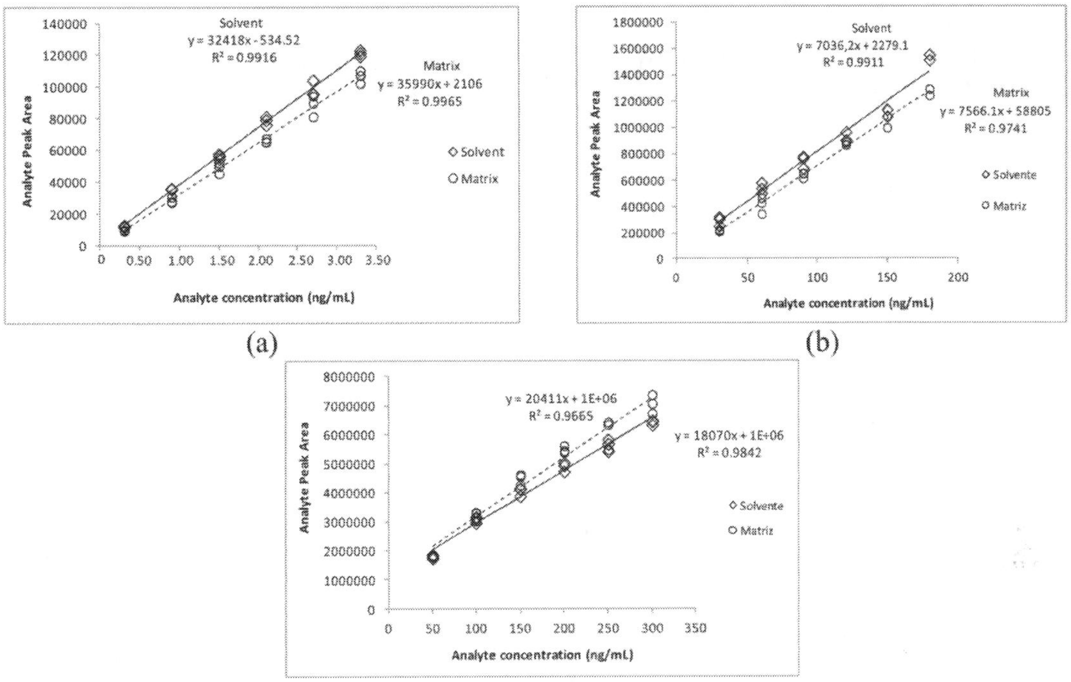

(a) (b) (c)

Figure 1. Calibration curves for: (a) chloramphenicol, (b) thiamphenicol (c) florfenicol in solvent and in the fish matrix in the ranges of 0.30 to 3.30 ng/mL, 30.0 to 180.0 ng/mL and 50.0 to 300.0 ng/mL, respectively.

Table 2 shows the comparisons of slopes and intersections of the calibration curves of chloramphenicol, thiamphenicol and florfenicol in solvent and in the matrix, respectively. By the t-test (Student), significant difference at 5% of probability between the slopes of the curves in solvent and in matrix was observed, indicating matrix effect for fish on the responses of chloramphenicol and florfenicol. For thiamphenicol there was significant difference at 5% probability in the slopes and intersections of the curves in solvent and in the fish matrix, indicating that there was matrix effect.

Table 2. Comparison of slopes and intersections of the chloramphenicol (0.30 to 3.30 ng/mL), thiamphenicol (30.0 to 180.0 ng/mL) and florfenicol (50.0 to 300.0 ng/mL) calibration curves in the matrix of fish and in the solvent.

	Results		
Statistics	CAP	TAP	FF
Comparison between intersections			
Ta	1.502	2.232	1.948
P	0.144	**<u>0.034</u>**	0.061
Comparison between slopes			
Tb	2.652	3.354	5.338
P	**<u>$p<0.05$</u>**	**<u>$p<0.05$</u>**	**<u>$p<0.05$</u>**

ta = t-statistic for contrasts between intersections, tb = t-statistic for contrasts between slopes, p = significance. Values in bold and underlined are significant at 5% probability.

4. Conclusion

Based on the results, matrix effect was observed for the three studied analytes chloramphenicol, thiamphenicol and florfenicol. Therefore, the calibration curves must be constructed using the matrix.

Acknowledgment

We acknowledge Conselho Nacional de Desenvolvimento Científico e Tecnológico (CNPq), Fundação de Amparo à Pesquisa do Estado de Minas Gerais (FAPEMIG) and Financiadora de Estudos e Projetos (FINEP) for the financial support.

References

[1] Eurachem 1998 The fitness for purpose of analytical methods. A laboratory guide to method validation and related topics 61

[2] BRASIL. Ministério da Agricultura Pecuária e Abastecimento (Mapa). Inst. Norm. nº 24. Guia para validação de métodos analíticos e controle de qualidade interna das análises de monitoramento do Plano Nacional de Resíduos e Contaminantes - PNCRC Animal. D.O.U. 2009

[3] Souza S V C 2007 Procedure for house validation of test methods: design and applicability in food analysis. Belo Horizonte – Faculty of Pharmacy UFMG (Thesis, DSc in Food Science) 296

[4] Souza S V C and Junqueira R G A 2005 A procedure to assess linearity by ordinary least squares method. *Anal. Chim. Acta* **552** 25-35

[5] Bruce P, Minkkinen P and Riekkola M L 1998 Practical method validation: validation sufficient for an analysis method *Mikrochim. Acta* **128** 93-106

[6] Thompsom M E, Ellison S L R and Wood R 2002 Harmonized guidelines for single-laboratory validation of methods of analysis (IUPAC Technical Report). *Pure and Appl. Chem.* **74** 835–855

Phase separation methodology for physicochemical studies of soils

C A Franchini[1], T L Oliveira[1], W G Teixeira[2], A Kuznetsov[1], C A Achete[1,3]

[1] Divisão de Metrologia de Materiais, Instituto Nacional de Metrologia, Qualidade e Tecnologia (INMETRO), Duque de Caxias, RJ 25250-020, Brazil
[2] Embrapa Solos, Rio de Janeiro, RJ 22460-000, Brazil
[3] Programa de Engenharia Metalúrgica e de Materiais (PEMM), Univ. Federal do Rio de Janeiro, Cx. Postal 68505, Rio de Janeiro, RJ 21945-970, Brazil

E-mail: okuznetsov@inmetro.gov.br

Abstract. A detailed study of a soil in every country is of a paramount importance, because it determines an entire economic strategy. The mineralogical properties of soils have been studied in the world for more than 100 years by various characterization techniques, X-ray diffraction being the most prominent. The main difficulty in most of employed techniques is the dominance of the majority phases in the response or the measured signal from the sample that becloud minority phases preventing their identification. The application of methods of phases' separation would provide the possibility to discern minority phases in soils. This work presents a phase separation method that employs a combination of two phenomena based on principles of fluid dynamics: flotation and sedimentation. Different characterization methods were used to analyse the produced soil samples. The methodology employed for separation of phases allowed the complete separation of clay phase from heavier mineral phases. This result makes it possible to discern minority mineral phases of soils that are difficult to detect. A more accurate determination of the mineralogical composition of a soil becomes feasible.

1. Introduction

Soils are composed of a mixture of numerous minerals, amorphous substances and organic complexes intimately associated in a variety of compositions in different areas of the planet. Inasmuch as the production of our food and the quality of our environment are related to the rational use of a soil, it is very important to thoroughly study its chemical mineralogy. Due to the great structural diversity, the soils' compositions are rarely identified and quantified in totality.

The X-ray diffraction is the standard technique for the identification of minerals. It allows reliable identification and quantification of mineral phases, especially when combined with such techniques as X-ray fluorescence and thermogravimetry. Over the last few years the transmission electron microscopy (TEM) has been playing an increasingly important role in this regard [1]. Even with the separation of phases, the identification of experimental diffraction peaks using the available databases does not allow a reliable identification of minerals. The use of computer modeling programs for powder X-ray diffraction patterns (e.g. via Rietveld method) increases the chances of identifying the different phases in a soil. To date, no attempts to combine the experimental procedures of phase separation of soils with computational methods of analysis of the phases within separated fractions were taken.

The separation of a series of mineral species, as singly as possible, prior to the application of various techniques for a soil's analysis, would be the key part to the identification of the exact nature of mineral soils. The methods for effective separation of phases that preserve their physical and chemical characteristics are rare and, in many cases, inefficient for detailed structural elucidation of mineral components.

Therefore, the aim of this study was to develop a methodology for a separation of mineralogical phases that enhances the capabilities of the techniques for a quantitative characterization of soils.

2. Experimental

2.1. Preparation of soil samples and phases separation

Soil samples from an Acrisol were collected on soil surface (0-20 cm depth) in the county of Mojú in the State of Pará in Brazil. This kind of soil is most representative of this region. Soil samples were initially crumbled and cleaned out of any unwanted wastes. The combination of two methods was used for the separation of phases: the flotation process coupled with the principles of fluid dynamics of particles. The terminal settling velocity, v, of an isolated particle with equivalent spherical diameter D_p moved by a constant force in the fluid of density ρ_f and viscosity μ_f is given by the empirical equation [2].

$$v = \frac{\mu_f}{D_p \cdot \rho_f} \cdot \left[\left(k_1 \frac{C_D Re^2}{24} \right)^{-1,20} + \left(\frac{C_D Re^2}{k_2} \right)^{-0,60} \right]^{-0,83} \quad (1)$$

where C_D is the particle's drag coefficient, Re is the Reynolds number, k_1 and k_2 are coefficients dependent on a particle´s shape. This equation allows estimating optimal speed for phase separation in a solvent with a given density. Upon this established methodology, it was possible to separate, with the help of a magnetic stirrer, the studied soil sample in 9 compositionally different solid fractions.

2.2. *Instrumentation*

The chemical composition of the sample was determined by X-ray fluorescence using the S4-Pioneer spectrometer from Bruker-AXS. The bottom fraction (fraction number 9) has been chosen then, in which one might expect the highest probability of dense minerals in order to elucidate the identity of these mineralogical phases.

The thermogravimetry (TG) coupled to mass spectrometry (MS) allowed the separation and identification of the released gases. Thermogravimetric analysis (TGA) data were recorded using a Universal V4.5 TA instrument coupled to a mass spectrometer.

X-ray diffraction analysis has been carried out in order to determine the crystalline phases present in the studied compounds. The powder diffraction patterns from fractions were collected on D8 Focus diffractometer from Bruker-AXS using Cu-Kα radiation (33 keV and 50 mA, Ni filter) with a 2θ step of 0.02° in the measurement 2θ range from 2° to 80°. The resulting patterns were fitted to the theoretical spectra using Rietveld structural refinement.

3. Results and discussion

The results of TG-MS are shown in Figure 1. The use of this technique in materials science and specifically in mineralogy has a long history. The analyzes of the integral soil samples performed herein are uncommon, mainly because of a soil being a mixture of mineral and organic matter, in which the organic component may mask characteristic peaks of strongly exothermic reactions of decomposition of certain minerals. The observed total mass loss of approximately 12% was partially (about 9%) attributed to dehydration and partially (~3%) originated from carbon compounds. A small percentage of the fragment ions with mass-to-charge ratio (m/z) of 46 indicates an eventual traces of nitrogen compounds. No ions having m/z=64 were observed, which otherwise would indicate the possible presence of sulfur compounds.

By looking at m/z=18 related to the desorbed water, one can see four regions of intense mass loss. The first region between 30° C and 200° C is related to the loss of adsorbed water in the soil. The temperature interval between 200° C and 360° C can be assigned to dehydroxylation of inorganic compounds [3]. The region between 380° C and 560° C is a characteristic region of dehydroxylation of clays of kaolinite type [4], whereas the highest temperature region is related to the decomposition of hydroxy-silicates.

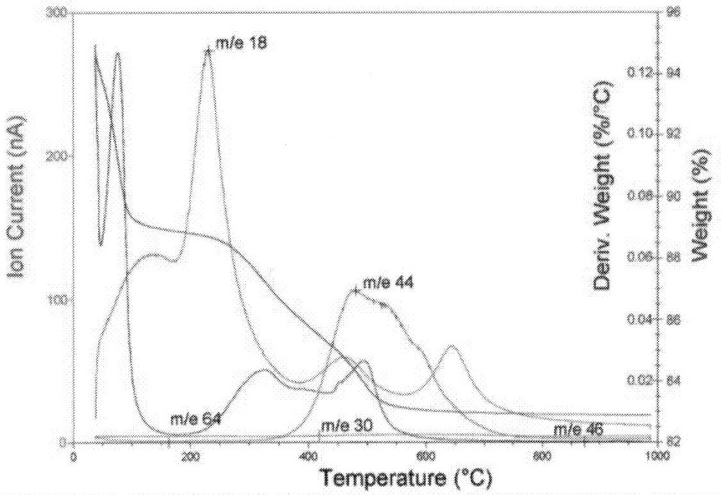

Figure 1. TG-MS curves of the soil sample. The blue line represents the weight loss and the red line shows its derivative. Green lines refer to the MS signal from the adsorbed molecular fragments identified by the respective m/z ratios.

The ions having m/z= 44, related to the desorption of CO_2, would be produced from decomposition of carbon composites. The thermal decomposition that took place below 500° C can be associated with a burning of organic carbon produced by living tissues, sap, microbes, fungi and animals. While a more stable carbon with the decomposition temperature range above 500° C would be of mineral origin. There is a big variety of complex chemical compounds of carbon; many have not been yet classified [5].

Ions with m/z=46 and 64, related to NO_2, and SO_2 fragments, respectively, were also examined. A small amount of nitrogen dioxide indicating nitrated compounds was detected while no MS signal was obtained for sulfur dioxide. This result does not discard the presence of sulfur-based compounds in the samples. Most probably, these substances are in very low quantities making their detection difficult.

Table 1 shows the results of XRF analysis. The results are semi-quantitative, showing only the relative amount of elements detected in the fraction 9 of the soil. This analysis when combined with XRD and TG-MS techniques, aims to assist in the elucidation of the possible phases.

Figure 2 shows the XRD pattern of the original soil sample, without separation. Two major phases, clay (61%) and quartz (39%) were observed, it being not possible to clearly distinguish any other mineral phase. The crystallographic phase of clay corresponds to kaolinite ($H_4Al_2SiO_9$).

Table 1. XRF analysis of the fraction 9 of the soil sample. Concentration of the elements is given in atomic %.

Element	Concent.	Element	Concent.	Element	Concent.	Element	Concent.
Si	48.2	**Au**	0.3	**Ru**	0.1	**Ba**	0.03
Al	29.8	**Zr**	0.2	**Cl**	0.1	**Mg**	0.02
Fe	12.5	**P**	0.2	**In**	0.08	**K**	0.01
Ti	6.5	**S**	0.2	**Bi**	0.07	**Pr**	0.01
Ca	0.4	**Ni**	0.1	**Zn**	0.05	**Th**	0.002
Cr	0.3	**Mn**	0.1	**Se**	0.05		
Cd	0.3	**U**	0.1	**La**	0.04		

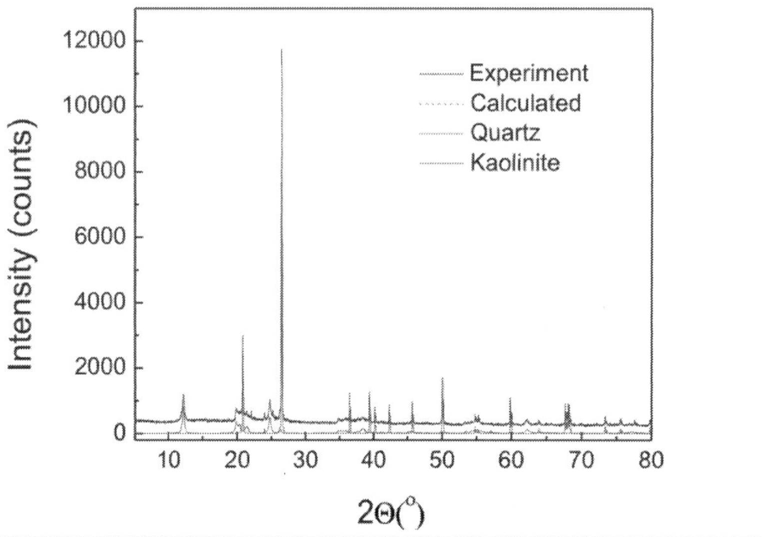

Figure 2. Experimental (black line) and theoretical (red line) XRD patterns of the initial soil sample. The green and the blue lines represent the calculated diffraction intensities produced by quartz and kaolinite phases.

The methodology for phases separation applied to the studied soil sample allowed us to separate it in nine fractions. Figure 3 shows the XRD pattern of the fraction 4, which composition consisted of more than 99% of clay phase (kaolinite) with a small amount of quartz (<1%), whereas the last fraction (number 9) shown in Figure 4 exhibits the complete withdrawal of the clay phase. Although the quartz phase dominates the composition of this fraction (75%), other mineral phases present in the soil become visible in the diffractograms, thus allowing their identification. To this end, we applied the Rietveld method which is based on the variation of instrumental, structural and microstructural parameters inserted into mathematical models that describe in the best possible way the whole experimental diffraction pattern. The excellent fit to the latter was obtained by modeling the majority phases.

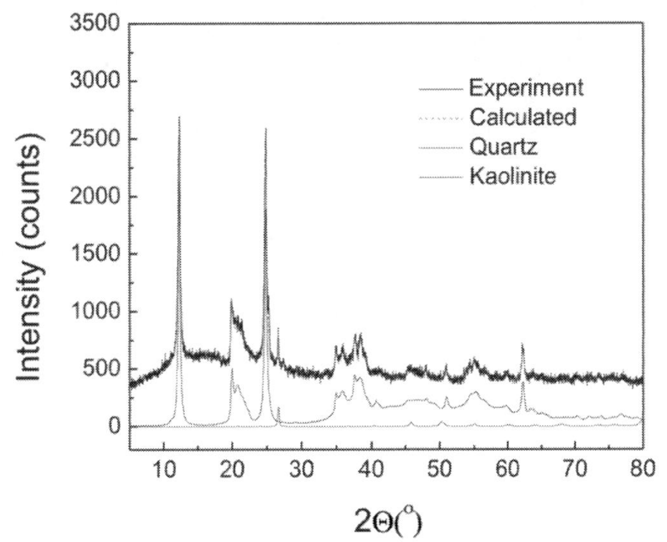

Figure 3. Experimental (black line) and theoretical (red line) XRD patterns of the fraction 4. The green and the blue lines represent the calculated diffraction intensities produced by quartz and kaolinite phases.

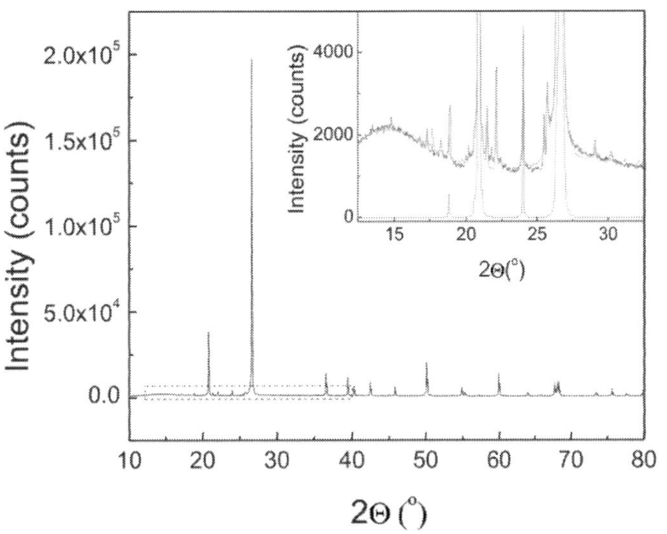

Figure 4. Experimental (black line) and calculated (red line) XRD patterns of the fraction 9. The inset shows the zoomed region of the diffraction pattern. The blue line in the inset denotes the calculated diffraction intensities produced by quartz. The diffraction intensities of the visible small peaks were calculated assuming the structural models of minority phases presented in the table 2.

Table 2 shows the result of possible phases present in the bottom fraction (fraction 9), applying this methodology of analysis. Nevertheless, some observed reflections remained not indexed with the supposed phases. This fact may indicate the presence of other minority phases in the bottom fraction.

Table 2. List of possible crystallographic phases found in the bottom fraction

Phase	Chemical Formula
Quartz	SiO_2
Pseudobrookite	MgO_5Ti_2
Olivine	$Fe_{1.249}Mg_{0.749}O_4Si$
Kalsilite	$KAlSiO_4$
Berlinite	AlO_4P
Prehnite	$Ca_2Al(AlSi_3O_{10})(OH)_2$
Gold	Au
Ferrocyanide	$Fe(CN)_2$
Lithium Aluminosilicate	$Li_{0.25}Al_{0.25}O_2Si_{0.75}$
Lithium Aluminosilicate	$Li_{0.33}Al_{0.33}O_2Si_{0.67}$

4. Conclusions

The methodology employed for separation of phases allowed the complete separation of clay phase from heavier mineral phases. This result makes it possible to discern minority mineral phases of soils that are hard or impossible to detect without an efficient separation. As a consequence, a more accurate determination of the mineralogical composition of a soil becomes feasible.

References

[1] Righi D, Elsass F, 1996 *Clays and Clay Min.* **44** 791
[2] Massarani G. *Fluidodinâmica em Sistemas Particulados*. Editora UFRJ, Rio de Janeiro, 1997
[3] Morgan D J, Warringtow S B, Warne S St J 1988 *Thermochim Act.* **135** 207
[4] Plante A F, Fernández J M, Leifeld J. 2009 *Geoderma* **153** 1
[5] Abd El-Tawab N A 2012 *Int. J. Conserv. Sci.* **3** 163

Absolute quantification of Bovine Viral Diarrhea Virus (BVDV) RNA by the digital PCR technique

R B Flatschart [1], D O Almeida [1], M B Heinemann [2], M N Medeiros [1], J M Granjeiro [1] and A V Folgueras-Flatschart [§1].

[1] INMETRO – Instituto Nacional de Metrologia, Qualidade e Tecnologia - Duque de Caxias-RJ, Brazil.
[2] EV/UFMG – Escola de Veterinária / Universidade Federal de Minas Gerais - Belo Horizonte-MG, Brazil.

[§]Email: avflatschart@inmetro.gov.br

Abstract. The quality control of cell lines used in research and industry is critical to ensure confidence in experimental results and to guarantee the safety of biopharmaceuticals to consumers. The BVDV is a common adventitious agent in many cell lines. We preliminarly evaluate the use of Digital Droplet PCR (ddPCR) for the detection and enumeration of genome copies of BVDV in cell culture and on FBS. The application of a commercial Real-Time PCR kit with the ddPCR technique was successful on different matrices. The technique allowed the absolute quantification of the genome without the use of calibration standards, suggesting its promising application on the development of reference materials for quantification of nucleic acids.

1. Introduction

With the uprise of therapeutic use of biopharmaceuticals produced in cell culture, the quality control of the producing cell lines assumed paramount importance to ensure its safety for users. The Bovine Viral Diarrhea Virus (BVDV) is an adventitious agent often associated with cell lines maintained in vitro [1] being propagated in culture during handling or introduced through the use of animal raw materials (fetal bovine serum, albumin, etc.) [2]. The BVDV infection on cells is often undetected [3], as sometimes it does not induce cytopathogenic changes.

Digital PCR is an emerging analytical technique used for the detection and quantification of specific sequences of nucleic acids. It can be performed without the use of any calibrant, having the potential to be considered as a primary method [4]. According to some National Metrology Institutes (NMI), when digital PCR is applied to the quantification of gene sequences or to the certification of DNA reference materials, the measurement values can be made traceable to the International System (SI), along with its associated uncertainty [5].

In this work we made a preliminary evaluation on the use of Digital Droplet PCR (ddPCR) for quantification of BVDV genome copy numbers on a commercial detection kit control, on cell culture supernatants used for virus propagation and on raw materials (fetal bovine serum) employed for the maintenance of cells in culture.

2. Materials and Methods

2.1. BVDV samples on different matrices
For virus on cell culture, supernatant of MDBK cells infected with the BVDV cytopathogenic strain NADL was collected. This material was produced at the Veterinary School of the Federal University of Minas Gerais - UFMG. For virus on fetal bovine serum (FBS), serum of naturally infected animal was used.

2.2. Viral RNA extraction
Total RNA from 500 μL supernatant of BVDV infected cell culture and from 500 μL FBS were extracted with Trizol, according to manufacturer's instructions (Life Technologies, USA). Concentration and overall purity of RNA was estimated, respectively, by UV spectrophotometry at 260 nm and by the ratio of readings at 260 nm and 280 nm.

2.3. Controls
Purified BVDV RNA from TaqMan BVDV and Xeno RNA Kit (Applied Biosystems, USA) was used as positive controls and Milli-Q water as negative control.

2.4. cDNA synthesis and ddPCR
For cDNA synthesis, RNA extracted from cell culture-produced BVDV (8 μL), from FBS (8 μL) and from positive control (2 μL) were processed according the instructions of the High Capacity kit (Applied Biosystems, USA). As negative controls, 8 μL of Milli-Q water was processed in parallel.

A 2 μL aliquot of the cDNA from the reverse transcription step of samples and controls were used for PCR amplification with the ddPCR Supermix for Probes kit (Bio-Rad, USA).

In all cases, the oligos used for cDNA synthesis, cDNA amplification and the FAM-labeled probes used for detection were from TaqMan BVDV kit (Applied Biosystems, USA).

3. Results and Discussion
The extraction, purification and cDNA synthesis scheme used for preparation of cDNA samples and controls proved to be valid. Also, the use of the High Capacity system with the TaqMan BVDV oligos along with ddPCR amplification with the same oligos allowed the detection of PCR products by the FAM probes included on the TaqMan BVDV system.

For the amplification step itself, given the composition differences among ddPCR Supermix and Real-Time PCR mixes, even an already standardized reaction would require further optimization. Thus, for the use of TaqMan BVDV oligos and probe on ddPCR, the optimum annealing temperature (60 °C) coincided with the one recommended by the manufacturer of the detection kit (Figure 1).

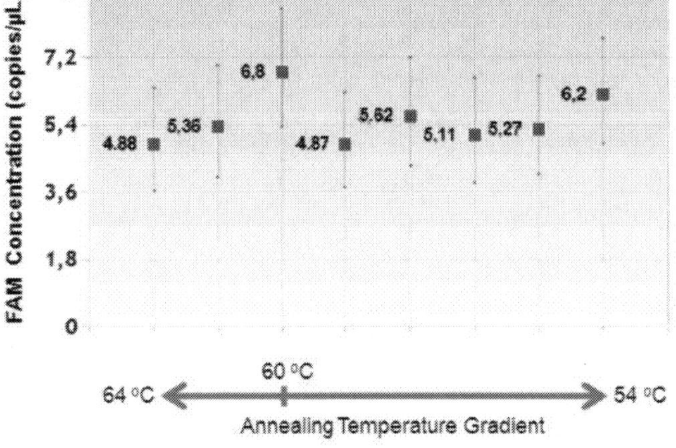

Figure 1 – Optimization of the ddPCR annealing temperature for BVDV cDNA using oligos and TaqMan probe from TaqMan BVDV. In a 54-64°C temperature gradient, 60°C was the optimum temperature for amplification. Annealing temperature versus number of FAM-labeled copies of BVDV cDNA/μL.

Once optimized, the ddPCR was able to amplify the virus-derived cDNA in samples from cells, fetal bovine serum and on the positive control. The negative control, derived from a RT reaction with water, gave no amplification signal. In addition to detect the presence of BVDV in different matrices, the ddPCR technique could quantify the specific viral nucleic acid present in each sample (Figure 2).

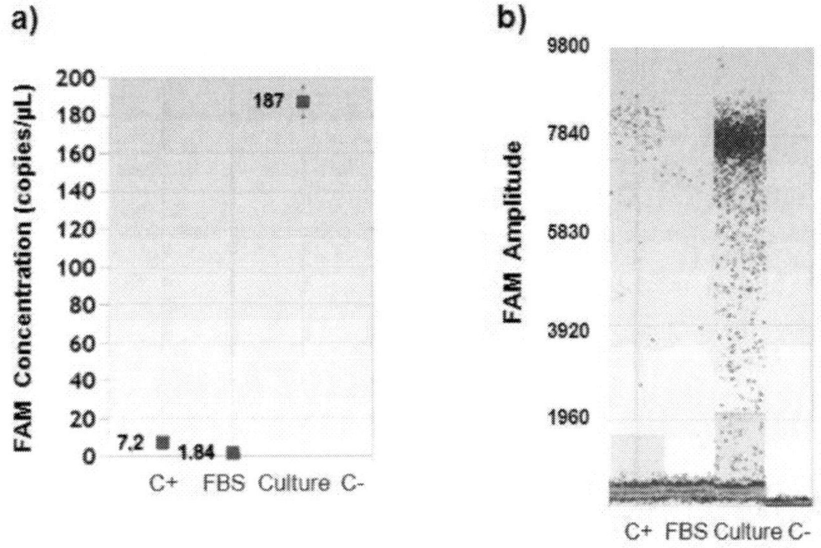

Figure 2 – Digital Droplet PCR quantification of BVDV cDNA synthesized from various matrices. Viral RNA samples from fetal bovine serum (FBS) and from cell culture (Culture) were quantified with BVDV specific oligos and probes. As positive control, cDNA was prepared from TaqMan BVDV control RNA. Water was used as negative control. In a) number of FAM-labeled.copies of BVDV cDNA/µL for tested samples and controls; in b) FAM fluorescence amplitude, which marks the difference between detection of BVDV and the lack of amplification signal for different samples and controls.

The quantification of BVDV RNA through the use of oligos and probe from TaqMan BVDV kit plus ddPCR Supermix proved to be suitable to test for adventitious agents in cell cultures and in raw material used on cell cultivation.

4. Conclusion

The experiments, while preliminary, shown the feasibility of adoption of commercial Real-Time PCR detection kits on a ddPCR scheme for de quantification of pathogens based on nucleic acid detection.

The ddPCR technique is easy to perform, with the advantage that absolute quantification of nucleic acid, expressed as copy number / µL does not require standards, not always available for this type of material, and overcome the need of any method or calibration like ΔΔCq.

With the experiment described here, the ddPCR showed to be an adequate tool on the development of reference materials for nucleic acid quantification on different matrices.

Acknowledgements

To Bio-Rad Laboratories do Brazil, in the persons of Carlos Martins and Fabiana Godoy, for its technical advising and for training on the use of QX-100 Digital Droplet PCR System.

References

[1] Cobo F, Stacey GN and Hunt C, in stem cell banks: approaches to standardization *Appl. Microbiol. Biotechnol.* **68** 456-66 Cabrera C, Nieto A, Montes R, Cortés JL, Catalina P, Barnie A and Concha A 2005 Microbiological control

[2] Pinheiro de Oliveira T F, Fonseca A A Jr, Camargos M F, de Oliveira A M, Pinto Cottorello A C, Souza Ados R, de Almeida I G and Heinemann M B 2013 Detection of contaminants in cell cultures, sera and trypsin *Biologicals* **41** 407-14

[3] Merten W 2002 Virus contaminations of cell cultures: a biotechnological view *Cytotech.* **39** 91-116

[4] Bhat S, Herrmann J, Armishaw P, Corbisier P and Emslie K 2009 Single molecule detection in nanofluidic digital array enables accurate measurement of DNA copy number *Anal Bioanal. Chem.* **394** 457-67

[5] Bhat S, Curach N, Mostyn T, Bains S, Griffiths R and Emslie K 2010 Comparison of methods for accurate quantification of DNA mass concentration with traceability to the International System of Units *Anal. Chem.* **82** 7185-192

Contributions to the revision of the *'Guide to the expression of uncertainty in measurement'*

G A Kyriazis

Instituto Nacional de Metrologia, Qualidade e Tecnologia
Av. Nossa Senhora das Graças, 50 – Duque de Caxias – 25250-020 – RJ - Brazil

gakyriazis@inmetro.gov.br

Abstract. Some inconsistencies of the current version of the *'Guide to the expression of uncertainty in measurement'* are discussed and suggestions to make this document consistent are commented. The paper is written taking into account the terminology of the third version of the *'International vocabulary of metrology'*.

1. Introduction

The Joint Committee for Guides in Metrology (JCGM) [1] was created in 1997 and is composed of representatives from the following organizations: Bureau International des Poids et Mesures (BIPM), International Electro-technical Commission (IEC), International Federation of Clinical Chemistry (IFCC), International Laboratory Accreditation Cooperation (ILAC), International Organization for Standardization (ISO), International Union of Pure and Applied Chemistry (IUPAC), International Union of Pure and Applied Physics (IUPAP) and International Organization of Legal Metrology (OIML). This committee is responsible for updating the *'International vocabulary of metrology'* (VIM) and the *'Guide to the expression of uncertainty in measurement'* (GUM) [2] (and its supplements). The VIM had its third version (VIM3), which was published in 2008 [3] (Portuguese-Brazilian translation in 2012 [4]), strongly influenced by the GUM uncertainty approach. Inconsistencies in the GUM text have been pointed out since its elaboration and publication in 1993. The concepts of Bayesian probability theory provide a consistent method for the quantitative reasoning under incomplete information. Suggestions to make the GUM consistent with such theory were proposed in 2003 [5], but at that time it was thought that it would be better to defer a revision of that document to a later time as a great effort had been done to disseminate its concepts. Nevertheless, in 2006, the JCGM decided on the elaboration of a consistent set of GUM supplements [6]. In 2012, the JCGM finally decided to revise the GUM to make it consistent both internally and with relation to its supplements [7].

The main inconsistencies of the current GUM are commented in section 2. The suggestions proposed in [5], which are endorsed by this author, are presented again here in section 3 (with small additions) for convenience of the reader.

The terminology of VIM3 is fully employed here. For convenience of the reader, the terms of that document are marked in italics the first time they appear in the text.

2. Inconsistencies in GUM

The initial step in an evaluation of *measurement uncertainty* (VIM3, 2.26) is to establish the *measurement model* (VIM3, 2.48). The latter, as defined in VIM3, is the 'mathematical relation among all *quantities* (VIM3, 1.1) known to be involved in a *measurement* (VIM3, 2.1)´. The generic form of the model is $h(Y, X_1, ..., X_n) = 0$, where Y, the *output quantity* (VIM3, 2.51), is the *measurand* (VIM3, 2.3), whose value should be inferred based on the information about the *input quantities* (VIM3, 2.50) $X_1, ..., X_n$. The latter can be: (a) the expected value(s) of the frequency distribution(s) of the (several sets of) *indications* (VIM3, 4.1), (b) the *corrections* (VIM3, 2.53) and (c) the *influence quantities* (VIM3, 2.52). In general, the measurement model can be expressed so that the output quantity becomes explicit, that is $Y = f(X_1, ..., X_n)$, where the function f is the *measurement function* (VIM3, 2.49). This measurement function is employed by the GUM and its supplements in the evaluation of measurement uncertainty.

Note that indications are neither input quantities as informed in VIM3, 2.50, Note 2 nor *measured values* (VIM3, 2.10) as alluded in VIM3, 2.28. Indications are ´*quantity values* (VIM3, 1.19) provided by a *measuring instrument* (VIM, 3.1) or a *measuring system* (VIM, 3.2)´. Once they are registered, they are to be considered as fixed and known. They are used to infer the expected value (assumed fixed and unknown) of the frequency distribution (assumed known) of the indications. The GUM Supplement 1 (GUMS1) [8] presents the inference results for two frequency distributions: one continuous (Gaussian) (GUMS1, 6.4.9) and one discrete (Poisson) (GUMS1, 6.4.11). It also presents several probability distributions for the values of the other input quantities (about which information cannot be obtained, or one does not desire it to be obtained, from indications), that can be used according to the information available about them.

The current GUM prescribes two methods of uncertainty evaluation: one that takes into account the availability of indications, *Type A evaluation* (VIM3, 2.28), and the other, *Type B evaluation* (VIM3, 2.29), which is applicable in the absence of those. The mathematical expressions used in Type A evaluation, such as those described in the current GUM, are derived from conventional statistics, whilst the GUM interprets the *combined standard uncertainty* (VIM3, 2.31) from a Bayesian viewpoint. This is an inconsistency that should be corrected in the future revision of the GUM. Observe that this inconsistency does not occur in GUMS1 where the classification in Type A and Type B evaluations is not necessary (GUMS1, 5.11.4). On the other hand, one may continue with the distinction between the two evaluation methods, if one wishes to do so, as they are in fact different.

VIM3 defines *measurement result* (VIM3, 2.9) as the ´set of quantity values being attributed to a measurand together with any other available relevant information´. The measurement result may be then expressed as (it is assumed here that the *definitional uncertainty* (VIM3, 2.27) is negligible in comparison with the other uncertainties involved):

- a probability distribution for the (essentially unique) *true value* (VIM3, 2.11) of the measurand based on all information available from the measurement. A numerical approximation of this distribution can be obtained by the method of propagation of distributions described in GUMS1;
- an interval of possible values (*coverage interval* (VIM3, 2.36)) within which it is believed that the (essentially unique) true value of the measurand is contained with a given *coverage probability* (VIM3, 2.37). GUMS1 describes how to calculate coverage intervals for given probabilities;
- a *measured value* (VIM3, 2.10) and its associated *standard uncertainty* (VIM3, 2.30). The measured value of the measurand is interpreted as the expected value of the probability distribution for the value of the measurand and the standard uncertainty associated with the measured value is interpreted as the standard deviation of this distribution. GUMS1 describes how to calculate the expected value and standard deviation of the numerical approximation of the probability distribution. If the model is linear or linearized, it is also possible to use the method described in GUM to obtain the expected value and variance of that distribution (note that the GUM does not allow one to specify completely the distribution) from the expected values and variances of the probability distributions for the input quantities.

Consider the measured value y and the combined standard uncertainty $u(y)$. It can be demonstrated that the minimum and maximum coverage probabilities of the interval $[y \pm 2u(y)]$ are 75% and 95%, respectively, for any distribution with expected value y and standard deviation $u(y)$ [5]. Therefore, the interval $[y \pm 2u(y)]$ covers a great fraction of the probability distribution represented by y and $u(y)$. As the GUM procedure does not yield a complete specification of the distribution represented by y and $u(y)$, the concept of minimum coverage probability of the interval $[y \pm 2u(y)]$ is in agreement with the GUM. In this case, $2u(y)$ is the *expanded measurement uncertainty* (VIM3, 2.35).

The current GUM prescribes two methods for the evaluation of the expanded measurement uncertainty. Though relegated to an appendix (GUM, Appendix G), such methods have systematically been adopted by the *metrology* (VIM3, 2.2) community when issuing *calibration* (VIM3, 2.39) certificates. The first one presupposes that the probability distribution for the value of the measurand can be approximated by a Gaussian. The second presupposes that the distribution can be approximated by a Student-*t* and prescribes the use of the Welch-Satterthwaite formula for the calculation of the corresponding degrees of freedom. The inconsistency of the last method is related to the fact that this formula is not compatible with quantities whose measurement uncertainties (associated with their measured values) were obtained from a Type B evaluation. This is an inconsistency that should be corrected in the future revision of the GUM. Observe that this inconsistency does not occur in GUMS1 where the numerical approximation of the probability distribution is known and the expanded measurement uncertainty can be directly calculated from it.

As informed in GUMS1, section 6.4.9.4, Note 2: ´In the Bayesian context of this Supplement, concepts such as the reliability, or the uncertainty, of an uncertainty are not necessary. Accordingly, the degrees of freedom in a Type A evaluation of uncertainty is no longer viewed as a measure of reliability, and the degrees of freedom in a Type B evaluation do not exist.´ The future version of the GUM should be correspondingly revised.

One drawback of the evaluation method based on the measurement function, as described in GUM and its supplements, is that prior information about the measurand cannot be used in a new evaluation of that same measurand (recalibration). On the other hand, the method allows for prior information about the expected value(s) of the frequency distribution(s) of the (several sets of) indications to be included in the new evaluation. It also permits the incorporation of the definitional uncertainty in the evaluation.

3. Steps of the future GUM

The steps of the evaluation of uncertainty in measurement that the author recommends be adopted in the revision of the GUM to make it consistent are set below. Such steps are in [5] and are presented here with small additions using the VIM3 terminology for convenience of the reader.

Step 1. Describe the *measurement method* (VIM3, 2.5) as a measurement function $Y = f(X_1, ..., X_n)$, where the input quantities $X_1, ..., X_n$ may have their own measurement functions and be evaluated by following the same steps described here. All input and output quantities are variables with probability distributions representing states of knowledge. Such distributions are obtained from the information available in the measurement as described in GUMS1.

Step 2. Determine the measured values $x_1, ..., x_n$ of the input quantities $X_1, ..., X_n$, respectively. All the measured values are interpreted as expected values of the probability distributions for the input quantities.

Step 3. Determine the standard uncertainties $u(x_1), ..., u(x_n)$ associated with the measured values of the input quantities $X_1, ..., X_n$, respectively. All the standard uncertainties are interpreted as standard deviations of the probability distributions for the input quantities.

Step 4. Evaluate if the probability distributions for each pair X_i and X_j of the input quantities are logically dependent and determine the correlation coefficients $r(x_i, x_j)$ for those pairs of the input variables that may be significantly correlated, for $i, j = 1, ..., n$, $i < j$.

Step 5. Determine the measured value y of the output quantity substituting all measured values x_1, ..., x_n of the input quantities in the measurement function $Y = f(X_1, ..., X_n)$. Thus, $y = f(x_1, ..., x_n)$.

Step 6. Determine the combined standard uncertainty $u(y)$ associated with the measured value y by propagating the standard uncertainties $u(x_1)$, ..., $u(x_n)$ and the correlation coefficients $r(x_1, x_2)$, ..., $r(x_{n-1}, x_n)$. A first-order Taylor series approximation of the measurement function $Y = f(X_1, ..., X_n)$ provides the following law of propagation of uncertainties:

$$u^2(y) = \sum_i c_i^2 u^2(x_i) + 2\sum_{i<j} c_i c_j\, u(x_i)u(x_j)r(x_i, x_j), \qquad (1)$$

where c_1, ..., c_n are the sensitivity coefficients or partial derivatives of Y with respect to X_1, ..., X_n evaluated at x_1, ..., x_n, respectively. The product $u(x_i)u(x_j)r(x_i, x_j)$ is equal to the covariance $u(x_i, x_j)$ for $i, j = 1, ..., n$, $i < j$.

Step 7. If it is necessary to express the measurement result as a coverage interval, multiply the combined standard uncertainty $u(y)$ by a *coverage factor* (VIM3, 2.38) k to obtain the coverage interval $[y \pm ku(y)]$. Generally, $k = 2$. Use values of k other than 2 in special cases only.

Step 8. Report the measurement result as the measured value y together with its associated combined standard uncertainty $u(y)$ or as the coverage interval $[y \pm ku(y)]$. Describe how y and $u(y)$ were obtained. Describe how k was chosen when k is not 2.

Note that if the evaluation is performed using only the method of propagation of uncertainties described in GUM, it will be no more necessary to report the coverage probability of the interval when the coverage factor is 2. In addition, it will not be needed anymore to use the Welch-Satterthwaite formula to calculate effective degrees of freedom.

References

[1] *Joint Committee for Guides in Metrology*, http://www.bipm.org/en/committees/jc/jcgm/ (accessed 13.11.13).

[2] BIPM, IEC, IFCC, ILAC, ISO, IUPAC, IUPAP and OIML, Evaluation of measurement data – Guide to the expression of uncertainty in measurement, GUM 1995 with minor corrections, Joint Committee for Guides in Metrology, JCGM 100, 2008, http://www.bipm.org/utils/common/documents/jcgm/JCGM_100_2008_E.pdf (accessed 13.11.13).

[3] BIPM, IEC, IFCC, ILAC, ISO, IUPAC, IUPAP and OIML, International vocabulary of metrology – Basic and general concepts and associated terms, JCGM 200, 2008, http://www.bipm.org/utils/common/documents/jcgm/JCGM_200_2008.pdf (accessed 13.11.13).

[4] BIPM, IEC, IFCC, ILAC, ISO, IUPAC, IUPAP and OIML, Vocabulário internacional de metrologia – Conceitos fundamentais e gerais e termos associados, 1ª edição luso-brasileira, 2012, http://www.inmetro.gov.br/inovacao/publicacoes.asp (accessed 13.11.13).

[5] Kacker R and Jones A 2003 *Metrologia* **40** 235-48.

[6] Bich W, Cox M G and Harris P M 2006 *Metrologia* **43** S161-S166.

[7] Bich W, Cox M G, Dybkaer R, Elster C, Estler W T, Hibbert B, Imai H, Kool W, Michotte C, Nielsen L, Pendrill L, Sidney S, van der Veen A M H and Wöger W 2012 *Metrologia* **49** 702

[8] BIPM, IEC, IFCC, ILAC, ISO, IUPAC, IUPAP and OIML, Evaluation of measurement data – Supplement 1 to the 'Guide to the expression of uncertainty in measurement', Propagation of distributions using a Monte Carlo method, Joint Committee for Guides in Metrology, JCGM 101, 2008, http://www.bipm.org/utils/common/documents/jcgm/JCGM_101_2008_E.pdf (accessed 13.11.13).

Homogeneity study of candidate reference material in fish matrix

J C Ulrich, J E S Sarkis and M A Hortellani [1]
Instituto de Pesquisas Energéticas e Nucleares – IPEN/CNEN-SP
Travessa R, 400 – Cidade Universitária – São Paulo, Brazil

E-mail: jculrich@ipen.br

Abstract. A material is perfectly homogeneous with respect to a given characteristic, or composition, if there is no difference between the values obtained from one part to another. Homogeneity is usually evaluated using analysis of variance (ANOVA). However, the requirement that populations of data to be processed must have a normal distribution and equal variances greatly limits the use of this statistical tool. A more suitable test for assessing the homogeneity of RMs, known as "sufficient homogeneity", was proposed by Fearn and Thompson. In this work, we evaluate the performance of the two statistical treatments for assessing homogeneity of methylmercury (MeHg) in candidate reference material of fish tissue.

1. Introduction

The homogeneity study is an integral part of planning the production of certified reference material (CRM), allowing the evaluation of one of the key uncertainty components in the certification model. Clearly, a high degree of homogeneity is anticipated during the preparation of reference materials. However, despite all efforts to ensure this, materials can still exhibit a degree of heterogeneity. The producer of the material thus has to ensure and demonstrate the level of homogeneity through the use of appropriate procedures to detect any impurities or interferences, caused by problems not detected during preparation.

According to ISO guide 35 [1], a material is perfectly homogeneous with respect to a particular characteristic or composition if there is no difference between the obtained values from one party to another. As a consequence, the amount of material necessary for homogeneity studies and the choice of appropriate measurement methods for determining the characteristics or chemical composition must be integrated from the beginning of experimental protocols.

A basic model for the evaluation of a homogeneity study comprises bottles (or parts) of the material and measured values that can be expressed by the relationship between the overall mean value of the measurements, error for homogeneity between bottles and random measurement error. The variances of these terms are represented by the variance between bottles and repeatability variance. Such evaluations are typically performed using one-way analysis of variance (ANOVA). However, this strategy can lead to erroneous conclusions in two cases: if the analytical method used is inaccurate, possible heterogeneities may not be detected and the test can incorrectly indicate that the material is homogeneous or that the variability in the results is significant; conversely, if the analytical method is highly sensitive, small differences the composition among bottles can be inappropriately interpreted as heterogeneity. Moreover, populations of data to be processed by ANOVA must have a normal distribution and equal variances, which limits the use of this statistical tool.

Given the limitations of ANOVAs, a new methodology to evaluate the homogeneity of reference materials was proposed by Fearn and Thompson [2]. Known as "sufficient homogeneity", this method (unlike ANOVA) imposes a limit to the analytical variance, designated s_{an}, in relation the "target standard deviation (or expected)", designated σ_p. This limitation requires that the ratio s_{an}/σ_p should be less than 0.5, if possible. Another difference from ANOVA is the calculation of the critical value for comparison, calculated from obtaining the values of allowable variance between-samples (s^2_{all}), the analytical variance (s^2_{an}) and factors (F1 and F2) extracted from a specific table to test "sufficient homogeneity" [2].

Due to the superior performance of the sufficient homogeneity test, it was incorporated into an international protocol [3] and has been widely cited in the literature [4,5].

The objective of the present study is to compare analysis by ANOVA and "sufficient homogeneity" in the assessment of homogeneity study for methylmercury (MeHg) in candidate reference material in a fish matrix.

2. Experimental

2.1 Planning the homogeneity study

Between-bottle homogeneity of MeHg was determined for the contents of 10 bottles. For evaluation of the data by ANOVA, three portions of approximately 0.5 g of sample material from each bottle were tested. Each portion was analysed three times, obtaining three average values for each bottle. Subsequently, "sufficient homogeneity" was evaluated using two more portions of 0.5 g each and measured in triplicate, producing two average values for each bottle.

2.2 *Sample preparation and determination of MeHg*

Determination of MeHg is based on the acid leaching with hydrochloric acid solution (HCl) 6 mol L^{-1} (by volume) and mercury separation of the organic and inorganic ion exchange resin (Dowex 1×8 100–200 mesh). The methodology was based on Horvat and May [6,7].

Once separated, MeHg was decomposed into inorganic Hg^{2+} by ultraviolet (UV) irradiation and the final solution was diluted to 30 g with demineralised water and inserted into the sample introduction system of an atomic absorption spectrophotometer (FS-SpectrAA220 Varian Australia Pty Ltd.). Methylmercury (such as mercury) is determined by atomic absorption spectrophotometry with cold vapor generation and flow injection (FIA-CV-AAS).

3. Results and discussion

3.1 Determination of methylmercury (MeHg) in candidate CRMs.

The portions from 10 bottles were analyzed and outliers were tested using the Grubbs test [8]. The analysis did not indicate the presence of any outliers.

3.1.1 One-way Analysis of Variance. Table 1 shows the results of ANOVA for MeHg.

Table 1. Data ANOVA of the homogeneity study for MeHg

Source of variation	SS	Degrees of freedom	MS	$F_{calculated}$	F_{critic}
Between groups	0.0950	9	0.0106	2.72	2.39
Within groups	0.0774	20	0.0039		
Total	0.1724	29			

As can be seen from table 1, $F_{calculated}$ is higher than the critical value F_{critic}, indicating that the sample is not homogeneous (i.e. the material has not passed the test for homogeneity). Comparing the calculated variances with the measured values (table 2) reveals that one of the assumptions of the ANOVA was not supported because the variances are different. This is further substantiated by dividing the highest value obtained at the lower variance giving a value of 223, as compared to recommended values of 3 or 4 [9]. The F test used is a hypothesis test based on sample variances. Thus, the fact that the variances are different is further evidence that the test does not support or reject the hypothesis of equality of MeHg values.

Table 2. Results of mean values, standard deviation and variance of the results of replicates analyzed for MeHg.

Bottle N^0.	Mean ($\mu g\ g^{-1}$)	Standard deviation ($\mu g\ g^{-1}$)	Variance	Replicate (n_j)
2	0.277	0.032	0.00104	3
9	0.205	0.033	0.00107	3
14	0.388	0.134	0.01789	3
26	0.241	0.017	0.00030	3
31	0.265	0.100	0.01001	3
36	0.317	0.074	0.00545	3
50	0.334	0.009	0.00008	3
55	0.216	0.022	0.00047	3
63	0.237	0.044	0.00195	3
78	0.223	0.021	0.00044	3

The difference in variance demonstrated by the ANOVA can be derived from the analytical method used in the measurements (FIA-CV-AAS), as performed on different days. This showed variability in the response signals of both the calibration curve and in the signals of the samples.

3.1.2 "Sufficient homogeneity" test. As recommended in the protocol [2], samples in duplicate were analyzed from 10 bottles under replicable conditions (results in Table 3). The items following table 3 are the equations used for the calculations. In item (a), the analytical variance (s^2_{an}) is obtained by dividing the sum of the squares of the differences (D^2) by 2m, where m is the number of bottles. Item (b) gives the analytical variance as the standard deviation. Item (c) is the value obtained for the estimation of variance between bottles (S^2_{sam}). The value of the allowable variance between the sample (s^2_{all}) calculated in Item (d), obtained from the value of the target standard deviation (σ_p) calculated by the Horwitz function [3,10]. The critical value for the homogeneity study, called "c" is calculated in

Item (e), being obtained by summing the results of multiplying specific factors [2] with the values of (s^2_{all}) and (s^2_{an}). The significance of this test is determined by the following criteria: if "$S^2_{sam} > c$" homogeneity test failed and this cannot be proven and "$S^2_{sam} < c$" homogeneity test was accepted and this is proven.

Table 3. Test of homogeneity between bottles, applied to the data of MeHg

Bottle N.	Result 1 (a)	Result 2 (b)	D = a - b	S = a + b	$D^2 = (a - b)^2$
2	0.296	0.295	0.001	0.591	0.000001
9	0.195	0.241	-0.046	0.436	0.002116
14	0.291	0.333	-0.042	0.624	0.001764
26	0.245	0.255	-0.010	0.500	0.000100
31	0.379	0.226	0.153	0.605	0.023409
36	0.382	0.332	0.050	0.714	0.002500
50	0.328	0.329	-0.001	0.657	0.000001
55	0.217	0.237	-0.020	0.454	0.000400
63	0.208	0.215	-0.007	0.423	0.000049
78	0.243	0.224	0.019	0.467	0.000361
				Sum =	0.030701

Item (a) $s^2_{an} = (\Sigma D^2)/2m = 0.030701/(2\ 10) = 0.001535\ \mu g\ g^{-1}$ (analytical variance)

Item (b) $s_{an} = \sqrt{s^2_{an}} = \sqrt{0.001535} = 0.039179\ \mu g\ g^{-1}$ (analytical variance expressed how standard deviation)

Item (c) S^2_{sam} = ((variance of the sums S=a+b)/2 – S^2_{an})/2 = (0.010686/2 – 0.001535)/2 = 0.001904

Item (d) $\sigma^2_{all} = (0.3\sigma_p)^2 = (0.3\ 0.05326)^2 = 0.000255\ \mu g\ kg^{-1}$.

σ_p (obtained by Horwitz function) = $0.02c^{0.8495} = 0.02(0.274\ 10^{-6})^{0.8495}/10^{-6} = 0.05326\ \mu g\ g^{-1}$ ("c" is average concentration of MeHg obtained, expressed as mass fraction).

Item (e) $c = F_1 s^2_{all} + F_2 s^2_{an} = 1.88\ 0.0002553 + 1.01\ 0.001535 = 0.002030$.

Using the criteria, $S^2_{sam} = 0.001904 < c = 0.002030$ the homogeneity is proved.

4. Conclusions

Tests for homogeneity of MeHg for candidate reference material in a fish matrix were carried out using two statistical tools.

The ANOVA test indicated that the material was not homogeneous. In contrast, the "sufficient homogeneity" test proposed by Fearn and Thompson indicated an acceptable level of homogeneity of the material.

The homogeneity study provides one of the components of uncertainty required to evaluate combined standard uncertainty in certified reference material, ensuring minimum levels of quality and acceptability.

References

[1] International Organization for Standardization (ISO); ISO Guide 35:2006, Reference Materials General and Statistical Principles for Certification, Geneva

[2] Fearn T and Thompson M 2001 *Analyst*, **126** 1414-17

[3] Thompson M, Ellison S L R and Wood R 2006 The International Harmonized Protocol for the Proficiency Testing of Analytical Chemistry Laboratories, *Pure Appl. Chem.* **78** 1 145–96

[4] Raposo F, Fernández-Cegrí V, De la Rubia M A, Borja R, Beltrán J, Cavinato C, Clinckspoor M, Demirer G, Diamadopoulos E, Frigon J C, Koubova J, Launay M, Méndez R, Menin G, Noguerol J, Uellehdahl H and West S 2010 *TrAC Trends in Analytical Chemistry* **9** 1082-91

[5] Semeraro A, de la Calle M B, Wysocka I, Quétel C, Linsinger T, Emteborg H, Cordeiro F, Verbist I, Vendelbo D and Taylor P *IMEP-27: total Cd, Pb and As and extractable Cd and Pb in mineral feed*, EUR 23776 EN-2009

[6] Horvat M, May K, Stoeppler M and Byrne A R 1988 *Applied Organometallic Chemistry* **2** 515-24

[7] May K, Stoeppler M and Reisinger K 1988 Studies in the ratio total mercury methylmercury in the aquatic food chain. In: *Carcinogenic and mutagenic metal compounds vol 2*, ed Merian E, Frei R W, Hardi W and Schlatter C H

[8] Roper P, Burke S, Lawn R, Barwick V and Walker R 2001 Applications of Reference Materials in Analytical Chemistry. Cambridge, UK Royal Society of Chemistry, Thomas Grahan House

[9] Vieira S 2006 *Análise de variância* (São Paulo, SP: Atlas)

[10] Albert R and Horwitz W, 1997 *Anal. Chem.* **69** 789-90

Estimation of uncertainty of a reference material for proficiency testing for the determination of total mercury in fish *in nature*

L V Santana[1], J E S Sarkis [1], J C Ulrich [1] and M A Hortellani[1]
[1] IPEN/CNEN Institute of Nuclear and Energy Research, São Paulo, Brazil

E-mail: santana-luciana@ig.com.br

Abstract: We provide an uncertainty estimates for homogeneity and stability studies of reference material used in proficiency test for determination of total mercury in fish fresh muscle tissue. Stability was estimated by linear regression and homogeneity by ANOVA. The results indicate that the reference material is both homogeneous and chemically stable over the short term. Total mercury concentration of the muscle tissue, with expanded uncertainty, was 0.294 ± 0.089 µg g^{-1}.

1. Introduction

Determination of property values of reference material requires the systematic assessment of uncertainties associated with studies of homogeneity and stability. In general these values are obtained using the Guide to expression of Uncertainty in Measurement (GUM) [1].

Characterization of a reference material is defined as the complete set of measurements and associated uncertainties needed to set the values of its properties, as opposed to assigning property values for batch material or for individual units. Several different approaches exist to characterize material from which the measurement methods are traceable to the references specified in the project planning. The main approaches are as follows [1, 2]:

1- Characterization by an independent method;

2- Characterization by multiple methods and/or different laboratories.

Homogeneity analysis is necessary to demonstrate that the material does not vary significantly in consistency and chemical content. Such analysis is performed between the sample containers and within each container to assess whether there is an excess of variability that might affect the specific property being evaluated [2, 3].

Stability analysis is carried out to verify transport conditions. Such analysis is typically performed by exposing reference samples to temperatures over short periods of time (3-6 months) that simulate a realistic range of transport conditions. The material is then evaluated for levels of degradation that may affect the state values of the evaluated property. The Stability study is considered ideal for proficiency testing samples from biological matrices [2, 3, 4].

In this paper we calculate short term uncertainties associated with homogeneity and stability study in order to obtain reference values for proficiency testing material for determination of total mercury content of fish fresh muscle tissue.

2. Methodology

2.1. Preparation of the material

Analysis was performed on two fish (589 g and 660 g) acquired from a retailer within Itaituba city, near the Tapajós River in the Amazon region of Brazil. The fish were prepared by removing the head, tail, intestines and skin with a stainless steel knife. Muscle tissue was triturated and filleted on a Walita® domestic multiprocessor, after which it was homogenized in a Walita® domestic blender (with "fouet" type attachments) for 10 minutes, resulting in 1.100 g of sample material [5]. The material was then divided into 86 sachets of 15 g each. Of this total, 43 sachets were randomly assigned to be irradiated in a cyclotron (Cyclone Applications ion beam 30), the Radiation Technology Center of IPEN, with 10.0 ± 1.05 kGy using a cobalt 60 source (^{60}Co) to temperature of 24 ± 4 °C for homogeneity and stability analysis.

2.2 Determination of total mercury

A 0.5 g sample was digested with 1 mL of HNO_3, H_2SO_4 and 2 ml of 1 mL $HClO_4$ on plate at 110 °C for 30 minutes. After digestion, the solution was diluted with 20 mL of water for analysis [6].

Total mercury was evaluated by atomic absorption spectrometry with cold vapor generation and flow injection (FIA-CV-AAS) using an atomic absorption spectrophotometer Varian model AA-220 FS, according to manufacturer's specifications. The reducing agent was $SnCl_2$ 25% (w / v) in HCl 25% (v / v) [6].

The certified reference material DORM-2 ("Dogfish") provided by the National Research Council of Canada (NRCC) was used for the validation of the analytical methodology. The value of total mercury concentration in the certified sample (with expanded uncertainty) is 4.64 ± 0.26 $\mu g\ g^{-1}$ [6].

2.3. Homogeneity and stability study

Homogeneity study was performed by randomly assigning ten sachets to temperature treatments of 23 ± 0.2 °C. Analysis of Variance (ANOVA) was used to evaluate heterogeneity between and within sachets, as described in the ISO Guide 35 [1, 3].

Stability analysis was performed by exposing 16 randomly selected sachets to 4 different temperatures ranges 5 °C, 23 °C, 40 °C and 60 °C for a period of 0, 7, 15, 30 and 45 days. Data was statistically evaluated by linear regression analysis [2, 3].

2.4 Estimation of uncertainty of the material

The uncertainty characterization u(char) was calculated in a spreadsheet according to the 'Guide to the expression of uncertainty of measurement' (GUM) [1-3]. Relative uncertainty of the homogeneity study was evaluated by the mean square analysis of variance (ANOVA-one way) according to Equation 1 [1]:

$$s_{bb} = s_A = \sqrt{\frac{MQ_{between} - MQ_{within}}{n_0}} \quad (1)$$

Where: $s_{bb} = s_A$ = uncertainty due homogeneity; $MQ_{between}$ = mean square between the sachets; MQ_{within} = mean square within the sachets and n_0 = number of sachets.

Relative uncertainty of the stability analysis was calculated by multiplying the value of the standard error of the slope of the regression curve for the time period (in months) according to equation 2 [2, 3].

$$u_{sts} = {}^{s}{}_{(b1)} \cdot t \quad (2)$$

Assuming the uncertainties associated with homogeneity and stability are independent, the overall uncertainty associated with the property values of the reference material, with coverage factor (k = 2), was calculated as follows [2]:

$$u_{MR} = k\sqrt{u_{car}^2 + u_{bb}^2 + u_{sts}^2} \qquad (3)$$

3. Results

The characterization of uncertainty u(char) gave a value of 0.033 μg^{-1} g (12%), obtained from a single measurement method (FIA CV-AAS). Error propagation was carried out by the Ishikawa model diagram, where each parameter was calculated by spreadsheet, developed in the laboratory for the determination of mercury in fish (test accredited in CGCRE / INMETRO second ABNT: ISO / IEC 17025: 2005). The results were submitted to Grubbs test for the presence of outliers. According to the ANOVA, the studied material was homogenous ($F_{calc}<F_{crit}$, Table 1). The Shapiro-Wilk test (p=0.05) indicated that the sample was drawn from a normal population.

Table 1-Average results ± standard deviation (SD), in μg g^{-1}, of homogeneity study at 23 ± 0.3 °C

Average ± SD	0.294 ± 0.009
RSD %	1.3
P-value	0.19
$W_{calc.}$	0.78
$F_{crit.}$	3.88
$F_{calc.}$	1.85
ubb, μg g^{-1}, (%)	0.015 (5 %)

Wcrit.(0.05;5)=0.762. F_{calc}: calculated; F_{crit}: critical, F value for α = 5 %; P-value>0.05; u_{bb}: uncertainty of homogeneity; sachets n° 56, 14, 21, 42, 3, n=30.

Likewise, the sample showed a considerable degree of stability under the experimental conditions imposed on the samples (Table 2).

Table 2- Average results ± standard deviation (SD), in μg g^{-1}, obtained for stability study over 0, 7, 15, 30 and 45 days

	Freezer (5 ± 0.2 °C)	Room (23 ± 0.3 °C)	Warm house (40 ± 1.3 °C)	Warm house (60 ± 1.2 °C)
Average ± SD	0.313±0.015	0.325± 0.025	0.375±0.038	0.380±0.029
RSD %	1.5	2.5	3.8	2.9
b1	0.0006	-0.0073	0.0024	-0.0003
s(b1)	0.00502	0.01325	0.00018	0.0013
|b1| < t(0,95,n-2)x s(b1)	0.02	0.02	0.03	0.03
usts, μg g^{-1}, (%)	0.010 (1 %)	0.026 (26 %)	0.001 (01 %)	0.003(03 %)

b_1: slope curve; s_b: standard deviation in b_1; u_{sts}: uncertainty of stability; t0, 95, n-2 x s(b1) value for factor t (Student's) for n-2 degrees of freedom to a level of confidence 95 %[2], n=48.

The statistical evaluation of the data obtained for the concentration of mercury in fish muscle tissue during the time intervals and temperature ranges showed no significant trend - | b_1 | is smaller than the value of the condition t0, 95-n 2 xs (b1) [2, 7]. Table 4 shows the value of total mercury concentration with an estimated expanded uncertainty for reference material according to equation 3:

$$u_{MR} = 2 \times \sqrt{(0.033)^2 + (0.015)^2 + (0.026)^2} = 0.089$$

Table 4- Estimation of the combined standard uncertainty in µg g^{-1}, for the element mercury

	[a]Average ± U (µg g^{-1})	U (%)
$U_{MR(Hg)}$	0.294 ± 0.089	30

[a]Mean obtained in the study of homogeneity, represented by the mean values of 30 replicates.

4. Conclusions

The material produced for proficiency tests showed high levels of homogeneity for Hg, both within and between sachets. The stability of the material was also robust in the short-term, demonstrating the efficacy of the irradiation process to transport the samples at room temperature. Proficiency testing was performed at 30 days and, for this reason, it was not deemed necessary to assess long-term stability.

The calculated uncertainties were compatible with the measurement technique used to estimate uncertainty of the material. Therefore, the material produced was considered fit for use in proficiency testing, providing an additional tool for control and quality assurance in the analysis of mercury content in fish.

Acknowledgements

We are grateful to IPEN-CNEN/SP and CNPq for financial support in the preparation of this work.

References

[1] BIPM, IEC, IFCC, ISO, IUPAC, IUPAP, OIML 2008 *Guide to the expression of uncertainty in measurement*, 1st edn. International Organization for Standardization (ISO), Geneva

[2] ISO, International Organization of Standardization 2006 *Certification of reference materials— general and statistical principles*. ISO Guide 35. 3rd edn, Geneva

[3] van der Veen A M H, Linsinger T P, Pauwels J 2001 *Accredit. Qual.* Assur. **6** 290

[4] van der Veen A M H, Linsinger T P J, Lamberty A, Pauwels J 2001 *Accredit. Qual. Assur.* **6** 257

[5] Aquino J S, Silva J A, Prado J P, Cavalheiro J M O 2008 *Rev. Inst. Adolfo Lutz*, 67(**3**):190-195

[6] Azevedo J S, J Sarkis E S, Hortellani M A, Ladle R J 2012 *Water, Air, & Soil Pollution*. **223** 3911-3922

[7] van der Veen A M H, Pauwels J 2000 *Accredit. Qual. Assur.* **5** 327

Mathematica tools for uncertainty analysis

M D Mikhailov, V Y Aibe
Inmetro, Fluid Dynamics Metrology, Rio de Janeiro, Brazil

E-mail: dinam@inmetro.gov.br

Abstract. Some functions of the author´s *Mathematica* 9 package are presented. Links are given to the author´s interactive demonstrations. *Mathematica* supports 143 types of statistical distribution covering not only metrology but almost all statistical areas. The Supplement 1 to the ISO GUM mentions that analytical methods are ideal ones, but assume they are applicable only in simple cases, and recommends Monte Carlo method. In contrast, this paper presents *Mathematica* functions that can explore a multitude of methods in order to find analytical, numerical, or statistical results using (usually) one million random variates.

1. Introduction

The GUM Guide [1] provides a well-known framework for assessment and evaluation of uncertainties based on the law of propagation of uncertainty. This guide has generated the appearance of a great number of computer systems and calculators dedicated in helping estimate the uncertainty of measurements. Typing the key "uncertainty calculator" in Google gives back about 2,460,000 results. Wikipedia alone describes 18 uncertainty propagation software systems [2].

The authors of the present work have previously devised an uncertainty calculus based on the GUM in Mathematica 6 [3]. An object $x\pm\Delta x$ called uncertainty number was defined. After assuming $\Delta x/x<<1$ the terms Δx were neglected in relational operations ($<$, $=$, $>$). Our Mathematica rules transform one expression of independent uncertainty numbers into single uncertainty number. Later on we were able to extend these rules to cover also fully dependent uncertainty numbers. Special cases of the rules we used in the interactive demonstrations [4, 5, 6, and 7].

The GUM approach [1] is only exact for linear models, since it is based on the first order Taylor expansion, neglecting higher order terms. Thus direct application of the GUM rules in nonlinear models can potentially give quite misleading results. In order to account for non-linear scenarios, GUM allows for techniques other than the law of propagation of uncertainty (ref. [1], section G.1.5). The Supplement 1 to the GUM [8] recommends propagation of distributions, which can be applied in non-linear problems. In this Supplement, it is mentioned that analytical methods are ideal, but only viable in simple cases. Because of this, the Supplement recommends the Monte-Carlo method.

In this paper we continue our previous work [3] by presenting a new *Mathematica* functions able to compute symbolically or numerically the propagation of statistical distributions, thereby extending our uncertainty calculus into the non-linear domain, at the same time avoiding the limitations present in the GUM approach. Interactive examples using this newly developed package are available on-line in Computable Document Format (CDF) files [9]. There users can interactively compare the exact uncertainties versus GUM uncertainties for **`Sin`** and **`Cos`** function [10]. The free *Mathematica* Player [11] is needed to evaluate the interactive tools [3, 4, 5, 6, 7, 9, 10].

Mathematica support 143 statistical distributions covered different research areas such as Actuarial

Journal of Physics: Conference Series **575** (2015) 012042 doi:10.1088/1742-6596/575/1/012042

Science, Finance, Metrology, Risk, Reliability etc. From these distributions, there are many ways to define new, user-derived distributions that behave just like any other built-in distribution. Recently, the authors have published a quick, interactive reference guide to the logical relationships obeyed by the statistical parameters of the built-in *Mathematica* distributions [12], and the percentile plots of 78 continuous statistical distributions [13].
All functions, input and output lines bellow are written in the standard for *Mathematica* Currier font.

2. Objectives
This paper aims is the comparison of results obtained by original *Mathematica* functions based on propagation of uncertainty and propagation of distributions.
Mathematica use **NormalDistribution[μ,σ]** for the Gaussian distribution with mean μ and standard deviation σ. Since almost all *Mathematica* built-in objects are full English names beginning with capital letters, we follow this convention for our package functions as well. The functions demonstrated are: **UncertainCalculus**, **GUM**, **CoverageInterval**, and **Measurand**.

3. UncertainCalculus
The first version of uncertain calculus is demonstrated at ENCIT 2008 [3]. The function **UncertainCalculus[*case*]** evaluate input line with expression including uncertain numbers like $\mu\pm\sigma$ (**PlusMinus[μ,σ]**). Similarly to **GUM** function ***case*** = 1 activate the rule for independent variables and ***case*** = 3 activate a rule for fully correlated variables. The ***case = 0*** deactivate **UncertainCalculus**.
The **UncertainCalculus[1]** activate the rule for independent variables.
Then the following input line gives the corresponding output lines.
c1 (μ1 ± σ1) + c2 (μ2 ± σ2)

$$(c_1\mu_1 + c_2\mu_2) \pm \sqrt{c_1^2\sigma_1^2 + c_2^2\sigma_2^2} \qquad \text{Eqs 1}$$

In case of detectable error, like **UncertainCalculus[2]** the warning message appears.
The **UncertainCalculus[3]** activate the rule for fully correlated variables.
Then the following input line gives the corresponding output lines.
c1 (μ1 ± σ1) + c2 (μ2 ± σ2)

$$(c_1\ \mu_1+c_2\ \mu_2)\pm(c_1\ \sigma_1+c_2\ \sigma_2) \qquad \text{Eqs 2}$$

For one variable independent and fully correlated variables rules give the same results.
Next examples evaluate two nonlinear functions
(μ ± σ)^2

$$\mu^2\pm 2\ \mu\ \sigma \qquad \text{Eqs 3}$$

Sin[μ±σ]

$$\mathrm{Sin}[\mu]\pm\sigma\ \mathrm{Cos}[\mu] \qquad \text{Eqs 4}$$

Finally, the **UncertainCalculus[0]** deactivate the uncertain calculus.

4. GUM
The function **GUM** implements the formulas given in [1].
GUM[expr,x==μ±σ,case] or **GUM[expr,{x1==μ1±σ1,...,xn==μn±σn},case]**
gives first order series approximation for expectation ±standard deviation of **expr**.
Mathematica use **lhs==rhs** to denote equation, while **lhs=rhs** to evaluate **rhs** and assign the result to **lhs**.
case = 1 or IdentityMatrix[n] (independent variables).
case = 2 nonlinear expr (more series terms included in case 1).
case = 3 or n x n constant matrix with elements 1 (fully correlated variables).
case = n x n symmetric matrix with main diagonal 1 and elements in the closed interval -1 to 1 .

```
GUM[c1 x1 + c2 x2, {x1 == μ1 ± σ1, x2 == μ2 ± σ2}, 1]
```

$$(c_1\mu_1 + c_2\mu_2) \pm \sqrt{c_1^2\sigma_1^2 + c_2^2\sigma_2^2}$$ Eqs 5

```
GUM[c1 x1 + c2 x2, {x1 == μ1 ± σ1, x2 == μ2 ± σ2}, 3]
```

$(c_1\ \mu_1+c_2\ \mu_2)\pm(c_1\ \sigma_1+c_2\ \sigma_2)$ Eqs 6

```
GUM[x^2, x == μ ± σ, 1]
```

$\mu^2\pm 2\ \mu\ \sigma$ Eqs 7

```
GUM[Sin[x], x == μ ± σ, 1]
```

$\text{Sin}[\mu]\ \pm\ \sigma\ \text{Cos}[\mu]$ Eqs 8

The uncertain calculus (Eqs 1 to Eqs 4), and GUM (Eqs 5 to Eqs 8) gives the same results.
The uncertain calculus could not use the correlation matrix **R**, where $0\leq\rho\leq 1$.

$$R = \begin{pmatrix} 1 & \rho \\ \rho & 1 \end{pmatrix}$$

Eqs 9

```
GUM[c1 x1 + c2 x2, {x1 == μ1 ± σ1, x2 == μ2 ± σ2}, R]
```

$(c_1\ \mu_1+c_2\ \mu_2)\pm\sqrt{(c_1^2\ \sigma_1^2+2\ c_1\ c_2\ \rho\ \sigma_1\ \sigma_2+\sigma_1^2\ \sigma_2^2)}$ Eqs 10

The solution given by Eqs 5 and Eqs 6 are special cases of Eqs 10 for $\rho=0$ and $\rho=1$, respectivelly.
For the non-linear measurand `x^2`, `Sin[x]`, and `Cos[x]` the results are

```
GUM[x^2, x == μ ± σ, 2]
```

$$\mu^2 \pm \sqrt{4\mu^2\sigma^2 + 2\sigma^2}$$ Eqs 11

```
FullSimplify[GUM[Sin[x], x == μ ± σ, 2]]
```

$$Sin[\mu] \pm \sqrt{\sigma^2 \cdot Cos[\mu]^2 - \tfrac{1}{4}\cdot\sigma^4\cdot(1+3\cdot Cos[2\mu])}$$ Eqs 12

```
FullSimplify[GUM[Cos[x], x == μ ± σ, 2]]
```

$$Cos[\mu] \pm \sqrt{\tfrac{1}{4}\cdot\sigma^4\cdot(-1+3\cdot Cos[2\mu])+\sigma^2\cdot Sin[\mu]^2}$$ Eqs 13

5. Uncertainty

The **Uncertainty** function propagate distributions to find the exact results analytically, numerically, or statistically.

Uncertainty[expr,case] in domain of symbols, rationals, or reals attempt to calculate the uncertainty (expectation ±sandard deviation) of expr.

Uncertainty[expr,case,n] in domain of reals calculate uncertainty of expr by using n pseudoradom variates.

The argument case specify statistical distributions of the variables as in the *Mathematica* function Expectation.

```
Uncertainty[c1 x1 + c2 x2,

{x1 ≈ NormalDistribution[μ1, σ1],

x2 ≈ NormalDistribution[μ2, σ2]}]
```

$$(c_1\mu_1 + c_2\mu_2) \pm \sqrt{c_1^2\sigma_1^2 + c_2^2\sigma_2^2}$$ Eqs 14

```
MatrixForm[Ξ = MultinormalCovarianceMatrix[{σ1, σ2}]]
```

$$\begin{pmatrix} \sigma_1^2 & \sigma_1\sigma_2 \\ \sigma_1\sigma_2 & \sigma_2^2 \end{pmatrix}$$

Eqs 15

```
PowerExpand[Simplify[Uncertainty[c1 x1 + c2 x2,
```

```
{x1, x2} ≈ MultinormalDistribution[{μ1, μ2}, Ξ]]]]
```

$(c_1\ \mu_1+c_2\ \mu_2)\pm(c_1\ \sigma_1+c_2\ \sigma_2)$ Eqs 16

```
PowerExpand[Simplify[Uncertainty[c1 x1 + c2 x2,
   {x1, x2} ≈ MultinormalDistribution[{μ1, μ2}, Ξ R]]]]
```

$(c_1\ \mu 1+c_2\ \mu_2)\pm\sqrt{(c_1^2\ \sigma_1^2+2\ c_1\ c_2\ \rho\ \sigma_1\ \sigma_2+c_2^2\ \sigma_2^2)}$ Eqs 17

Since the measurand is linear the GUM solutions Eqs 5, Eqs 6, and Eqs10 coincide with Eqs14, Eqs16, and Eqs17.

For the non-linear measurand **x^2** , **Sin[x]**, and **Cos[x]** the results are

```
Uncertainty[x^2, x ≈ NormalDistribution[μ, σ]] /.
Sqrt[a_] Sqrt[b_] :> Sqrt[Expand[a b]]
```

$(\mu^2+\sigma^2)\pm\sqrt{4\mu^2\sigma^2+2\sigma^4}$ Eqs 18

The exact analitical solutions for **Sin[x]** and **Cos[x]** needs more time measured in second by the function **Timing**.

```
Timing[Uncertainty[Sin[x], x ≈ NormalDistribution[μ, σ]] /.
   Sqrt[x_] :> Sqrt[FullSimplify[ExpToTrig[x]]] /.
  (x_ + Exp[y_]) :> Exp[y] (1 + Exp[-y] x)]
```

$\left\{213.986572, e^{-\frac{\sigma^2}{2}}\cdot Sin[\mu]\pm\frac{1}{\sqrt{2}}\cdot\sqrt{(1-e^{-\sigma^2})(1+e^{-\sigma^2}\cdot Cos[2\mu])}\right\}$ Eqs 19

```
Timing[Uncertainty[Cos[x], x ≈ NormalDistribution[μ, σ]] /.
 Sqrt[x_] :> Sqrt[FullSimplify[ExpToTrig[x]]] /.
  (x_ + Exp[y_]) :> Exp[y] (1 + Exp[-y] x)]
```

$\left\{142.647314, e^{-\frac{\sigma^2}{2}}\cdot Cos[\mu]\pm\frac{1}{\sqrt{2}}\cdot\sqrt{(1-e^{-\sigma^2})(1-e^{-\sigma^2}\cdot Cos[2\mu])}\right\}$ Eqs 20

The exact results for **x^2** , **Sin[x]**, and **Cos[x]** obtained by **Uncertainty** function (Eqs 18, Eqs 19, and Eqs 20) correspond to (Eqs 11, Eqs 12 and Eqs 13) obtained by **GUM** function.

Finally for a nonlinear measurand we generate 5 times 1 million random variates and extract the mean and standard deviation from the obtained data. The speed of computation is because *Mathematica* use a package array that permitted to do manipulations at once.

```
Timing[TableForm[Table[Uncertainty[x1^2 + x2^3,
   {x1 ≈ NormalDistribution[10, 0.1],
    x2 ≈ NormalDistribution[5,0.2]}, 10^6], {5}]]]
```

```
           225.627±15.2012
           225.632±15.1982
{2.059213, 225.579±15.1832},
           225.598±15.1781
           225.6±15.1789
```

Eqs 21

The exact numerical result is:

```
Uncertainty[x1^2 + x2^3,
{x1 ≈ NormalDistribution[10, 0.1],
```

```
x2 ≈ NormalDistribution[5, 0.2]}]
```

225.61 ± 15.1803 Eqs 22

6. CoverageInterval

CoverageInterval[expr,case,q] evaluate **CoverageInterval[expr,case,(1-q)/2,(1+q)/2]**
CoverageInterval[expr,case,q,n] evaluate **CoverageInterval[expr,case,(1-q)/2,(1+q)/2,n]**
CoverageInterval[expr,case,{q1,q2}] attempt to calculate coverage interval from **q1** to **q2**,
where **0<q1<q2<1**and **q1+q2==1**
CoverageInterval[expr, case, {q1,q2}, n] calculate coverage interval from **q1** to **q2** by using n pseudoradom variates.
The argument case specifies statistical distributions of the variables as in the *Mathematica* function Expectation.
The function **ToInterval** transform the uncertainty given by Eqs 17 as:
ToInterval[Eqs17]
Interval[{$c_1\ \mu_1+c_2\ \mu_2-\sqrt{(c_1^2\ \sigma_1^2\ +2\ c_1\ c_2\ \rho\ \sigma_1\ \sigma_2+c_2^2\ \sigma_2^2)}$,
$c_1\ \mu_1+c_2\ \mu_2+\sqrt{(c_1^2\ \sigma_1^2\ +2\ c_1\ c_2\ \rho\ \sigma_1\ \sigma_2+c_2^2\ \sigma_2^2)}$}], Eqs 23
For the same problem the function **CoverageInterval** gives

```
PowerExpand[Simplify[
  CoverageInterval[c1 x1 + c2 x2,
   {x1, x2} ≈ MultinormalDistribution[{μ1, μ2}, Σ R], q], 0 < q < 1]]
```

Interval[{$c_1\ \mu 1+c_2\ \mu_2-\sqrt{2}\ c_1^2\ \sigma_1^2\ +2\ c_1\ c_2\ \rho\ \sigma_1\ \sigma_2+c_2^2\ \sigma_2^2)$ InverseErfc[1-q],
$c_1\ \mu 1+c_2\ \mu_2-\sqrt{2}\ \sqrt{(c_1^2\ \sigma_1^2\ +2\ c_1\ c_2\ \rho\ \sigma_1\ \sigma_2+c_2^2\ \sigma_2^2)}$ InverseErfc[1+q]}], Eqs 24
We like to find such a parameter q that transform Eqs 24 in Eqs 23

```
FullSimplify[Solve[Eqs23[[1, 2]] == Eqs24[[1, 2]], q]][[1, 1]]
```

q→Erf[1/√2]
The obtained probability q is exact and could be computed numerically with any number of digits

```
N[%, 12]
```

q → 0.682689492137 Eqs 25
The following example shows that exact numerical solution spends more time that random variate method with a million variate.

```
Timing[CoverageInterval[
x^2, x ≈ NormalDistribution[10, 0.1], 0.95]]
```

{7.503648, Interval[{96.1185, 103.958}]} Eqs 26

```
Timing[CoverageInterval[
x^2, x ≈ NormalDistribution[10, 0.1], 0.95, 10^6]]
```

{1.357209, Interval[{96.1248, 103.972}]}, Eqs 27

7. Feasible Region

In the domain of symbols, the function **Uncertainty** obtain exact solution for uncertainty of the measurand **Exp[x^2]**, when **x** is normally distributed with mean μ, and standard deviation σ.
The exact solution is [15]:

$$\frac{e^{\frac{2\mu^2}{1-2\sigma^2}}}{\sqrt{1-2\sigma^2}} \pm \sqrt{\frac{e^{\frac{2\mu^2}{1-4\sigma^2}}}{\sqrt{1-4\sigma^2}} - \frac{e^{\frac{2\mu^2}{1-2\sigma^2}}}{1-2\sigma^2}}$$

For the value used in [14] μ=**0**, and σ=**1** the obtained solution becomes $-i\pm\sqrt{1+i\sqrt{3}}$, where $i=\sqrt{-1}$.
For this reason "Monte Carlo techniques is computationally problematic" [14]
The full analysis of the exact solution is not object of the present notebook. In addition to the well-known restriction σ>**0**, we take two more restrictions. Than we **Reduce** function finds the following feasible region:

```
Reduce[{σ>0,1-2σ2>0,1-4 σ2>0},σ],
0<σ<1/2,
```

Eqs 27

In the domain of **Rationals**, the function **Uncertainty** needs about 2 second to obtain uncertainty of **Exp[x^2]**, when **x** is normally distributed with mean **1**, and standard deviation **1/10**.

```
Timing[Uncertainty[Exp[x^2], x ≈ NormalDistribution[1, 1/10]]]
```

$$\left\{1.562500, \frac{5}{7}\sqrt{2}\cdot e^{\frac{50}{49}} \pm \sqrt{-\frac{50\cdot e^{\frac{100}{49}}}{49} + \frac{5\cdot e^{\frac{25}{12}}}{2\sqrt{6}}}\right\}$$

Eqs 28

Since the result is also in the domain of **Rationals** it could be transformed in **Reals** with arbitrary number of digits.
Next line show 20 digits:

```
2.8024933887978839525 ± 0.58552102828175616088
```

```
Timing[TableForm[
  Table[{Uncertainty[Exp[x^2],
    x ≈ NormalDistribution[1., .1], 10^6]}, {5}]]]
```

```
           2.80168±0.58508
           2.80251±0.58524
{0.468750, 2.80227±0.585468}
           2.80199±0.586149
           2.80262±0.585053
```

Eqs 29

For the values μ=**0**, and σ=**1** used by Bruce Christianson and Maurice Cox the solution is complex number and **Uncertainty** function return the input line after 327 seconds.
Finally we use different number of random variates to obtain completely non convergent results

"random variates"	"uncertainty"
1000	110.54±1880.48
10000	417.604±12989.1
100000	3182.44±431251
1000000	183460.±1.24656×10^8

Eqs 30

For **UnitSetep** function the feasible regions of the results are built-in *Mathematica.*

```
Uncertainty[UnitStep[x - a], x ≈ UniformDistribution[{2, m}]]
```

Journal of Physics: Conference Series **575** (2015) 012042 doi:10.1088/1742-6596/575/1/012042

$$\left(\begin{cases} 1 & a \; 2 \;\&\&\; m \; 2 \\ \frac{\cdot a \; m}{\cdot 2 \; m} & a \; 2 \;\&\&\; a \cdot m \; 0 \\ 0 & \text{True} \end{cases}\right) \; \sqrt{\begin{cases} \frac{2 a \cdot a^{2} \cdot 2 m \; a m}{(\cdot 2 m)^{2}} & a \; 2 \;\&\&\; a \cdot m \; 0 \\ 0 & \text{True} \end{cases}} \qquad \text{Eqs 31}$$

8. Acknowledgments

The authors are thankful to the Inmetro - National Institute of Metrology, Quality and Technology for the financial support to this research through a CNPq Research Fellowship - Edital MCT/ CNPq / Inmetro nº 059/2010 - PROMETRO Processo Nº 563061/2010-3. This work is a contribution of the INMETRO and is not subject to copyright in the Brazil.

9. References

[1] JCGM 100:2008 - GUM 1995 with minor corrections
http://www.bipm.org/en/publications/guides/gum.html
[2]. Uncertainty propagation software
http://en.wikipedia.org/wiki/List_of_uncertainty_propagation_software
[3]. V Y Aibe and M D Mikhailov, Uncertainty Calculus in Metrology, Proceedings of ENCIT 2008, 12th Brazilian Congress of Thermal Engineering and Sciences, Belo Horizonte, MG, November 10-14, 2008
http://repositorios.inmetro.gov.br/handle/10926/448
[4]. V Y Aibe, M D Mikhailov,
http://demonstrations.wolfram.com/SolvingMatrixSystemsWithRealIntervalOrUncertainElements/
[5]. V Y Aibe, M D Mikhailov,
http://demonstrations.wolfram.com/SolvingALinearSystemWithUncertainCoefficients/
[6]. V Y Aibe, M D Mikhailov,
http://demonstrations.wolfram.com/AreaPerimeterAndDiagonalOfARectangleWithUncertainSides/
[7]. V Y Aibe, M D Mikhailov,
http://demonstrations.wolfram.com/FitInterpolationOrPolynomialInterpolationInUncertainCalculus/
[8]. JCGM 101:2008 Supplement 1 to the GUM
http://www.sci.utah.edu/~kpotter/Library/Papers/jcgm:2008:PDMC/
[9]. V Y Aibe and M D Mikhailov, Uncertainty and UncertaintyGUM *Mathematica* functions,
http://repositorios.inmetro.gov.br/handle/10926/1898
[10]. M D Mikhailov and V Y Aibe.
http://demonstrations.wolfram.com/GUMVersus
ExactUncertaintyOfSinXAndCosXWhenXIsNormallyDistrib/
[11]. M D Mikhailov and V Y Aibe.
. http://www.wolfram.com/cdf-player/
[12]. M D Mikhailov and V Y Aibe.
http://demonstrations.wolfram.com/LogicalExpressionForAssumptionsOnParametersInMathematicasSym/
[13]. M D Mikhailov, V Y Aibe, and I Yasuda
http://demonstrations.wolfram.com/PercentilesOfContinuousProbabilityDistributions
[14]. Bruce Christianson and Maurice Cox, Automatic propagation of uncertainties,
In H. M. Bücker, G. Corliss, P. Hovland, U. Naumann, and B. Norris, editors,
Automatic Differentiation: Applications, Theory, and Implementations,
Lecture Notes in Computational Science and Engineering, pages 47-58, Springer, 2005.
[15] I Yasuda, M D Mikhailov and V Y Aibe, New *Mathematica* Uncertainty Analysis Tools, Proceedings of Flomeko 2013, 17th International Flow Measurement Conference, Paris - France, November 24-26, 2013.

Proposal of Scope of Clinical Assays of Safety and Effectiveness of Cosmetic Products

L S Almeida[1,4]

[1] National Institute of Metrology, Quality and Technology - Inmetro, General Coordination for Accreditation. Av. Nossa Senhora das Graças, 50, Duque de Caxias-RJ, CEP 25250-020, Brazil.

lsalmeida@inmetro.gov.br

Abstract. The demand for accreditation for clinical assays involving cosmetic products has led the Division of Laboratory Accreditation (Dicla) to study the possibility of implementing the General Coordination for Accreditation (Cgcre) which is a specific accreditation program for Good Clinical Practice (GCP). This work represents the very beginning of such a study thus conveying a proposed model of scope for clinical assays on safety and effectiveness.

1. Introduction

1.1. The Mission and Function of Inmetro Cgcre/Dicla
The Inmetro is a federal agency whose mission is to "provide confidence to the Brazilian Society regarding measurements and products through metrology and conformity assessment, thus promoting harmonization of consumption relationships innovation and competitiveness of the country" [1]. Among the powers and duties of Inmetro the accreditation activities stand out, structured under Cgcre (General Coordination for Accreditation), which is in charge of accreditation of conformity assessment bodies recognized by the Brazilian government [2].

In this context, it is noteworthy that the laboratory accreditation by Cgcre is performed by the Division of Laboratory Accreditation (Dicla), which operates in the following modes of accreditation/ recognition: i) accreditation of testing and calibration laboratories, following the requirements of ABNT NBR ISO / IEC 17025:2005; ii) accreditation of clinical laboratories in accordance with ABNT NBR NM ISO 15189:2008; iii) accreditation of producers of reference materials in accordance with ABNT ISO Guide 34:2012; iv) accreditation of proficiency testing providers, through ABNT NBR ISO IEC 17043:2010 and v) recognition of conformity to the principles of Good Laboratory Practice (GLP) according to the definitions explained in NIT-Dicla-035: Principles of Good Practice Laboratory .

1.2. Conformity Assessment Bodies (CABs)
In order to promote the CABs as well as the testing facilities, which are recognized under GLP, Cgcre provides, at the site of Inmetro the scope of each accredited CAB and test facility that has recognition of conformity with the GLP.

[4] To whom any correspondence should be addressed.

Journal of Physics: Conference Series **575** (2015) 012043 doi:10.1088/1742-6596/575/1/012043

The scope of each CAB is a formal document, which sets Cgcre tests, calibrations, certified reference materials and proficiency testing of accredited providers, as well as the tests that were recognized by Cgcre.

1.3. The Current Demand for Accreditation of Testing on Cosmetics
Currently, Dicla has faced a demand of requests for accreditation for tests involving cosmetic products.

Perhaps this situation can be best explained by the fact that the companies have increasingly looked for recognition for the services they provide, coupled with the high rate of use of cosmetic products in Brazil. In 2011, Brazil was ranked as the third largest consumer of cosmetics [3].

Just as occurs with other products that promote health and the well - being of consumers, cosmetic products may occasionally experience adverse reactions in users due to individual factors or improper use of the product [4-5].

Therefore, the National Health Surveillance Agency (Anvisa) establishes several regulations so as to standardize the registration of cosmetics in the country.

This work is aimed at studying the main Standards and Guidelines established by Anvisa, regarding the regulation of cosmetic products in order to begin a study to verify the feasibility of establishing a program at Cgcre/Dicla accreditation / recognition of a clinical assay for those products.

2. Materials and Methods
The development of the work consisted of 4 distinct and subsequent phases.

The first phase consisted in the study and bibliographic surveys of the Standards and Guidelines established by Anvisa.

In the second phase the tests required by Anvisa that Dicla does not yet have were verified.

The third phase consisted of a query to the sites of organizations from different body country accreditors.

In the fourth and final stage, a model of scope for clinical trials of safety and effectiveness of cosmetic products was suggested.

3. Results and Discussions

3.1. 1st phase: Study and Surveys of Anvisa Standards
A study of the Standards requested by Anvisa was carried out in order to establish criteria concerning the registration of products in Brazil.

3.1.1. RDC Resolution 211, July, 14th, 2005
The classification of cosmetics is established in this resolution, which was defined in terms of the likelihood of unwanted effects due to inappropriate use of the product, its formulation, purpose of use, areas of the body as intended and precautions to be observed when using it [6]. The products were classified as "Product Grade 1 and Product Grade 2" [6].

According to this resolution, required physic- chemical and microbiological tests are required on raw materials and finished products for Risk Level 1 products.

For Risk Level 2 products are required in addition to those listed for the Risk Level 1, the pre-clinical safety assays and the clinical assays of safety and effectiveness in the finished products are also required.

3.1.2. RDC Resolution 38, March, 21th, 2001
This resolution describes the criteria and procedures for the registration of new categories of cosmetic products intended for children use [7].

In conformity with resolution, the cosmetic products for children are classified as Product Grade 2.

The relevant information of this resolution is the specification of clinical assays that are required by ANVISA for registration of children cosmetic product.

3.1.3. RDC Resolution 237, August, 22th, 2002
This resolution approves the technical regulations for sunscreens in cosmetics.

According to this resolution, ANVISA states specifically what are the tests to be used for sunscreens for determining the Sun Protection Factor (SPF) as well as the use of the terms "water resistant" or "very water resistant" on product labeling [8]. All assays are classified as effectiveness assays.

3.1.4. Guide for Safety Evaluation of Cosmetic Products
This Guide ranks the trials into two groups: studies of compatibility and acceptability studies [9]. It also lists the clinical safety trials according to the classification of the study.

Besides the description of these tests, this guide provides ANVISA clinical trials to be conducted in cosmetics when the manufacturer indicates to consumers that the product has the following security attributes: Tested in terms of Dermatology and Ophthalmology, Hypoallergenic. It also describes the tests to be performed when the product meets certain safety standards, according to the target audience, such as children or pregnant women or sensitive skin people [9].

3.1.5. Technical Chamber of Cosmetics (CATEC)
In addition to the standards cited, there are some opinions of CATEC of Anvisa establishing the need for proof of efficacy and safety for certain products. For example: Products containing Vitamin C, Dimethylaminoethanol (DMAE) and Fostatidilcolin; products indicated for dark circles; products for cellulite and stretch marks and fat, among others [10].

3.2. 2nd phase: Testing verification required by Anvisa
In this phase it was realized that Dicla does not only have a program for BPC accreditation / recognition for clinical assays, among the tests required by Anvisa for registration of cosmetic products in Brazil.

3.3. 3rd phase: Consultation on websites of accreditation bodies
A consultation was held on the websites of the following accreditation organizations to check if any of them has the accreditation program / recognition in BPC: IPAQ (Portugal), UKAS (United Kingdom), NATA (Australia), TVR (Netherlands), A2LA (USA) A Class (USA), EMA (Mexico) and CNCA (China).

Of all the bodies consulted, none has the accreditation program / recognition for BPC.

3.4. 4th phase: Proposal of Scope of Clinical Assays of Effectiveness and Safety of Cosmetic Products
Taking into consideration the consulted documents as well as the standard formatting of the Dicla Scope, two tables were devised, 1 and 2 below, one model of scope for clinical assays and another model for safety and effectiveness of cosmetic products.

Table 1. Model of Scope for Clinical Assays of Safety for Cosmetic Products

Classification of Clinical Assay/Cosmetic Product	Classification of study/Assay description	Norm and/or procedure
Safety Assays Chemical "peeling"	Compatibility Study Evaluation of Dermal Sensitization	POP 540 Rev.01, 2013
Safety Assays Products for acne prone skin	Acceptability Study Evaluation of acnegenicity	IT 98 Rev.00, 2012

Table 2. Model of Scope for Clinical Assays of Effectiveness for Cosmetic Products

Classification of Clinical Assay/Cosmetic Product	Assay description	Norm and/or procedure
Effectiveness Assays Sunblock / anti-solar sunscreen	Evaluation of the dry solar sunscreen factor	International Sun Protection Factor (SPF) Test Method, Colipa 2006
Effectiveness Assays Kids sunscreen	Evaluation of the determination of "extremely resistant to the water"	Norm FDA 2012 – Sunscreen drug products for over-the-counter human use, Code of Federal Regulations, Title 21, Vol.5, Sec.352.76

4. Conclusions

After all the studies we came to the conclusion that this work can be viewed as a initial study regarding the possibility of the implementation of an accreditation program/recognition BPC under Cgcre/Dicla, and If it materializes it surely will be of great value to the Brazilian society.

References

[1] Inmetro. O que é Inmetro [on line]. Rio de Janeiro, Brazil; 2013. [captured 11 may 2013] Available in: http://www.inmetro.gov.br/inmetro/oque.asp
[2] Brazil. Presidency of the Republic of Brazil. Law 1245, of 14 december of 2011.
[3] Chiari B G et al 2012 Int J Res CosmetSci **2** (2) 14
[4] Rogiers V 1999 Atla **27** 4
[5] Santos H 2008 Cosm Toil **20** 2.

[6] Brazil. Resolution 211, July, 14 th, 2005, Establish the Definition and the Classification of Personal Care Products, Cosmetics and Perfumes, in conformity with Annex I and II this Resolution and give others recomendations. Union Official Journal. Brasília, 18 jul. 2005.

[7] Brazil. Resolution 38, March, 21th, 2001, Approve the Technical Regulamentation of Cosmetic Products for Kids.

[8] Brazil. Resolution 237, August, 22th, 2002, Approve the Technical Regulamentation about Sunblocks in Cosmetic Products. Union Official Journal. Brasília, 23 agos. 2002.

[9] Brazil. Guide for Avaluation of Safety of Cosmetic Products. Brasília: Anvisa, 2012.

[10] Anvisa. Technical Chamber [on line]. Brasília, Brazil; 2013. [captured 22 may 2013] Available in:
http://portal.anvisa.gov.br/wps/content/Anvisa+Portal/Anvisa/Inicio/Cosmeticos/Assuntos+de+Interesse/Camara+Tecnica

Journal of Physics: Conference Series **575** (2015) 012044 doi:10.1088/1742-6596/575/1/012044

Review on how proficiency testing needs in Brazil are supplied by accredited providers by Cgcre

M H Moura[1] and R M H Borges
National Institute of Metrology, Quality and Technology (INMETRO), Duque de Caxias, RJ, Brazil

E-mail: mhmoura@inmetro.gov.br

Abstract. Proficiency testing schemes are an important tool to quality assurance in measurement as well as a tool to harmonization of multilateral recognition arrangements for accreditation. The General Coordination for Accreditation (Cgcre) of INMETRO developed a new program to accredit proficiency testing providers according with the International Standard ISO/IEC 17043. This work presents a review on needs for proficiency testing schemes in Brazil and assesses how these needs are supplied by accredited providers.

1. Introduction

Laboratory accreditation is a means of determining the technical competence of laboratories to perform specific types of testing and calibration, providing the formal recognition to competent laboratories, thus a ready means for customers to identify and select reliable testing, measurement and calibration services able to meet their needs. According with the international standard ISO/IEC 17025 accredited testing and calibration laboratories shall establish quality control procedures for monitoring the validity of tests and calibrations undertaken. This monitoring shall be planned and reviewed and may include participation in interlaboratory comparison or proficiency testing schemes [1].

The need for ongoing confidence in laboratory performance is not only essential for laboratories and their customers but also for other interested parties, such as regulators, laboratory accreditation bodies and other organizations that specify requirements for laboratories [2], which can be demonstrated through laboratory participation in a proficiency testing scheme. Besides, the International Standard ISO/IEC 17011 [3] and the document published by the International Laboratory Accreditation Cooperation (ILAC), ILAC P9, establish as a requirement the accreditation bodies shall consider the participation and the performance of laboratories in proficiency testing schemes [4].

Proficiency testing providers are organizations that develop and operate proficiency testing schemes [2] and their competence can be evaluated based on ISO/IEC 17043 published in 2010 replacing the ISO Guide 43:1997 and ILAC-G13:08/2007 document [5].

In line with the ILAC resolutions [6, 7], accreditation bodies have implemented programs to guarantee the accreditation of proficiency testing providers in different regional cooperations for accreditation.

[1] To whom any correspondence should be addressed.

1.1. National review
In Brazil, the General Coordination for Accreditation (Cgcre) of the National Institute of Metrology, Quality and Technology (INMETRO) is the legally recognized accreditation body to perform the activities of accreditation of calibration and testing laboratories, proficiency testing providers and other conformity assessment bodies [8].

According to statistical data from March 2013, there are 836 testing and calibration laboratories accredited by Cgcre, which stimulates the demand for proficiency testing schemes in Brazil since the Cgcre policy requires regular participation of its laboratories in proficiency testing activities [9].
In order to encourage the provision of proficiency testing programs in Brazil by organizations with recognized competence Cgcre developed the modality of accreditation of proficiency testing providers. It was established initially a pilot project aimed at learning this new modality [10], as requirements of ILAC-G13:08 guide, prevailing at the time.

This paper presents how the program for accreditation of proficiency testing providers was developed by Cgcre and evaluates the meeting the demand in Brazil for proficiency testing schemes by accredited providers. In the context of a larger study still under development, which includes this work, it will explore the characteristics and demands of the proficiency testing schemes, regulatory requirements, criteria for selection of programs, studies of homogeneity and stability of proficiency testing items and statistical methods among other aspects, in order to subsidize Cgcre in improving its process.

2. Basis for the accreditation program of proficiency testing providers by Cgcre

2.1. Overview
The Brazilian Accreditation Body (Cgcre) decided to establish a proficiency testing provider (PTP) accreditation program based on the experience of the United Kingdom Accreditation Service (UKAS). In 2006, it was published in Accreditation and Quality Assurance an article about the pilot accreditation program of reference materials producer accreditation (RMP), involving five participant organizations. The accreditation criteria used by UKAS was ISO Guide 34 in combination with ISO/IEC 17025. Considering this paper, Cgcre established a similar approach to implement a pilot program for PTP accreditation based on ILAC G13.

Table 1 shows the steps followed to implement the PTP accreditation in Brazil.

Table 1. Steps followed to implement the pilot program for PTP accreditation in Brazil by Cgcre.

Period	Activities
2006	Study of technical feasibility.
2008	Training course to assessors and the Cgcre's staff including presentations from international experts in PTP accreditation.
2009	Establishment of documentation for the implementation of the pilot program in Brazil
	I Seminar opened to the public to more than 170 people with presentations done by representatives of ILAC and APLAC, also training Brazilian professionals, assessors and Cgcre´s staff. Revised documents were made available in Cgcre´s website and the PTP were invited to apply for accreditation, till October 2009. Received applications were reviewed by a group of Cgcre´s assessors and professionals. Twelve out of fourteen applications were accepted in the pilot accreditation program.

Journal of Physics: Conference Series **575** (2015) 012044 doi:10.1088/1742-6596/575/1/012044

	Document review of PTP by trained assessors, including statisticians.
2010/2011	Dates for the assessments were scheduled and the assessment visits were carried out in 12 PTP facilities, from October 2010 to April 2011. Delivery of certificates for granting accreditation to 12 providers in September 2011 at the 2nd Workshop on Accreditation of PTP.

2.2. Implementation of the accreditation program in the Cgcre's process

With the experience and improvement process in the pilot project, the new modality was offered to all interested PT providers following the Cgcre's process in 2011. Up to now, there was 1 new request for initial accreditation and 3 requests for the extension of the scope of accreditation.

3. Analysis of needs of proficiency testing schemes in Brazil

3.1. Needs for PT schemes

Testing and calibration laboratories accredited by Cgcre represent much of the specific needs for proficiency testing schemes in Brazil in their different fields of activities. A survey with accredited laboratories until March 2013 for each field of activity (for testing laboratories) and each group of calibration services, shown in figure 1 and figure 2 was performed.

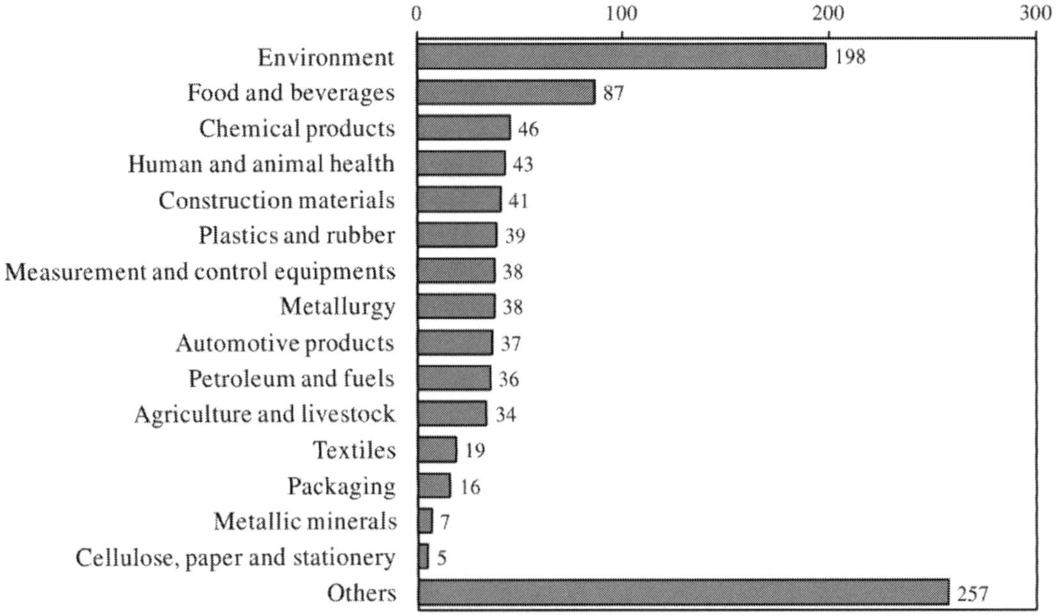

Figure 1. Number of testing laboratories accredited by Cgcre according to fields of activity.

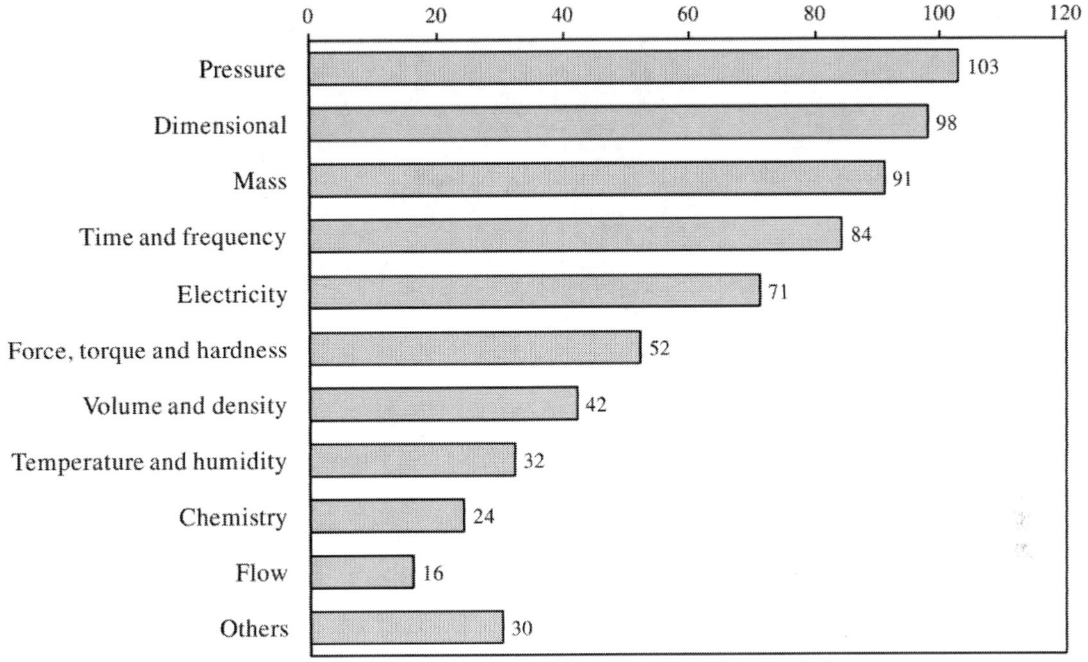

Figure 2. Number of calibration laboratories accredited by Cgcre according to groups of calibration service.

3.2. Offering of proficiency testing schemes

Based on the scopes of accredited PTP, the available schemes were categorized for the same fields of activity and groups of calibration services. Additionally, a survey was conducted with the accredited PT providers, identifying 11 (92%) of the 12 PT providers are planning to request extension of their scopes of accreditation for other schemes over the next three years, including different proficiency testing items and statistical approaches.

Figure 3 and Figure 4 show the number of accredited and non-accredited proficiency testing schemes (with possible extension of accreditation scope) offered by PT providers accredited by Cgcre.

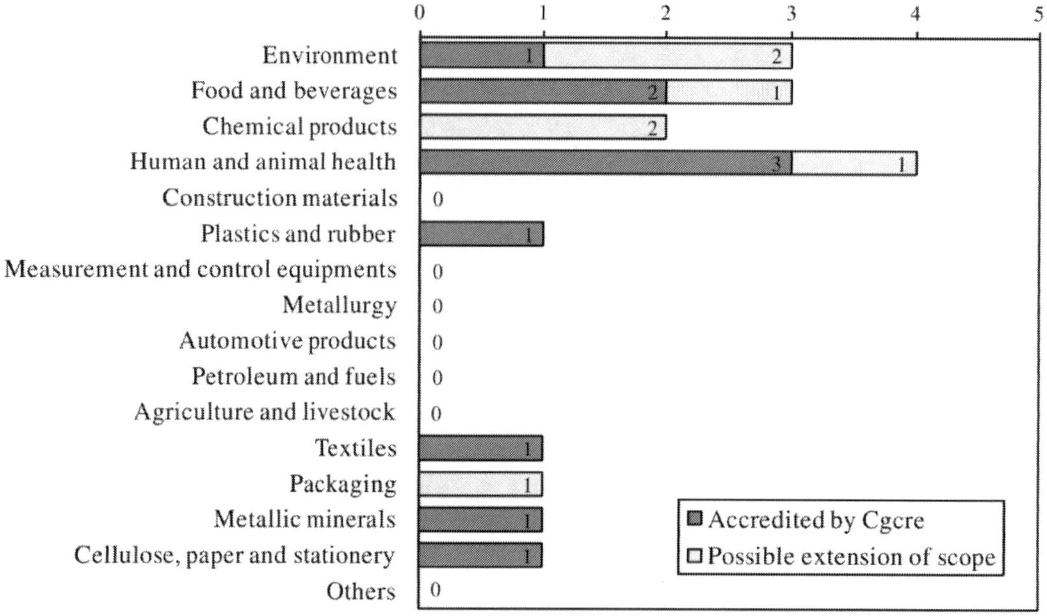

Figure 3. Number of proficiency testing schemes for testing laboratories according to fields of activity.

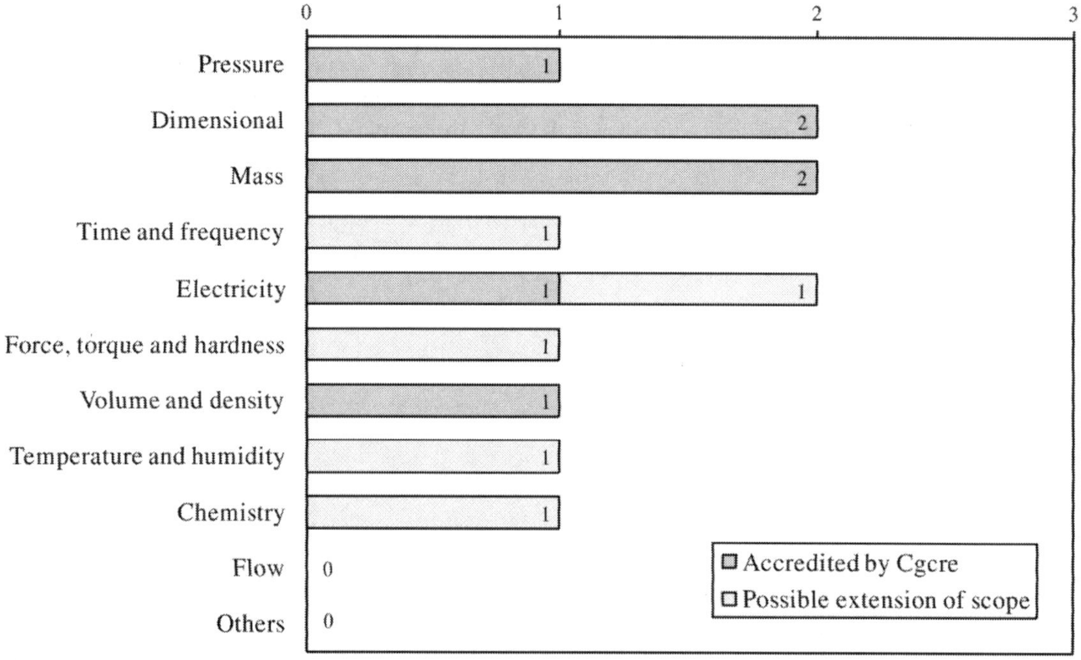

Figure 4. Number of proficiency testing schemes for calibration laboratories according to groups of calibration service.

3.3. Comparing the offering and needs of proficiency testing schemes
Reviewing the scopes of accreditation of different proficiency testing providers it was possible to identify accredited proficiency testing schemes for 7 (44%) of the 16 fields of activities, while 5 (45%) of the 11 groups of calibration services with accredited calibration schemes. Adding to the possible extensions of accreditation scopes, these values increase to 9 (56%) and 9 (86%), respectively.

3.4. Comparing the offering and needs of proficiency testing schemes
The laboratories that intend to obtain accreditation shall demonstrate their competence to perform tests and calibrations for which they seek accreditation, through satisfactory participation in proficiency testing activities. The standard NIT-DICLA-026 details the requirements for participating in proficiency testing activities. Cgcre provides information regarding proficiency testing schemes available in Brazil and abroad, in addition to documents and webpages on the matter. Cgcre established in its policy criteria for selection of PT providers [9], which in addition to the programs offered by accredited providers, accepts, for example, proficiency testing schemes offered by:

- Cgcre by itself through its Metrological Reliability Sector of Dicla (Secme) or its technical advisory committees. These specific sector organizes proficiency testing programs (measurement audits[2]) to 8 (73 %) of the 11 groups of calibration services;
- INMETRO - the National Metrology Institute;
- Proficiency testing providers registered in the International Database of Proficiency Testing Schemes (EPTIS), www.eptis.bam.de. In April 2013, there were 125 Brazilian PT schemes registered in this specific database and 9 (75 %) of the 12 PT providers accredited by Cgcre have registered schemes in EPTIS.

In case the laboratory proves that there are no proficiency testing scheme available and if the laboratory does not have access to interlaboratory comparisons mentioned in Cgcre's policy, the laboratory can demonstrate its competence by satisfactorily participating in an interlaboratory comparison whose purpose is to compare the results of two or more laboratories, organized by itself [9].

4. Conclusion
The pilot program established by Cgcre and the review of the accreditation process of proficiency testing providers adopted showed to be effective, considering that the pilot project resulted in the accreditation of the first providers in Brazil and in the dissemination of the accreditation process to assessors and Cgcre's staff. The results demonstrate that these proficiency testing providers have provision a significant portion of the fields of activity and groups of calibration services, representing the needs of testing and calibration laboratories accredited by Cgcre. However, there are gaps in the provision of proficiency testing schemes, for example, 7 (44 %) of the fields of activity are not supplied by accredited proficiency testing schemes and also no perspective to the extension of accreditation scope in the short term by accredited PT providers .

The results obtained allow Cgcre to establish preview of the possible scope of accreditation by PT providers to gauge the needs of its process, for example, prospecting and/or training of assessors/ experts in the different fields detected. In addition, the data obtained in this work still informs proficiency testing providers or other interested organizations to potential areas where there are needs for the development of proficiency testing schemes.

[2] Measurement audits are interlaboratory comparisons organized by Secme, used by the assessment team to assess the technical competence of laboratories to perform calibrations, including internal calibrations.

5. References

[1] Associação Brasileira de Normas Técnicas (ABNT) 2005. ABNT NBR ISO/IEC 17025. Rio de Janeiro, ABNT
[2] Associação Brasileira de Normas Técnicas (ABNT) 2011. ABNT NBR ISO/IEC 17043. Rio de Janeiro, ABNT
[3] Associação Brasileira de Normas Técnicas (ABNT) 2005. ABNT NBR ISO/IEC 17011. Rio de Janeiro, ABNT
[4] International Laboratory Accreditation Cooperation (ILAC) 2010. ILAC-P9:11. Silverwater, ILAC
[5] Gun-Munro J 2012. Accred Qual Assur. 17 363 - 70
[6] International Laboratory Accreditation Cooperation (ILAC) 2005. Adopted resolutions of the 9th GA. Auckland, ILAC
[7] International Laboratory Accreditation Cooperation (ILAC) 2009. Adopted resolutions of the 13th GA. Vancouver, ILAC
[8] BRASIL. Decreto nº 6.275, de 28 de novembro de 2007
[9] Instituto Nacional de Metrologia, Qualidade e Tecnologia (INMETRO) 2011. NIT-DICLA-026_08. Rio de Janeiro, INMETRO
[10] Borges R M H and Moura S 2011. IAAC Regional Talk. 12 13 - 5

VII Brazilian Congress on Metrology (Metrologia 2013) IOP Publishing
Journal of Physics: Conference Series **575** (2015) 012045 doi:10.1088/1742-6596/575/1/012045

Reliable data acquisition for inspection systems

V P Silva, D S Silva, D R Boccardo, R C S Machado and L F R C Carmo

Inmetro — National Institute of Metrology, Quality and Technology, Rio de Janeiro, Brazil

E-mail: {vpraia-prometro, dssilva, drboccardo, rcmachado, lfrust}@inmetro.gov.br

Abstract. In Brazil, the road transportation of dangerous goods is subject to regulatory control, which is conducted by Inmetro, the National Institute of Metrology, Quality and Technology. Currently, the process of monitoring such inspections is done manually, leaving the possibility of some inconsistencies: inspections held outside the authorized local inspection, incorrectly, incompletely, or even be held. Thus, in order to increase the reliability of such inspections, it is needed a closer monitoring by the regulatory agency. One approach towards that, is the implementation of an automated process in which evidences are collected in a reliable way during the inspections, enabling further analyses. This work employs security mechanisms on a portable device to ensure the confidence of the evidences collected during an inspection, paving the way for later more robust analyzes.

1. Introduction

In Brazil, the National Institute of Metrology, Quality and Technology (Inmetro) is responsible for the conformity assessment program for inspection of vehicles and equipments that transports dangerous goods road [1]. Until very recently, the inspection in such vehicles and equipments where was made in carriers or shippers of dangerous goods, which made it difficult to trace inspections records, and consequently, with little or no monitoring by the government authorities. Recently, via Regulation 91 of 2009 [2], Inmetro began a new phase in this program. This regulation dictates, among other things, that the entire inspection should be conducted on an accredited inspection site.

This regulation represented a vast improvement in the dangerous good transportation. However, there are evidences that a large number of inspections is still conducted in improperly ways: such inspections are held outside the authorized local inspection, incorrectly, incompletely, or even are not held. In addition, one cannot be confident on the correctness of the data generated in such inspections. Such scenario suggest that it may be necessary a closer monitoring by regulatory agency to monitor and verify the reliability of inspections. In the present paper, we describe a model developed at Inmetro that aims at monitoring inspections on vehicles and equipment that transport dangerous goods [3]. Our model is strongly based on an *a posteriori* analysis of the information generated by the inspection equipments. Any inconsistency in the analysis leads to suspension or even cancellation of accreditation of an inspection body.

In this model, the main idea is the acquisition of reliable evidences indicating the proper execution of the inspection by accredited inspection bodies. These evidences are composed by data and images recorded at key points of the inspection process. Each evidence is linked to time information — instant record generation — and space — the geographical location where the record was generated. Finally, the record time and its spatial position are digitally signed in order to trace the equipment origin, including the time and coordinates of the records. This work presents the implementation of a mobile device that will be used to collect records of a inspection in a reliable way. More specifically, this work focuses on mechanisms employed to ensure security properties of the records, so guarantying the reliability of a inspection.

This work is structured as follows. Section 2 discusses the inspection system to perform an automated and reliable inspection for dangerous good transportation. It also specifies some requirements for the evidences collection in such systems. Section 3 presents our implementation in a mobile device responsible for the data acquisition and how we addressed reliable issues. Section 4 presents our concluding remarks.

2. Proposed Model for Enhancing the Reliability of Inspection Systems

Historically, in Brazil, all inspections were done on checklist paper and the checking analysis conducted by sampling. However, this kind of checking is highly prone to be forged, leading to inaccurate records, then misleading checking analysis.

The following model overcomes such a problem and it is based on a post checking analysis. The core idea is a trustful generation, by the inspection body, of some evidences of the inspection, with the subsequent analysis of those by Inmetro. The evidences generated consist of a set of records and images, and their respective geographic and temporal information, all digitally signed, guaranteeing the authenticity and integrity of them. Inmetro is responsible for reviewing all the evidences of each inspection, searching for any inconsistencies that may compromise its reliability to ensure that it was performed correctly. Any deviations committed by inspection bodies will lead to punishments such as suspension or cancellation of the accreditation of the inspection body.

The main advantage of this model is that it offers very low impact to current inspection processes conducted in accredited inspection bodies. The only differences are the following tasks:

- evidences generation (photographic and records) at key points of the inspection process;
- evidences submission to Inmetro — along with additional information.

Hence, the inspection process will suffer little change. The proposed model for enhancing the reliability of inspection systems, represented in Figure 1, consist of three main parts: evidence acquisition subsystem, communication subsystem, and post analysis checking subsystem.

Evidence Acquisition Subsystem. The evidences of conducting an inspection will be a set of images and records associated with geographic and temporal information. The records represent the data inserted throughout the inspection process. These evidences will be generated by GPS-equipped devices with photo capability. Moreover, all evidences will be digitally signed by the equipment, ensuring the authenticity and integrity of such evidences. Such evidences are said traceable, because the moment and place of its execution can be identified.

Communication System. Once completed an inspection (or set of inspections) and produced the necessary data related to this inspection, it becomes necessary to consolidate such data in the form of a structured file and forward it to Inmetro for later analysis.

Post Checking Subsystem. Every inspection conducted by an inspection body shall be communicated to Inmetro, along with its set of evidences. The Inmetro will receive data from an inspection and review them, seeking for inconsistencies by making use of the records and images, and their associated spatial and temporal information.

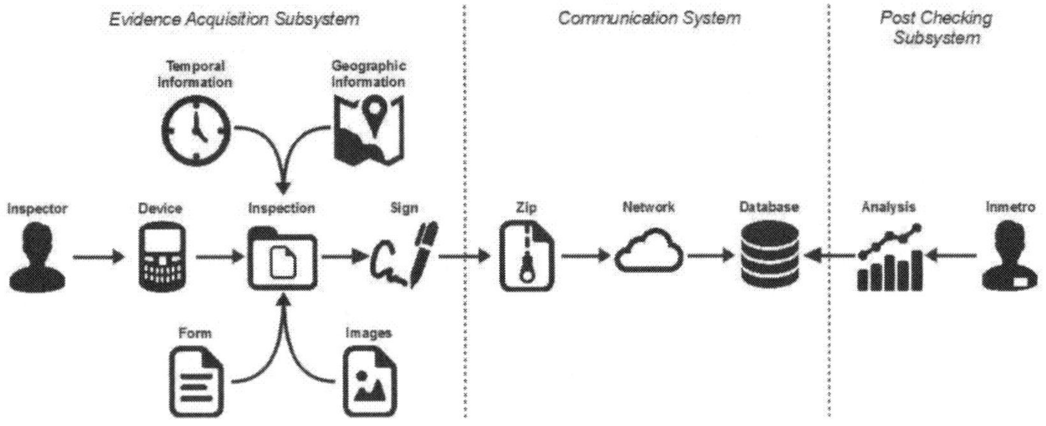

Figure 1. Proposed model for Inspection Systems.

3. Evidence Acquisition Subsystem

In this work, we focus on the evidence acquisition subsystem since it is the core of the model and it is prone to Man-At-The-End attacks. It is the core of the model since it provides the evidences in their origin; and it is prone to Man-At-The-End attacks because an adversary with physical access to the device is capable of compromising it, by inspecting or tampering with the hardware or software it contains.

The evidence acquisition subsystem, in our model, will be embedded in a mobile device that will be used by accredited inspection bodies for each vehicle or equipment inspection performed. The embedded software contains forms for each kind of inspection to be filled by the inspector, and security mechanisms to guarantee authenticity and integrity of the evidences, and to deter Man-At-The-End attacks. The security requirements for this equipment are discussed in the following:

- Authenticity of the evidence. The device must provide a means to verify the origin of the inspection evidence, ensuring that they were generated by a trusted device.
- Integrity of the evidence. The device should provide assurance that the inspection evidence remained the same (intact) from its generation. This aspect concerns the preservation of the evidence generated during an inspection, making it difficult to change without the consent of the responsible body.
- Tamperproofing of the embedded software. The device should contain mechanisms to keep always in the expected state even before tampering or surveillance. This aspect aims to provide device defenses against modification or monitoring functions of the embedded software.
- Confidentiality of secret information. The device must have mechanisms to protect the confidentiality of secret information, for example, cryptographic keys used in the device. An attacker on the possession of such keys could compromise the information security aspects of giving legitimacy to a given inspection that, in fact, is untrue.

3.1. Prototype Implementation

A prototype was developed in the Android platform since it provides greater ease of implementation, as well as greater popularity, which ensure greater portability in the case of prototype approval. During development and testing were collected results which endorsed the feasibility of this inspection system model, however some problems occur in the use of the

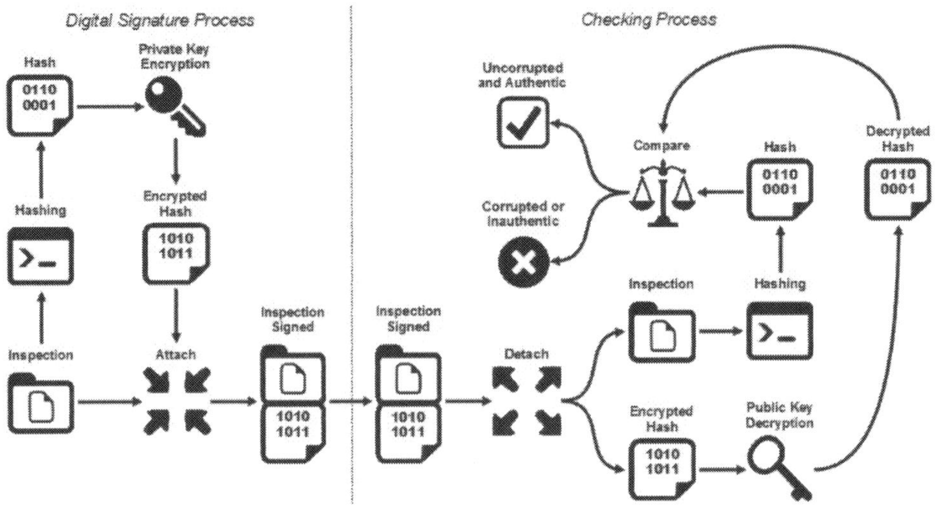

Figure 2. Digital signature process carried by the handheld device, and checking process to validate the digital signature.

Android platform. The favorable results to the inspection system were the various inconsistencies that could be observed by analyzing the inspection records generated during the test, which confirmed the effectiveness of our model. The unfavorable results were related to aspects of usability and security. The Android platform besides presenting vulnerability to various forms of attacks, did not allow the development of effective security mechanisms as planned. Other problems were the fragility and difficulty of using devices equipped with Android, such as smartphones and tablets, in the environment of an inspection body.

3.2. Final version

In order to overcome the prototype issues, we adopted a mobile device with the Windows Mobile Operating System. This device presents robustness and ease of use more suitable to the environment where they are carried out the inspections, and provides a more effective implementation of security mechanisms. An important aspect refers to the environment in which the device is operated, where there is no form of control over who or how it is operating. So, then we detail the techniques that were applied aiming to guarantee aspects of information security.

Digital signature. This technique deals with the integrity and authenticity of the evidences of an inspection. The implementation of this technique is performed by calculating the cryptographic digest (MD5) of the inspection with its correlated time and spatial information. After, it encrypts using asymmetric encryption (RSA). The result of this operation is the digital signature. More details about digital signatures can be found on [4].

In order to verify the authenticity and integrity of the inspection, the cryptographic summary is calculated and compared with the digital signature decryption. If the comparison result is equal it is checked the authenticity and integrity of such inspection. Otherwise, it means that the data was corrupted or is not authentic. Figure 2 illustrates the technique: the digital signature performed in the mobile device, and the checking procedure carried by Inmetro.

Obfuscation of secret information. The implementation of this technique is to make changes in the code of a software, preserving the same functionality, so that it shows greater difficulty

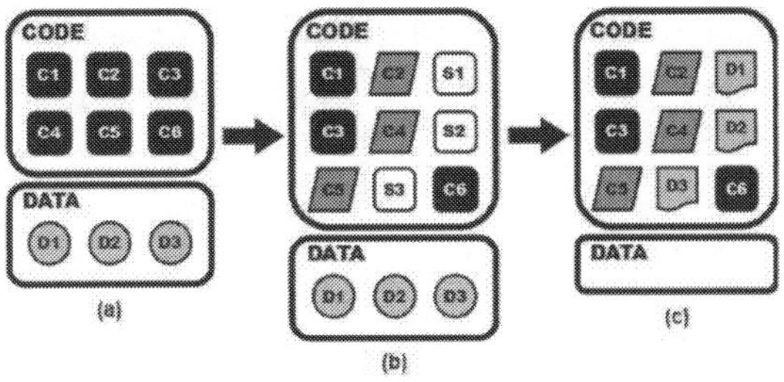

Figure 3. Code obfuscation technique: (a) non-ofuscated code, (b) obfuscated code after applying control obfuscations techniques and (c) after camouflaging the data.

on reverse engineering.

A software is commonly divided into two segments: (i) code segment, containing the instructions to be executed, and (ii) data segment, where data used by the software (constants and variables) are stored. However, this provision facilitates reverse engineering.

So, we used the obfuscation of secret information technique, described in details in [5], which camouflages the data of the data segment in the code segment, leading to incorrect interpretation of these data by reverse engineering tools. Such a technique uses obfuscation control techniques (call, return and return false) [6] in order to create sections in the code segment, capable of storing data. Using this technique to camouflage secret information, such as cryptographic keys, is critical to ensure authenticity and integrity security aspects. Figure 3 illustrates this technique and its steps: (a) non-ofuscated code, (b) obfuscated code after applying control obfuscations techniques and (c) after camouflaging the data. Observe that, in (a) the secret information is clearly visible to an reverse engineer attacker in the data segment, (b) we apply control obfuscation techniques to create sections in the code segment, capable of storing data, and in (c) the secret information is interleaved with other instructions, making it harder to reverse engineering.

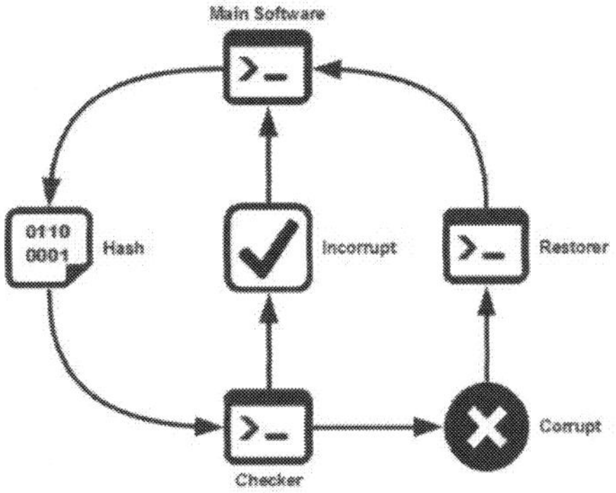

Figure 4. Tamperproofing technique: Self-repairing corrupted software.

Tamperproofing. This technique deals with the tamperproofing of the embedded software. It works in two steps: first by looking up some corruption in the software running in memory and, in the second, taking a response action, for example, self-repairing the corrupted software[7]. This technique was implemented in software, and it runs concurrently, always looking for some corruption of the software. It performs the comparison by calculating the cryptographic digest (hash) of an authentic (unmodified) embedded software and the current embedded software. If the comparison result is equal the embedded software is not corrupted and the device keeps running. Otherwise, the modified embedded software is replaced with the unmodified one and the device is restarted. Figure 4 illustrates this technique.

To give greater robustness to the confidentiality of the secret data and the integrity of the embedded software running on the device, some physical restrictions were applied in the device to enhance the security. For example, the USB communication ports were disabled by hard locks and the shortcut keys of the operating system were disabled by register manipulation.

4. Conclusions

Our proposed model for the evidence acquisition subsystem is critical to aggregate trust of an inspection system since it provides reliable acquisition of evidences in their origin. Such reliable acquisition is obtained by the employment of security mechanisms. Once the evidences are acquired, they are used for a more effective and efficient post checking analyses by Inmetro, looking for clues that indicate inconsistencies in the inspections.

References

[1] Decree n. 96.044 , May 18 1988
[2] Regulation INMETRO n. 91, March 31 2009
[3] Machado R C, Boccardo D R, Carmo L F R C, Prado C B, Nascimento T M, Ribeiro L C and Oliveira T D 2011 Sistema de acompanhamento de inspeções de produtos perigosos *VI Congresso Brasileiro de Metrologia, Natal. Anais do VI Congresso Brasileiro de Metrologia*
[4] Stallings W 2005 *Cryptography and Network Security: Principles and Practice* (Prentice Hall) ed 4
[5] Costa R O, Pirmez L, Boccardo D R, Carmo L F R C and Machado R C S 2012 TinyObf: Code Obfuscation Framework for Wireless Sensor Networks *Int. Conf. on Wireless Networks (ICWN), 2012, Las Vegas. Proc. of the 2012 Int. Conf. on Wireless Networks. Las Vegas: CSREA, 2012. v. 1. p. 68-74*
[6] Lakhotia A, Boccardo D R, Singh A and Manacero Jr A 2011 Context-sensitive analysis without calling-context *Higher-Order and Symbolic Computation, v. 23, p. 275-313*
[7] Collberg C and J Nagra 2010 *Surreptitious Software: Obfuscation, Watermarking, and Tamperproofing for Software Protection* (Addison Wesley)

Evaluation of critical indicators in the process of acquiring supplies and services LAC-UFPE

V F Caetano[1], C V Ferreira[2], M J dos Santos[3] and F A Honorato[1]

[1]Departament of Chemical Engineering, University Federal of Pernambuco, PE, Brazil
[2] Laboratory of Fuels of University Federal of Pernambuco, PE, Brazil
[3]Departament of Mechanical Engineering, University Federal of Pernambuco, PE, Brazil

E-mail: fah@ufpe.br

Abstract. In laboratories linked to public universities and accredited by the NBR ISO/IEC 17025, to meet efficiently item 4.6 (procurement of supplies and services) is a challenge that can be accomplished by programming based on historical purchases and services. In this study, we evaluated the critical procurement items to meet the quality management system of the LAC-UFPE: reagents, certified reference material, of equipment parts, maintenance and calibration of equipment and instruments. It was found that the most critical item is the certified reference material, the purchase or repair of which must be expedited within 125 days prior to the receipt to occur within the desired period.

1. Introduction

The acquisition of supplies and services, whether in public service or private enterprise, requires planning and management, as poor management of this process has a direct impact on quality, time, cost and risk activities. For testing and calibration laboratories that, in addition to having to meet the requirements of certificates in NBR ISO/IEC 17025[1], develop their activities under public management, such as the federal universities, the challenge is greater. Besides attending to item 4.6 of the standard that requires compliance with internal deadlines (maintenance plans and calibration, intermediate checks and others) and external (e.g., customer satisfaction during the process of selling analytical services or calibration), the rules established by Law No. 8666[2] of the federal constitution for bidding and procurement of public administration must be met. Thus, it is essential for these laboratories meet the time requirements of the acquisition process, i.e., from request until receipt.

For laboratories accredited to ISO/IEC 17025, the understanding of this process is crucial to avoid running out of supplies or unavailability of equipment that may cause noncompliance.

In laboratories that are linked to public entities, procurement processes must undergo the bidding process, which is governed by Law No. 8.666. Although this procedure is necessary so that public resources are spent efficiently, at the same time they lengthen the procurement processes that may bring negative consequences to the quality system. In this work, the process of acquisition of supplies and services for the Laboratory of Fuel, University Federal of Pernambuco (UFPE-LAC) was analyzed. We evaluated the stages of acquisition of certain items in order to establish indicators improve the buying process, in order to identify and act at the most critical steps and ensure compliance with the deadlines set by the system laboratory management system.

Journal of Physics: Conference Series **575** (2015) 012046 doi:10.1088/1742-6596/575/1/012046

2. Methodology

2.1. Data acquisition
A number of historical processes (42) for the acquisition of supplies and services for LAC-UFPE were used, representing the years 2010 to 2012. These represented items most important for the laboratory: reagents, certified reference material (CRM), equipment parts, calibration of equipment and instruments and preventive maintenance of equipment and instruments.

2.2. Process of acquiring supplies and services
The acquisition process starts from the request of a member of the lab to the immediate supervisor. From there, the request goes to the commercial sector of LAC-UFPE, where estimates are requested from suppliers. These estimates constitute input for purchasing decisions by the Foundation for Development Support (UFPE–FADE), which is the administrative entity responsible for the acquisition process of the laboratory. During the acquisition process, FADE evaluates each company's compliance with the regulations. Once the process is complete, FADE generates an order to supply and/or to service. Once the order has been released, the supplier is supposed to deliver the supplies or perform the services by the deadline specified in the order. Upon delivery of supplies or performance of services, FADE generates a payment, while at the same time the commercial sector of LAC-UFPE evaluates the performance of the supplier using a specific form. After this step, the acquisition process is finished.

2.3. Calculation of the time intervals
The acquisition process was divided into four phases. For each phase, we calculated the respective time intervals: one for the internal process, which basically depends on the LAC-UFPE procedures (i.e. (1)) and the other three related to external agents (i.e. (2), (3) and (4)). The four ranges are:

- (1): Δt_1 = (date of request for quotes from suppliers) - (date of request made by a member of the laboratory).
  This represents the time interval between the day someone made the request to the commercial sector of LAC-UFPE and the day on which the sector requested proposals from suppliers.
- (2): Δt_2 = (date of entry protocol of the acquisition process in the foundation) - (date of request for quotes to suppliers).
  This represents the time interval between the day the commercial sector of the LAC-UFPE requested proposals from suppliers and the day on which the request was sent to FADE.
- (3): Δt_3 = (date of release order / delivery by the foundation) - (date of entry protocol of the acquisition process in the foundation);
  This is the time interval in which the FADE evaluates the performance of each company with the regulations and generates an order to supply and / or service to the chosen supplier.
- (4): Δt_4 = (date of receipt of application) - (date of release order / delivery by the foundation).
  This is the time in which the supplier carries out the request after authorization by the FADE.

In the first time interval, the commercial sector of LAC-UFPE has a greater responsibility as it is in a position to identify potential suppliers and write up the formal request using the technical language required.

In the second interval, the swing phase depends on how quickly the suppliers submit proposals to the LAC-UFPE commercial sector, so that these can be forwarded to the purchasing department of FADE, so that there is continuity in the acquisition process.

In the third time interval, the internal procedures are recorded at FADE. At this stage, the LAC-UFPE commercial sector awaits the completion of the acquisition procedure and the release of the request for a contractor to be authorized to provide the material or perform the service.

In the fourth interval, the supplier is largely responsible, because the foundation has authorized the delivery of the service or supply of material. Thus the time involved depends on when the provider can perform the service or deliver the material. In some cases, because of the length of time between the original request from LAC and the authorization by FADE, the supplier no longer has the material required on hand.

3. Results

3.1. Acquisition of reagents

All stages showed a short time interval in the acquisition of reagents (figure 1) because of the large number of reagent suppliers on the market. Because it is a procurement item with high demand (which forces a greater internal control, causing the Δt_1 to be lower), there are a greater number of offers (which favors the reduction of Δt_2). The incidence of many suppliers leads to a faster processing of the process within the foundation (reduction Δt_3).

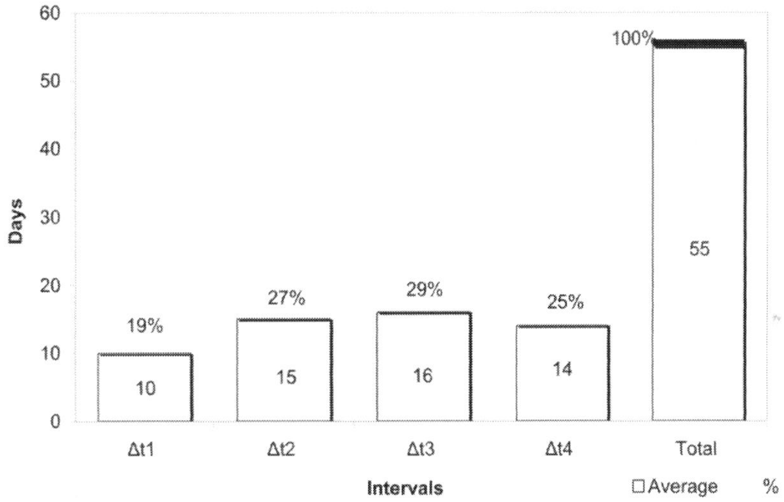

Figure 1. Evaluation of the process of provision of reagents.

3.2. Acquisition of certified reference material

In this case (figure 2), the critical point is the low inventory of these products available on the market and the small number of suppliers for some of the certified reference materials (CRM's). As many of the CRM's used in the laboratory are imported, national or international suppliers that offer this type of product may have reduced inventories, which make delivery time slow (higher Δt_4). As there is little availability of certain types of CRM's on the market, the acquisition process becomes slower (higher Δt_3) due to the low number of suppliers participating in the process. In this case, there was a period of 125 days for the request to be met.

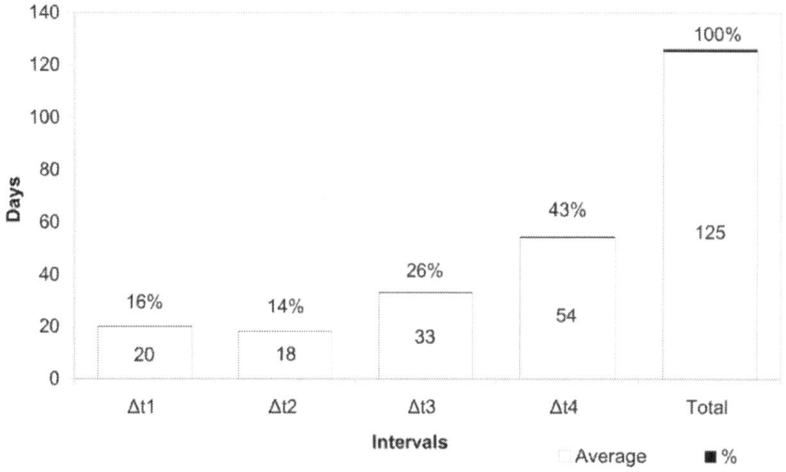

Figure 2. Evaluation of the process providing the reference material.

3.3. Acquisition of pieces of equipment
As equipment parts are mostly standardized and supplied by a single representative, this explains the low value of Δt_2 (figure 3). The predominant range was Δt_4, owing to low stocks of specific imported parts.

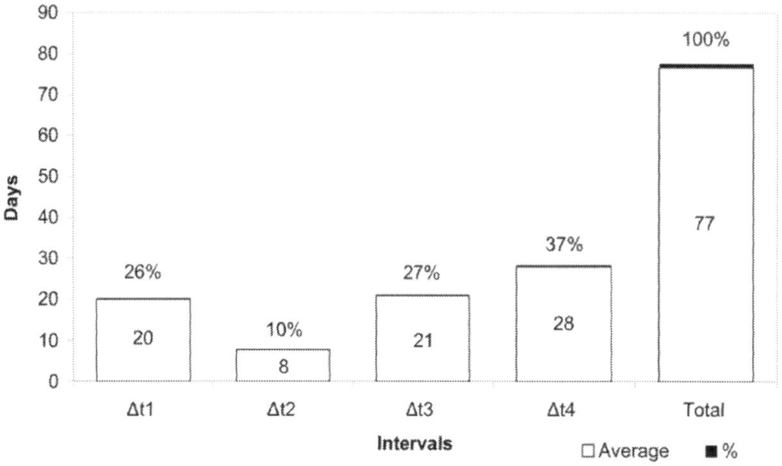

Figure 3. Evaluation of the supply process of equipment parts.

3.4. Acquisition of calibration service equipment and instruments
Because LAC-UFPE is an accredited laboratory calibrations of equipment and instruments should be carried out by laboratories belonging to the Brazilian Calibration Network (RBC). Thus, there is a restriction on the number of suppliers. In addition, as services are provided through scheduled visits, how few certified suppliers scheduling it takes a long time to schedule a visitation. All these factors contribute to a high value for Δt_4 (figure 4).

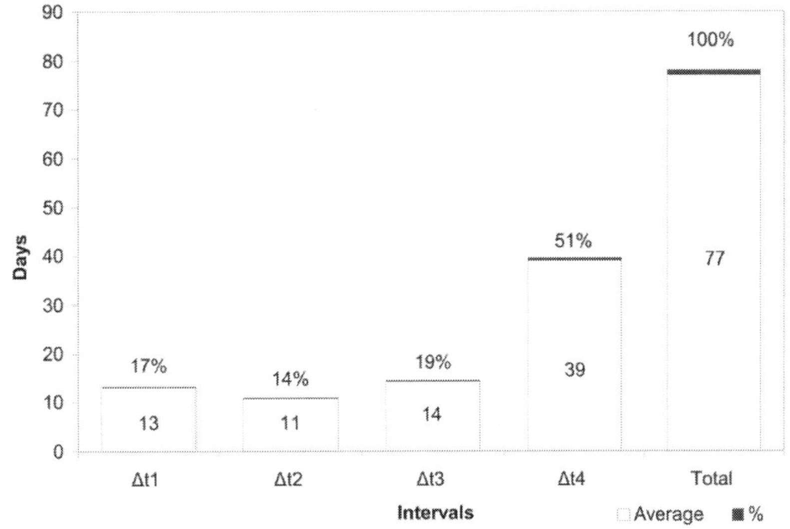

Figure 4. Evaluation of the process of the calibration service.

3.5. Acquisition of preventive maintenance service of equipment and instruments

Unlike what occurs in maintenance services (which has a smaller number of companies able to offer the service) and the supply of parts, most of which can be acquired by waiver of bidding, the acquisition of certain equipment maintenance services requires the participation of a larger number of suppliers. But many of them do not possess the necessary qualification (figure 5), which increases the process time (greater Δt_3). Moreover, as the service is also subject to the availability of technical maintenance, some processes have specific deadline for service completion (most Δt_4).

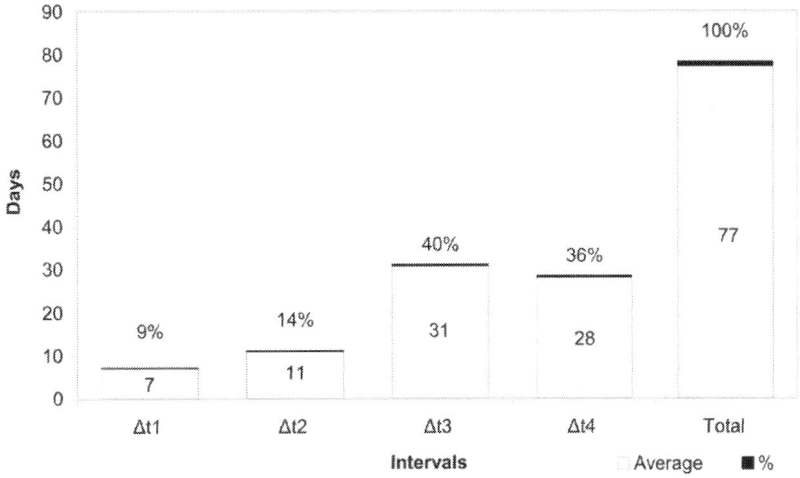

Figure 5. Evaluation of the process of preventive maintenance of equipment and instruments.

From the comparison between the five (5) items evaluated, we observed a lower overall acquisition time for the purchase of reagents (55 days). This is explained due to a high demand and ample supply for this item. On the other hand, for the purchasing CRM's, which have less demand and less supply, the overall acquisition time was 125 days, demanding greater control by the laboratory. For other items, the total acquisition time was the same (77 days). This can be explained by their similar values of demand, supply and technical skills.

4. Conclusions and final considerations

From the indicators raised, it was observed that the most critical item for purchase in the LAC-UFPE is the certified reference material which requires careful attention to planning for the purchase of this item.

In general, it is observed that the most critical intervals are 3 and 4, where there are bureaucratic procedures that must be followed and the FADE deadlines established for the suppliers have a great impact. In these intervals, the laboratory can act giving greater technical assistance to FADE for the indication and evaluation of suppliers; and in interval 4, the commercial sector can intensify the action on the part of the suppliers to comply with the deadlines that have been set. Intervals 1 and 2 can also be optimized to contribute to the decrease in average total time of the process by reviewing internal procedures, checking the availability of financial resources with senior management, paying greater attention to procurement planning, and in the searching for new suppliers.

Based on the results obtained, this work provided to members of the LAC-UFPE a greater understanding of the acquisition process and, consequently, information for more consistent prior planning. Thus, compliance with deadlines for maintenance or even intermediate checks for testing may be warranted. In addition, the commercial sector will have available an indicator that permits monitoring the procurement process, their most critical ranges and maintenance.

References

[1] BRAZILIAN TECHNICAL STANDARDS ASSOCIATION. **NBR 17025**: General requirements for the competence of testing and calibration laboratories. Rio de Janeiro, 2006.

[2] BRAZIL, Law n. 8,666, of June 21, 1993. Regulates article. 37, XXI, of the Federal Constitution, establishing rules for bidding and contracts for Public Administration and other measures. **Official Gazette**, Brasília, DF 22 June 1993. p. 8269.

Prospection for metrological control in medical scales and sphygmomanometers in the state of santa catarina – Brazil

M R Paiva, O Pohlmann Filho and A Soratto
National Institute of Metrology, Quality and Technology – Inmetro
Av. Nossa Senhora das Graças, 50, Xerém, Duque de Caxias, Rio de Janeiro, Brazil

E-mail: ansoratto@imetro.sc.gov.br

Abstract. This study presents a method to identify the universe of establishments in the health area where the use of measuring instruments subject to metrological control and periodic verification are mandatory. Through research of external databases and complemented with field visits, it was possible to estimate the number of scales and clinical sphygmomanometers that are currently in use by these establishments in the state of Santa Catarina. The results are useful for developing strategies in order to prospect, plan and execute the mandatory metrological verification of instruments in the health area in Santa Catarina.

1. Introduction

In the health area, the National Institute of Metrology , Quality and Technology - INMETRO develops regulations and delegates powers to a network of competent state institutes to carry out field activities such as mandatory verification of measuring instruments [1,2]. Among these instruments we highlight the clinical scales for weighing adults, pediatric scales and sphygmomanometers, also known as blood pressure cuffs.

The lack of metrological control or verification as well as the inadequate frequency of verification of these instruments causes several measurement errors, leading to misdiagnosis and even improper application of medication or procedures that endanger the health and lives of people.

Several studies demonstrate the need to invest efforts in metrological control, and we draw special attention to the sphygmomanometers [3-7].

It is also important to stress the lack of awareness and initiative of almost all professional or holders of these instruments, which do not look for INMETRO or one of their delegates institutes in the country to carry out the mandatory verification of their instruments.

Therefore, the Metrology Institute of Santa Catarina - Imetro/SC, the delegated Institute of INMETRO in the state of Santa Catarina, started visiting health establishments in order to verify the compliance of the above mentioned measuring instruments.

However, this initiative has brought great challenges, starting with the need to find out and estimate the quantitative of measuring instruments subject to verification.

This study aims to identify the potential demand for the mandatory verification of clinical scales and sphygmomanometers in the state of Santa Catarina and proposes ways to cope with the demand.

2. Method for the prospection of instruments in the health area

The study to identify the potential demand for the mandatory verification of clinical scales and sphygmomanometers in the state of Santa Catarina has been divided in two steps:

- Step 1 – Estimate the universe of health establishments in the state of Santa Catarina.
- Step 2 – Estimate the number of clinical scales and sphygmomanometers in the identified establishments.

Data collection was conducted through visits to health establishments which addresses were obtained from the National Register of Health Establishments (CNES) of the Ministry of Health in Brazil, and from the database of the Annual Social Information (RAIS) of the Ministry of Labour and Employment in Brazil (MTE).

The CNES information is constantly updated and is available in the Internet [8].

On the other hand, information from RAIS used in this research was picked up from the 2011 database. RAIS is now available to INMETRO as a result from a memo of understanding celebrated by INMETRO and MTE.

Research from RAIS was done using the Monitoring System of Coverage - SIMCS, developed by INMETRO/Cored [9,10].

2.1. Step 1 - Estimate the universe of health establishments in the state of Santa Catarina.

For the purpose of legal metrological control, the following are considered health area establishments: hospitals, clinics, doctors' offices, health clinics, pharmacies, including manipulation ones, fitness gyms (bodybuilding, swimming ...).

The survey began with research from CNES, which showed that there are 11.457 health establishments that actually operated in Santa Catarina in 2012. It includes hospitals, clinics, doctors' offices and public health posts.

Estimate of pharmacies, including manipulation ones and fitness gyms, was obtained from SCE (updated with information from RAIS 2011) that showed the existence of 3.417 pharmacies, 412 manipulation pharmacies and 983 fitness gyms in Santa Catarina.

Therefore, as per figure table 1, the conclusion of this initial estimation was that there are approximately 16.269 health establishments with scales and/or sphygmomanometers subject to mandatory metrological verification.

Table 1. Estimation of establishments in the health area subjected to mandatory verification in Santa Catarina.

Hospitals, clinics, doctors' offices, public health posts (CNES 2012)	Pharmacies, fitness gyms (RAIS 2011)	Total
11,457	4,812	16,269

2.2. Step 2 - Estimate the number of clinical scales and sphygmomanometers in the identified establishments

During 2012 and early 2013, visits were carried out in the health area establishments chosen from a sample of eight randomly selected municipalities of Santa Catarina: Tubarão, Criciúma, Araranguá, Içara, Morro da Fumaça, Imbituba, Braço do Norte e Orleans.

For each of these municipalities were generated lists of health area establishments (with its addresses) extracted from the database of CNES and RAIS. The number of establishments visited by municipality was defined considering to how long each field agent stays in each municipality during their regular visits to other establishments. In other words, the filed agents were invited to visit some additional establishments (the health area ones) during their regular visits to the mentioned municipalities.

In total, 813 visits in hospitals, clinics, doctor's offices, nursing activities, physiotherapy, health centers, pharmacies and gyms were visited.

From this total, scales and sphygmomanometers were found in 516 establishments. In the majority of them there were no check verification tag or its validity was expired.

Table 2 presents the percentages of correct CNES and RAIS in terms of establishments which were actually operating and with scales and/or sphygmomanometers in use.

Table 2. Number of establishments with clinical scales e sphygmomanometers

Database	Number of establishments that were visited	Number of establishments were instruments were found
CNES 2012	446	311 (69,7 %)
RAIS 2011	367	205 (55,9 %)
TOTAL	813	516 (63,5 %)

The percentage of correct information from CNES was 69.7%. In this database, 9.9% of establishments were closed or with address errors and 20.4% of establishments had no scales or sphygmomanometers in use.

The percentage of correct information from RAIS was 55.9%. In this database, 40.6% of establishments closed or with address errors and 4.4% of establishments had no scales or sphygmomanometers in use.

Table 3 shows the number of clinical scales (adult and pediatric) and sphygmomanometers found and the amount of these instruments per type of establishment.

Table 3. Number of scales and sphygmomanometers per establishment

Type of establishments	Hospitals, clinics, doctors' offices, public health posts (CNES 2012)	Pharmacies, fitness gyms (RAIS 2011)	Total
Number of establishments were instruments were found	311	205	516
Number of Clinical Scales	544	213	757
Scales per establishment	1.75	1.04	1.47
Number of sphygmomanometers	1,000	164	1,164
Sphygmomanometers per establishment	3.20	0.80	2.26

From the sample of 516 existing establishments, 1,164 sphygmomanometers were found and 757 clinical scales were found, representing on average, 2.26 sphygmomanometers and 1.47 scales per actually operating establishment.

Such trend is assumed to be similar in the other municipalities and allows a projection of the number of clinical scales and sphygmomanometers in use throughout the state of Santa Catarina , as shown in Table 4 .

To calculate the numbers of measurement instruments were considered 69.7% of establishments from CNES and 55.9% of establishments from RAIS, as these were the percentages of establishments found that were actually operating and with measuring instruments in use.

The research identified an estimated universe of 10,666 establishments in the health area in Santa Catarina, with approximately 16,763 clinical scales and 27,699 sphygmomanometers in use and therefore subject to legal metrological control of Imetro/SC.

Table 4. Estimation of clinical scales and Sphygmomanometers currently in use in Santa Catarina.

Type of establishments	Establishments with scales and/or sphygmo.	Estimation of clinical scales	Estimation of sphygmomanometers
Hospitals, clinics, doctors' offices, public health posts (CNES 2012)	7,986 (69,7% do CNES)	13,976	25,555
Pharmacies, fitness gyms (RAIS 2011)	2,680 (55,7% da RAIS)	2,787	2,144
Total	10,666	16,763	27,699

It is important to mention the significant number of establishments that were closed or with address errors as well the report from our field agents that confirms the lack of awareness of the health professional as far as the legislation and benefits of verification of the instruments are concerned. More than forty medical offices visited did not allow the access of the fiscal agent, claiming not to have scales or sphygmomanometer. Such information seems no to be true. Countermeasures for such facts need to be developed.

3. Results of the research

The results are useful for developing strategies in order to prospect, plan and execute the mandatory metrological verification of instruments in the health area in Santa Catarina.

The number of establishments and instruments that were identified as a consequence of this study can be used to foresee the equipment required to meet the existing demand for each municipality.

The prior identification of the addresses, through external databases CNES and RAIS, represents a major breakthrough against the usual practice of the fields agents. It is now possible to look for new establishments with appropriate tools.

This previous knowledge of addresses to be visited enables logistics decisions that prioritize efficiency.

4. Conclusions

The identification of the universe and the location establishments in the health area, and the estimated number of scales and sphygmomanometers in use, enable us to:

- Define the required number of fiscal agents;
- Improve logistics operations;
- Establish planning with realistic goals;
- Estimate the costs of implementing of mandatory verification in this the health area.

It is also important to stress that the methodology of research can be applied in any other state in the country. The states that have got the same level of Santa Catarina in the health area will probably have the same results once the number of scales and sphygmomanometers in health establishment will probably be similar.

The experience of looking for information in the external environment and the development of innovative tools such as the Monitoring System of Coverage - SIMCS developed by INMETRO/Cored, are important initiatives to improve the planning and execution of the legal metrological verification in the country.

Acknowledgements

Inmetro/Cored , Imetro/SC and the fiscal agents Ms Walkiria Aparecida Costa and Ms Janine May.

References

[1] International Vocabulary of Terms in Legal Metrology 2007 OIML available on <http://www.oiml.org/en/publications/vocabularies>

[2] Principles of Metrological Supervision 2004 OIML international document available on <http://www.oiml.org>

[3] Instituto Nacional de Metrologia, Qualidade e Tecnologia 1997 Esfigmomanômetros Disponível em <http://www.inmetro.gov.br/consumidor/p rodutos/esfigmo.asp>.

[4] Serafim T S, Toma G A, Gusmão J L, Colósimo F C, Silva S S B E e Pierin A M G 2012 *Acta Paul Enferm*. vol **25**(6) pp 940-6

[5] Padilha H e Oliveira M S 2011 In: *6º Congresso Brasileiro de Metrologia Anais*. Rio Grande do Norte

[6] Monteiro E C e Lessa M L 2005 *ENGEVISTA* vol **7**, n. 2, pp 51-60

[7] Palota L, Cordella M P, Oliveira S M and Cesarino C B 2004 *Arq Ciênc Saúde* abr-jun; vol **11**(2)

[8] CNES. Cadastro Nacional de Estabelecimentos de Saúde 2012 available on <http://cnes.datasus.gov.br/>

[9] Soratto A N, Pohlmann Filho O, Paiva M R 2011 In: *6º Congresso Brasileiro de Metrologia*, Anais. Rio Grande do Norte

[10] Soratto A N, Pohlmann Filho O, Paiva M R, Giordani R B, and Bringhenti C Development of a system to increase the legal metrological control of measuring instruments in Brazil Submitted to CAFMET 2014

AUTHOR INDEX

AUTHOR INDEX

Institute of Physics
Dirac House, Temple Back
Bristol BS1 6BE UK